LIGHT AND COLOR

LIGHT AND COLOR

R. DANIEL OVERHEIM
DAVID L. WAGNER

Edinboro State College

JOHN WILEY & SONS, INC.

New York Chichester Brisbane Toronto Singapore

Cover: Photograph of reflected light source on a sheet of metal foil. The foil is lit from behind. Light rays are reflected towards the camera at the same angle, capturing the reflection. Colors come from gels over the light source, providing color only to the highlight areas. Photographed by Paul Silverman, designed by Ann Marie Renzi.

Library of Congress Cataloging in Publication Data:

Overheim, R. Daniel.
Light and Color.

Bibliography: p. 259
Includes index
1. Color. 2. Optics. 3. Light. I. Wagner, David L.

II. Title.
QC495.088 535 81-21955
ISBN 0-471-08348-8 AACR2

Printed in the United States of America

10 9 8 7 6 5 4 3 2 1

PREFACE

This book grew out of a general education course entitled *The Nature of Light and Color* taught by the physics department at Edinboro State College. The intended audience consists of nonscience majors from virtually all areas of the college. Their science and mathematics background is assumed to be minimal. In the text, therefore, scientific concepts are fully explained as they are introduced, and the mathematical level is kept to rudimentary algebra.

From experience we have found that students from virtually every academic area of the college take the course on which this text is based. However, we have also found that majors in art, photography, television, and theater take the course in very large numbers. From our discussions with students and faculty in these disciplines it is clear that this book could form the basis for a course that was intended specifically for any of these students. It is appropriate both in topical content and in technical level for this audience.

This is not a traditional optics text. The choice of topics and their order of presentation here is quite different from existing texts on light. Our main emphasis is on color – the properties of light and materials that give rise to it, how it is described and analyzed, how we see it, and how it is produced in nature. Only when this material has been fully discussed do we turn, in the later chapters of the book, to more traditional topics such as geometrical and wave optics.

We begin in Chapter 1 with a discussion of the physical nature of light. This is approached from a historical perspective for purely personal reasons. We feel that whenever possible an attempt should be made not only to present the basic principles of science, but also to examine how those principles came to be accepted. Science is, after all, not revealed truth but rather an ongoing process that continually deepens our understanding of nature. Thus, in Chapter 1 we discuss some of the controversies concerning the physical nature of light and how these were resolved. We do not, at this point, go into any details of geometrical optics, and, although wave concepts are developed more fully, traditional wave optics is also delayed to later in the book.

Chapter 2 discusses illumination units, light sources (continuum and bright line), reflection, transmission, absorption, and primary colors (both additive and subtractive). A few simple examples of the application of these concepts are given.

Chapter 3 deals with the *description* of color, a science known as colorimetry. Three basic systems are described: the CIE, Munsell, and Ostwald.

Chapter 4 outlines the development of theories of color vision, describes a number of the basic features of color vision, and points out some of the problems and controversies involving modern color vision theories.

Chapter 5 discusses why objects appear the way they do. Appearance, of course, involves more than just color. Thus it is necessary to distinguish carefully between specular and diffuse reflection, and between selective nonselective reflection. In this chapter the properties of transparent and opaque colorants are also discussed.

Chapters 6 and 7 are essentially a traditional treatment of geometrical optics and simple optical instruments, while Chapter 8 deals with wave optics.

Chapter 9, the concluding chapter, describes how some of the more spectacular natural color phenomena arise. Such topics as rainbows, halos, the blue sky, the color of oil spots and soap bubbles, mirages, the aurora, and the blue-gray of distant mountains are discussed.

Each chapter is followed with a number of problems, questions or, in some cases, things to do. These exercises are intended to reinforce concepts introduced in the chapter. In most cases, especially those involving numerical concepts, it is essential that students work through all the problems we have provided if they are to understand the basic principles involved.

There are four appendices. The first, on the fundamentals of the spectrophotometer, is included for students who wish to understand exactly how transmission and reflection spectra are obtained. The difference between forward and reverse optics is also discussed.

Appendix B includes nomographs for calculating CIE tristimulus values and instructions for their use. Appendix C is a table of values for the sine function, and Appendix D is an annotated bibliography.

An instructor's manual with additional discussion of some of the topics covered in the book, as well as some that have not been included, is available. This manual contains suggestions for demonstrating many of the features of color vision.

We thank the individuals who reviewed our manuscript as it was in preparation and who provided many helpful and constructive suggestions. We also thank Susan Weigers and Debbie Kirk for their patience in typing repeated versions of the manuscript.

R. Daniel Overheim
David L. Wagner

CONTENTS

LIGHT AND COLOR

1 THE PHYSICAL NATURE OF LIGHT

Sight is a wonderful gift. It allows us to understand a great deal about the world around us and fills our lives with richness and beauty. Today we understand that this miracle of vision results from the combination of two distinct things. The first is light, a physical entity that has its own unique properties and that exists whether or not anyone is around to see it. The second is the eye, which is sensitive to light and, in combination with the brain, allows us to see. Since one of the main tasks of this book is to develop an understanding of how vision, especially color vision, takes place, we need to know something about both the physical nature of light and the operation of the eye-brain system. The subject of this chapter is the physical nature of light.

As we shall see, light can be understood on several different levels. Thus, in one set of circumstances we speak of light rays while under different conditions we refer to light waves or even to particles of light. These apparently contradictory terms indicate that light is a unique physical entity that cannot be compared to any other simple physical phenomenon. For convenience, the study of light is often broken down into broad categories dealing with the behavior of light under different circumstances. Thus geometrical optics describes light as if it traveled in straight lines and obeyed simple geometrical laws upon reflection from or passage through various materials. On the other hand, wave optics deals with phenomena that can best be described as the motion of waves, while quantum optics deals with situations in which the particle nature of light becomes important. To gain some appreciation for these different ways of describing light and to understand the circumstances in which each is appropriate, let us review the earliest recorded ideas concerning the nature of light.

1.1 EARLY IDEAS

Actually, the clear distinction between light and the eye has not always been understood. The ancient Greeks, although they developed some fairly sophisticated optical theories, tended to regard light and vision as one and the same. Pythagoras (~550 B.C.) and Democritus (~400 B.C.) thought vision resulted from "images" that traveled from objects to the eye. On the other hand, Euclid (326 B.C.) and Hero (40 A.D.) held that "visual rays" traveled from the eye to the object being seen. The rays were thought to "see" whatever they encountered. It is not clear what role these early Greeks assigned to light sources, such as the sun or a fire. They were certainly aware that such a source was necessary for vision to occur.

A rather interesting theory was put forward by Aristotle (~350 B.C.), perhaps the best-known Greek philosopher-scientist. He felt that transparent substances such as air, water, and glass were only "potentially" transparent. They became "actually" transparent in the presence of a fiery object, which we would call a light source. Once the medium was made actually transparent by the presence of fire, the "colors" of the objects in the world could travel from the objects to our eyes allowing us to see.

It was not until around 1000 A.D. that the Arab scholar Alhazen finally drew a clear distinction between light as a physical entity and the eye as a detector of light. With this crucial point cleared up it was possible for the scholastics of the Middle Ages to focus their attention on light as a separate physical entity and investigate its properties independently of vision. By the early seventeenth century men such as Johannes Kepler, Willebrod Snell, and René Descartes had developed the science of geometrical optics to nearly its present-day form.

Knowledge of the rules of geometrical optics allows the reliable construction of optical devices such as lenses, microscopes, and telescopes, but does not tell us a great deal about what light actually is. Historically, the laws of geometrical optics were fairly well known long before the true physical nature of light was understood. Later, in Chapter 6, we shall consider these geometrical laws in detail. At this point in our discussion, however, we are more interested in other problems such as what is the speed of light, what accounts for color, and what exactly is light? The first of these questions was answered in the late seventeenth century, and much progress was made on the second. But the complete answer to the third question had to await the twentieth century.

1.2 THE SPEED OF LIGHT

From earliest times there had been a difference of opinion about the speed of light. The Greeks were divided on the question, some proposing a finite speed, others arguing that light traveled infinitely fast. By the middle of the seventeenth century the two giants of physics, René Descartes and Isaac Newton, were still divided on the question, primarily because of their basic philosophies of nature.

Descartes, a French mathematician and natural philosopher, viewed the universe as completely filled with matter with no gaps at all. He pictured light as a disturbance that traveled infinitely fast from one place to another. For the English natural philosopher Newton, on the other hand, the universe was primarily empty space, with matter occupying only a small fraction of the total volume. Newton viewed light as a stream of tiny corpuscles or particles traveling at a large but finite speed. Until late in the seventeenth century, no direct observational evidence was available to decide the issue. Early in the seventeenth century Galileo Galilei (1564–1642), an Italian natural philosopher considered by many as the father of modern physics, had made an interesting attempt to measure the speed of light using basically a time of flight approach. This kind of approach uses the definition of speed as the basis for measurement.

The concept of speed involves the two more fundamental ideas of distance and time. When we say that something is traveling at a certain speed, we mean that it is moving in such a way that it travels a certain distance in a specified period of time. If a ball is traveling 100 ft/sec this means that in one second the ball travels 100 feet. For an object traveling at constant speed, there is a simple relationship between the speed of the object (V), and the time (t) required for the object to travel a given distance (d). That is

$$V = \frac{d}{t} \tag{1}$$

For example, suppose that an object travels 450 feet in 9 seconds. Then

$$d = 450 \text{ feet}$$

$$t = 9 \text{ seconds}$$

and, using eq. (1)

$$V = \frac{450 \text{ ft}}{9 \text{ sec}} = 50 \text{ ft/sec}$$

What Galileo proposed to do was to measure the time required for light to travel a definite distance. He did this by stationing himself on one hilltop and an assistant on a distant hilltop. Working at night with lanterns, Galileo and his assistant attempted to flash light signals back and forth between the two hilltops, and measure the resulting round trip travel time for the signals. After discounting the reaction times involved in opening and closing the lanterns, Galileo found that essentially no time was left to be attributed to the travel time of the light. He therefore concluded that light either traveled infinitely fast or at least at a very high speed. Galileo's method was conceptually sound but it relied on very crude measuring techniques. The first success with resolving the question of the speed of light came about more indirectly.

In the early 1670s a young Danish astronomer named Olaus Roemer was working as an assistant to the prominent French astronomer Jean Picard. One of

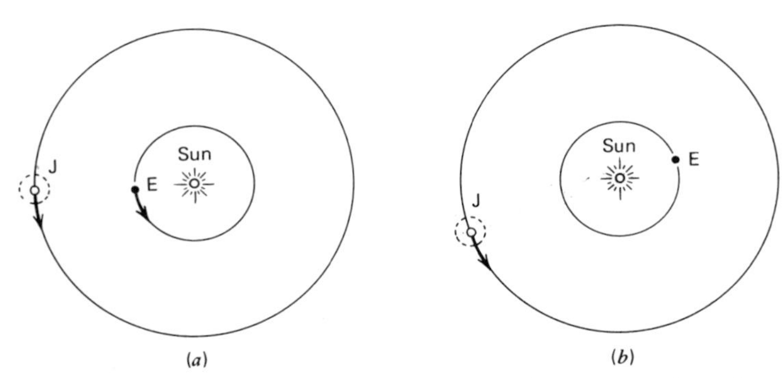

Figure 1.1 (*a*) Jupiter and earth at closest approach. (*b*) Jupiter and earth at maximum separation.

Roemer's tasks was to undertake detailed observations of the moons of Jupiter, which were easily visible through a telescope. The larger moons of Jupiter complete their orbits around Jupiter in a fairly short time, only 7.2 days for Ganymede for example. Thus, if observations are made when Jupiter and Earth are closest to each other (Figure 1.1*a*), it is possible to measure the orbital periods of the moons with great precision. This is usually done by measuring the time between two successive passes of any given moon behind Jupiter (i.e., between two successive eclipses of that particular moon by Jupiter). It should therefore be possible to predict well into the future the exact times when the various moons of Jupiter will be eclipsed by the planet. Much to Roemer's surprise, the predictions proved to be wrong, and in a rather peculiar way. As the earth got farther away from Jupiter (going toward the position shown in Figure 1.1*b*), the eclipses fell increasingly behind schedule. Then, as the earth moved back nearer the giant planet, the eclipses occurred progressively nearer to the correct times. Roemer concluded that the effect was due to the *finite velocity of light.* The time lags were produced by the light having to travel progessively farther to reach the earth. To fully comprehend this result, keep in mind that we see an event occur *when light reaches our eyes.* If light travels at a finite speed, the actual event will have occurred *before* we see it, by however much time is required for the light to travel to our eyes. In the case of the moons of Jupiter, if light travels at a finite speed, the light announcing each eclipse will take progressively longer to reach the earth as the earth gets farther from Jupiter. This is essentially the effect Roemer was witnessing. Roemer reported his findings in 1676 to the Academy of Sciences in Paris. He did not, as is often stated, actually make a calculation of the speed of light. He was more interested in having proved that light did have a finite speed than in what the speed actually was. He also lacked reliable values for key astronomical distances that would have been necessary to make an accurate calculation.

Clever methods have since been used to measure accurately the speed of light. The modern value is very nearly 300,000,000 m/sec (3×10^8 m/sec) or

186,000 miles/sec. It is easy to see why Galileo was unsuccessful. If his hills had been 15 km apart, the round trip for the light would have been 0.0001 sec (10^{-4}sec), a very short time indeed.

1.3 THE ORIGIN OF COLOR

By the seventeenth century, as we have seen, a great deal of progress had been made toward understanding the optical properties of light. The origin of color, however, remained a mystery. Most theories attributed color to some sort of *modification* of light that was thought to occur when light interacted with matter. Light, in its purest form, was regarded as essentially colorless. But, for example, as light from the setting sun passed through the many additional miles of earth's atmosphere the light was thought to be modified by the atmosphere into a red-orange color. Or, if light reflected from a green surface, the surface was supposed to have modified the light so that it became green. Exactly how this modification was carried out was not clearly explained.

The first real progress in understanding color was made by Isaac Newton during the years 1665–1666. During the years 1665–1666 a great plague swept London, and Newton, who was studying at Cambridge University, sought refuge at his mother's farm in the countryside. Here he turned his attention to a number of scientific problems, including the nature of light. Newton's optical experiments were so brilliantly clear in their conception and design that his book, *Opticks,* is still used in some classes today as a supplementary text. Of particular interest to us are Newton's experiments with color.

Newton began with a narrow beam of white light, which he produced by allowing sunlight to pass through a small hole in an otherwise darkened window. The beam was then allowed to fall on a glass prism (Figure 1.2). The light coming out the other side of the prism was observed to be spread out into a spectrum of colors from red through violet. This was a well-known effect at the time, but Newton pushed his investigations further. He used a narrow slit to block off most of the colors coming through the prism and selected one particular color, which he allowed to pass through a second prism (Figure 1.3). He observed that the second prism *had no effect* on the color of the light. From this he concluded

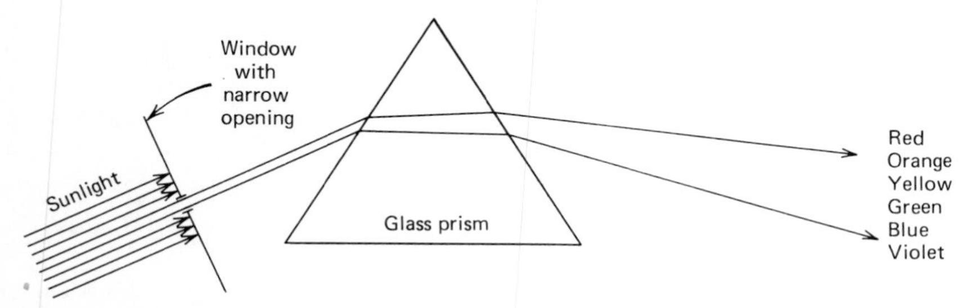

Figure 1.2 A prism separating the white light of the sun into the spectrum of colors.

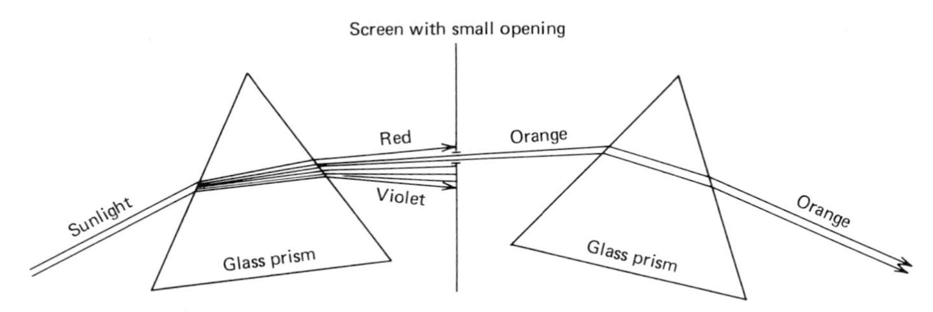

13 **Figure 1.3** A second prism has no effect on the pure orange light.

that colors were *not* the result of the prism actually changing the light. Instead, he theorized that normal white light already contained all the colors of the spectrum, and the prism only served to *separate* these colors from each other. To test this idea Newton used a second inverted prism (Figure 1.4), which he hoped would put all the colors back together again. Sure enough, when he tried this experiment he found the original beam of white light emerging from the second prism. Newton also did a number of experiments with colored objects. He placed objects of various colors directly in the path of the colored light coming through the prism. Not surprisingly, he found that a green object, for example, would efficiently reflect green light but would appear nearly black when illuminated by red or blue light. Thus it became clear that objects have color because they *selectively reflect* certain colors of the spectrum while absorbing light of other colors. This was a tremendous conceptual advance which led, over 100 years later, to Thomas Young's theory of color vision.

1.4 WAVES

During the seventeenth century there were two main kinds of theories about light. In one camp, which included Newton, were those who believed that light consisted of a stream of particles. On the other side were those who believed light was a propagating disturbance of some kind, much like a sound wave.

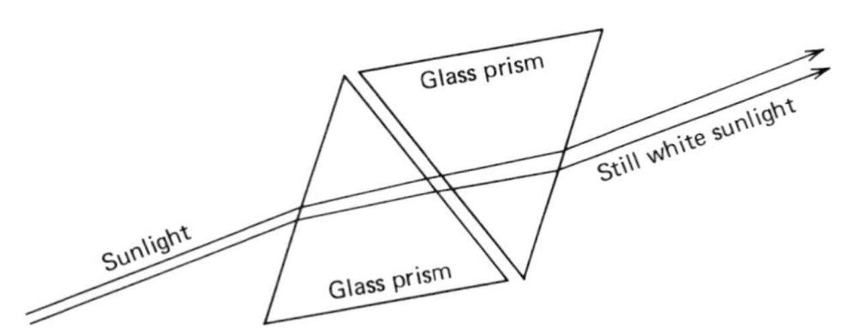

Figure 1.4 Inverted second prism puts the colors back together again, resulting in white light.

Before discussing the details of these two theories, it would be useful to learn a little about the general properties of waves. As a practical matter, until the twentieth century, when physicists discussed waves they meant mechanical waves. We now know that other types of waves exist, but the terminology used to describe these other waves is the same as for mechanical waves. Thus we begin with mechanical waves.

Basically a mechanical wave is a periodic *disturbance* that travels at a definite speed through a material medium. The medium itself undergoes relatively little motion, whereas the disturbance may travel a considerable distance. In order to help clarify this idea of a disturbance traveling in a material medium, let us imagine holding a very long rope tied to a post. Now imagine a flip of the wrist that sends a pulse down the rope (this is a *very* long rope, so the pulse doesn't reflect from the post and come back until after we've gone home). The pulse constitutes a disturbance that travels along the rope. Notice that the rope itself is not transported, but remains attached to your hand. If, instead of a single flip of the wrist, we move our hand up and down in a particular periodic fashion (like a weight oscillating at the end of a spring) then we create a periodic disturbance on the rope such as that shown in Figure 1.5. Notice that the disturbance along the rope repeats itself at regular intervals. This repeat distance is called the *wavelength* of the wave and is usually designated by the Greek symbol λ (lambda). Next notice that the hand goes up and down a definite number of times each second. This number of wavelengths produced per second is called the *frequency* of the wave and is designated by the symbol f. Frequencies are measured in oscillations per second (sometimes called cycles per second). This unit is now called the Hertz and is abbreviated by the notation Hz. Thus a frequency of 10 oscillations per second is referred to as 10 Hz. Related to the frequency is the *period* of the wave, designated by T, which is the time required to produce one wavelength. By definition the frequency and period are related by

$$T = \frac{1}{f} \tag{2}$$

where T is in seconds and f is in Hz.

The wave also travels at a definite *speed.* This speed can be calculated by observing that during one period, T, the entire wave shown in Figure 1.5 moves

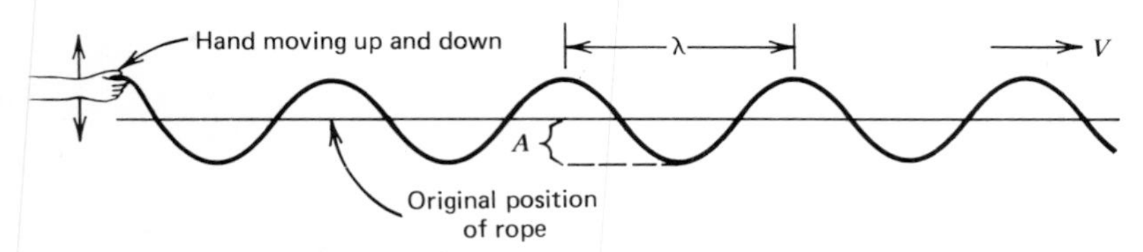

Figure 1.5 Wave on a rope showing wavelength (λ) and amplitude (A).

one full wavelength, λ, to the right. Thus, recalling eq. (1), the speed of the wave, V, is given by

$$V = \frac{\lambda}{T}$$

Or, in view of eq. (2),

$$V = \lambda f \tag{3}$$

Equation (3) is a basic equation that applies to waves of all kinds. We shall use this result several times later in the book.

Since eq. (3) is so important, let us consider a few concrete examples.

Example 1

In air, sound waves travel at a speed of about 335 meters/sec. The human ear can hear frequencies ranging from 20 to about 20,000 Hz. What are the corresponding wavelengths for these sound waves?

We begin with

$$V = \lambda f$$

and rearrange the equation algebraically:

$$\lambda = \frac{V}{f}$$

Then for $V = 335$ m/sec and $f = 20$ Hz

$$\lambda = \frac{335 \text{ m/sec}}{200 \text{ Hz}} = 16.75 \text{ m}$$

For $V = 335$ m/sec and $f = 20{,}000$ Hz

$$\lambda = \frac{335 \text{ m/sec}}{20{,}000 \text{ Hz}} = 0.01675 \text{ m} = 1.675 \text{ cm} \quad \blacksquare$$

Example 2

The end of a rope is moved up and down at a frequency of 3 Hz. The resulting waves on the rope have a wavelength of 9 feet (ft). What is the speed of the waves?

Beginning with

$$V = \lambda f$$

we have $f = 3$ Hz and $\lambda = 9$ ft.

$$V = (3 \text{ Hz})(9 \text{ ft}) = 27 \text{ ft/sec} \quad \blacksquare$$

Another basic property of all waves is the *amplitude, A*. The amplitude is a measure of the size of the disturbance. For the wave on a rope, the amplitude is just the maximum distance that the rope is displaced from its equilibrium position as the wave goes by (Figure 1.5). The amplitude is also a measure of how much energy the wave carries, or the *intensity* of the wave, *I*. In general, the intensity and amplitude are related by

$$I \propto A^2 \tag{4}$$

The concept of intensity is more fully discussed in Chapter 2.

Waves, broadly speaking, come in two basic varieties. Waves on a rope illustrate what are called *transverse* waves. Notice that in the example of the rope in Figure 1.5, the wave travels to the right but the rope itself moves up and down – *transverse* to the direction that the wave is moving. This is what defines a tranverse wave. *Transverse mechanical waves usually require a solid medium for their existence.* You cannot create a transverse mechanical wave in air for example.

The other kind of wave is known as a *longitudinal* wave. With this kind of wave, the disturbance causes the medium to be displaced back and forth in the *same* direction that the wave is traveling. Sound is a good example of a longitudinal wave. As the diaphram of a speaker moves in and out it alternately creates regions of compression and rarefraction that propagate through the air (Figure 1.6). The air molecules themselves move back and forth along the *same line* as the sound wave is traveling. Longitudinal mechanical waves can exist in solid, liquid, or gaseous media.

Waves generally exhibit another important property. They have the ability to *intefere* with each other. For instance, if two or more waves pass through the same point within a medium at a given time, they will combine together in a particular way. This is best illustrated by considering two waves of identical

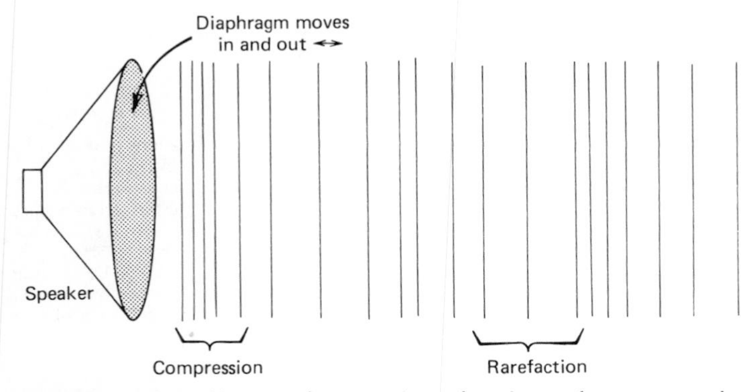

Figure 1.6 Sound produced by speaker consists of regions of compressed and rarefied air that propagate to the right. Air molecules oscillate back and forth in the same direction that the sound travels.

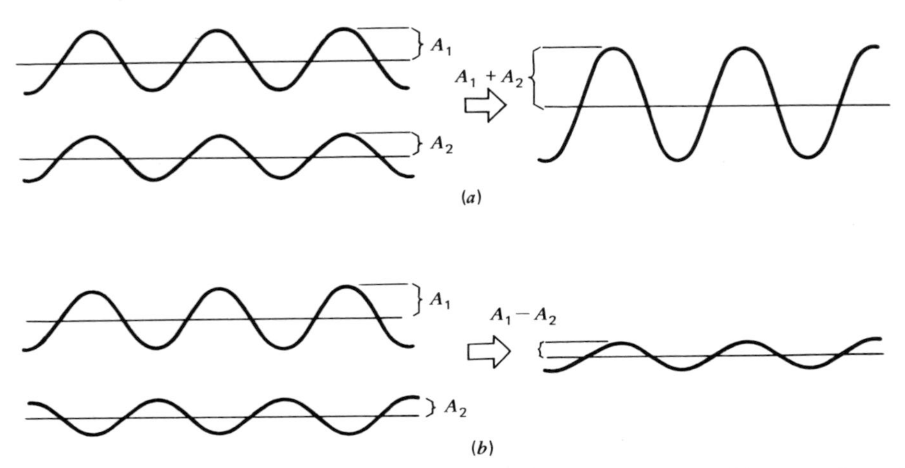

Figure 1.7 (*a*) Two waves interfering constructively to give a wave of increased amplitude. (*b*) Two waves interfering destructively to give a wave of decreased amplitude.

wavelengths but slightly different amplitudes. The two extreme possibilities are that the two waves arrive at a given point either completely in phase (Figure 1.7*a*) or completely out of phase (Figure 1.7*b*). By "in phase" we mean that the peaks of one wave will align with the peaks of the second wave, whereas "out of phase" means that the peaks of one wave will align with the valleys of the second wave. If the waves are in phase, they will reinforce each other and combine to produce a single wave with an amplitude equal to the *sum* of the individual wave amplitudes. If the waves are out of phase they will oppose each other and combine to produce a wave that has an amplitude equal to the difference of the individual wave amplitudes. This effect can be demonstrated quite strikingly by playing a pure note from a signal generator over two loud-

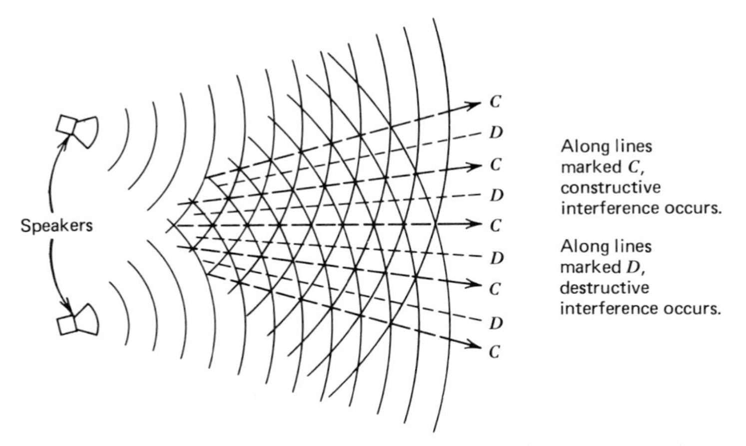

Figure 1.8 Interference between sound waves from two speakers.

speakers spaced a few feet apart (Figure 1.8). At certain places the waves from each speaker will arrive in phase, and reinforcement (*con*structive interference) will occur resulting in a loud sound. At other places the waves will arrive out of phase, and opposition (*de*structive interference) will occur, resulting in hardly any sound at all. Of course, in most places the waves will be neither exactly in phase nor exactly out of phase. In this case something intermediate between complete reinforcement and complete cancellation will result.

Another property that waves universally exhibit is the phenomenon of diffraction. That is, waves spread around corners to an extent that depends on their wavelength. This is why you can hear someone call to you from around a corner.

1.5 THE WAVE-PARTICLE CONTROVERSY

Let us now take up the controversy that arose in the seventeenth century between those who believed that light was a wavelike disturbance and those, like Newton, who believed that light consisted of a stream of particles. By Newton's time several optical phenomena had been discovered. In addition to reflection and refraction (Figure 1.9) scientists knew about diffraction and polarization. Diffraction (see Figure 1.10) is the slight bending of light that occurs when light passes very close to an edge. Polarization can most easily be understood by wearing a pair of polaroid sun glasses. (In Newton's time, naturally occurring materials were used to make these observations.) Suppose you look at the surface of a calm lake or pond. Normally you see a great deal of reflected glare. But if you put on polaroid glasses the glare disappears. On the other hand, if you rotate your head sideways 90° (Figure 1.11), the glare returns. Apparently, there is some characteristic of light that permits light reflected from a lake surface to pass through polaroid glasses when they are oriented in one way, but blocks the light when the glasses are rotated 90°. We say that the light reflected from the lake is polarized. The arguments between the wave theorists and particle theorists came down to providing convincing explanations for these, and other, effects.

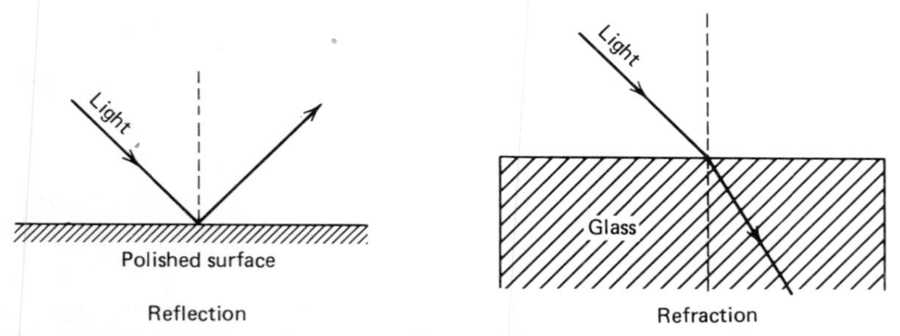

Figure 1.9 Light can reflect from a surface (reflection). When light passes from one transparent medium to another, it changes direction (refracts).

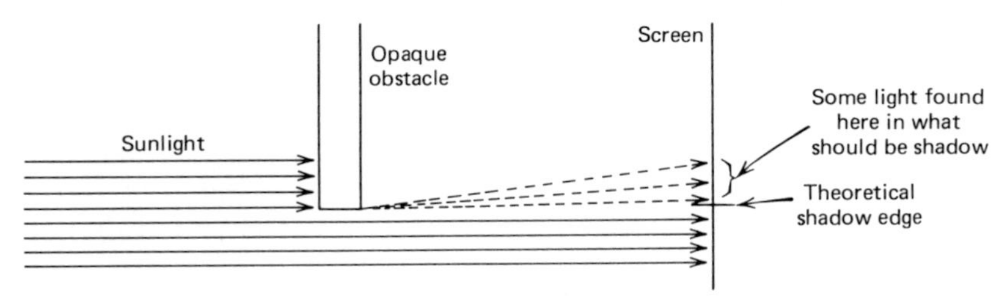

Figure 1.10 Diffraction is the slight bending of light around corners.

Newton's particle theory of light was able to account plausibly for all the known phenomena. For example, Newton explained the reflection of light as basically an elastic collision between the particles of light and a smooth mirror surface, much as a billiard ball rebounds from a table cushion. Likewise, Newton explained that refraction was caused by a slight impulse exerted on the particles of light as they crossed a boundary between two transparent media (Figure 1.19 and problem 2). Diffraction was supposed to be the result of a small attractive force between the light particles and any edge that they passed close to (Figure 1.10). And the colors in white light presumably were the result of the various different sizes of the particles of light. Newton was not unsympathetic to the wave theory of light, for he realized that it too could explain most of the known optical phenomena. The one major stumbling block was polarization. Newton's particle theory explained this by postulating that the particles of light were somewhat elongated. As the particles traveled, the elongated axis would be oriented crosswise to the direction of motion (Figure 1.12). Normal light had the

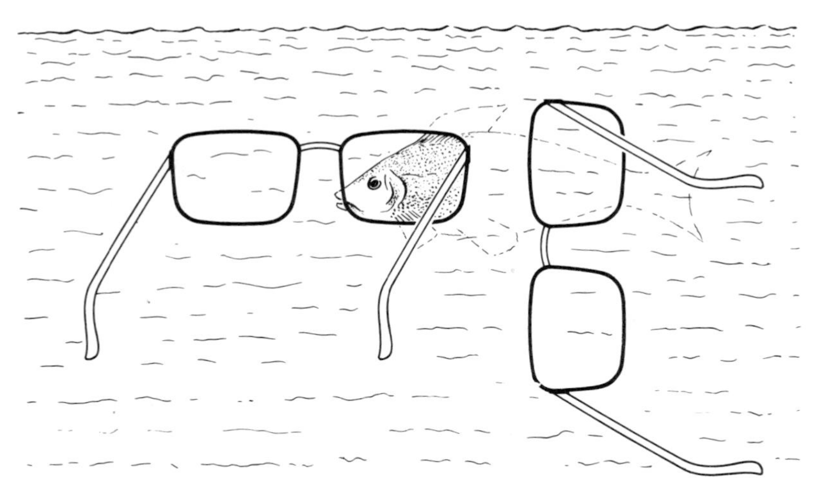

Figure 1.11 Polaroid sunglasses in two orientations. Notice that with the glasses on the left the glare from the water is eliminated and the fish is clearly visible. On the right, with the glasses rotated, the glare is still present.

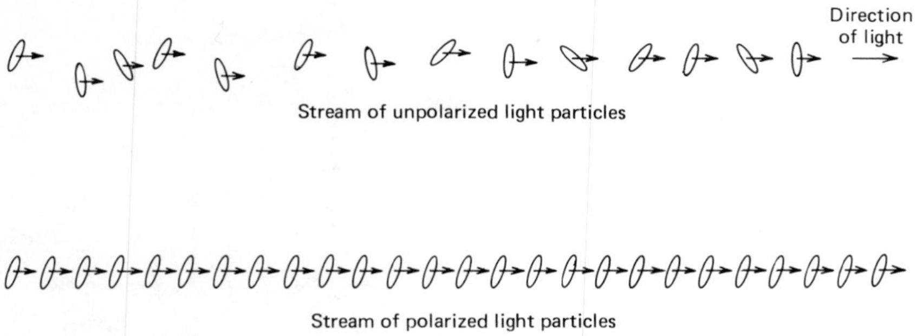

Direction of light

Stream of unpolarized light particles

Stream of polarized light particles

Figure 1.12 Newton's elongated light particles.

particles' long axes oriented randomly around the line defining the direction of propagation. However, polarized light was thought to consist of particles all oriented in the same way. The wave theorists had great difficulty in explaining polarization because they regarded light as a *longitudinal* wave. They insisted on a longitudinal wave because they thought that light was a mechanical wave traveling through some kind of tenuous material medium. From their point of view a solid medium and, thus, transverse waves seemed impossible. It's hard to disagree in view of the obvious nonrigid nature of air and of the space containing the solar system, both of which propagate light very nicely. Longitudinal waves simply cannot exhibit polarization. And thus Newton's particle theory triumphed for the time being.

1.6 THE TRANSVERSE WAVE THEORY OF LIGHT

The particle theory of light dominated physics for over 100 years. But in the early nineteenth century Thomas Young undertook a series of experiments that began to undermine the particle theory. In his most decisive experiment light passed through two very narrow slits, then striking a screen beyond (Figure 1.13). On the screen he observed alternating lines of light and dark, not just the two images of the slits one might expect. This reminds us very much of the two loudspeakers discussed earlier (see Figure 1.8). Young concluded that he was observing interference and that light must be a wave. Since he was mindful of the polarization problem, he proposed that light was a transverse wave, laying aside for a moment the question of the nature of the light medium. Many scholars vociferously opposed Young's idea, but Augustin Jean Fresnel, a young French physicist, followed up Young's suggestion. Fresnel developed a detailed mathematical theory of transverse light waves that was so spectacularly successful that the wave nature of light could no longer be denied. It was even possible to calculate the wavelengths of visible light. They turned out to range continuously from 700 nm (red) to 400 nm (blue), the different wavelengths representing the different colors of the visible spectrum. A nanometer (nm) is one billionth of a meter, so that light happens to have rather short wavelengths. The

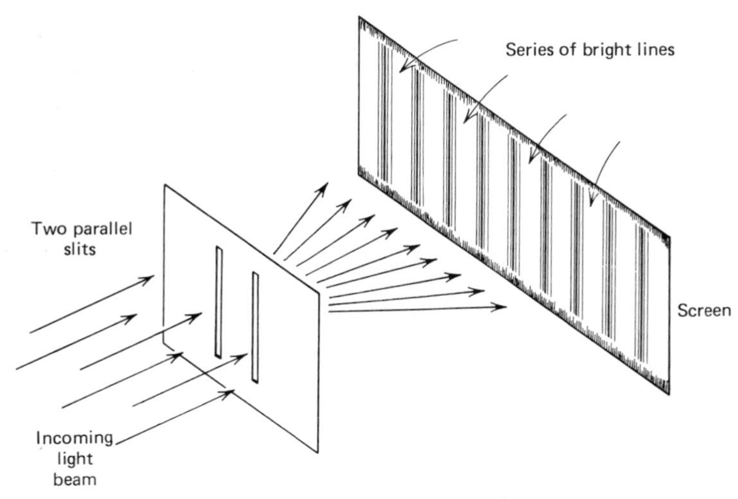

Figure 1.13 Young's double slit experiment. A beam of light of a single color is incident on a pair of parallel slits. On the far side of the slits is a screen. On the screen is seen a *series* of parallel bright lines instead of just the image of the two slits.

thickness of a pin is thus several hundred times the wavelength of visible light. The sort of medium that supported the propagation of light remained a mystery.

Physicists spent a great deal of time and effort speculating about the light medium. They felt that it had to be rigid to support transverse waves and yet somehow had to allow normal objects to pass through it unimpeded. Such a medium would be very strange indeed.

1.7 ELECTRIC AND MAGNETIC FIELDS

During the early nineteenth century while many physicists were struggling with the problem of the light medium, others were attempting to unravel the secrets of electricity and magnetism. Many basic phenomena involving electrically charged objects were known. For example, there seemed to be two basic kinds of charge called positive (+) and negative (−). Two like charges (both + or both −) were always found to repel each other according to a definite mathematical law. Unlike charges (one + and one −) always attracted each other. Similarly, magnets were found to interact with each other according to another mathematical law. One aspect of these phenomena that bothered some physicists was the "action at a distance" nature of the interactions. There seemed to be no direct physical link between two interacting charges. Michael Faraday, a young British physicist, was particularly troubled by the action at a distance view of electrical and magnetic interactions. He therefore invented a new concept called the line of force. These lines of force (which we now refer to as fields, or field lines) were seen as radiating out from charges or magnetic objects (Figure

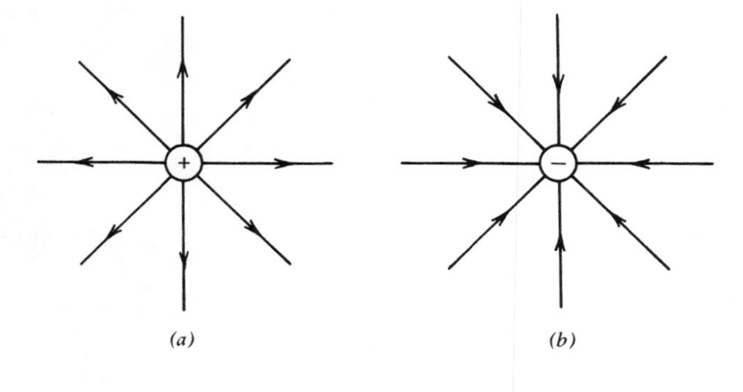

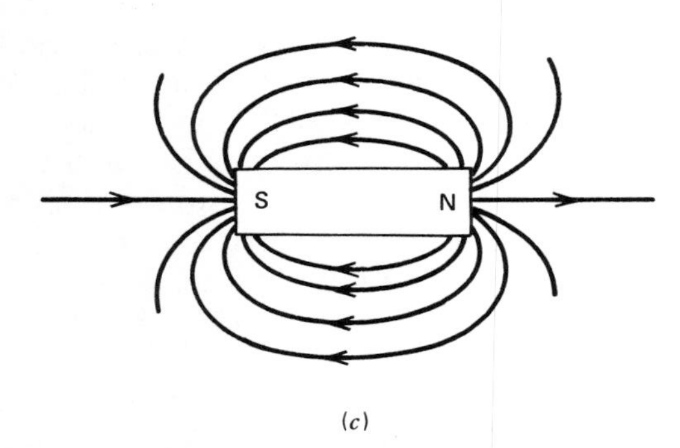

Figure 1.14 Electric and magnetic lines of force. (*a*) Electric lines of force emanating from a positive charge. (*b*) Electric lines of force radiating into a negative charge. (*c*) Magnetic lines of force around a bar magnet.

1.14), and were apparently thought by Faraday to represent tensions in a weak mechanical medium that pervaded all space. The important new point is that the lines of force became an intermediate agent between two interacting charges, thus removing the need for direct action at a distance.

An electrical interaction between two charges could now be visualized in the following way (Figure 1.15). One charge sets up lines of force all around itself. If a second charge is brought into the vicinity of the first, the existing lines of force from the first charge will produce a force on the second charge. Thus the two charges do not interact directly, but rather though their respective lines of force.

Faraday was primarily concerned with static situations. However, an interesting question arises. Suppose a charge is suddenly moved from one place to another. Do its lines of force follow along instantly, rearranging themselves

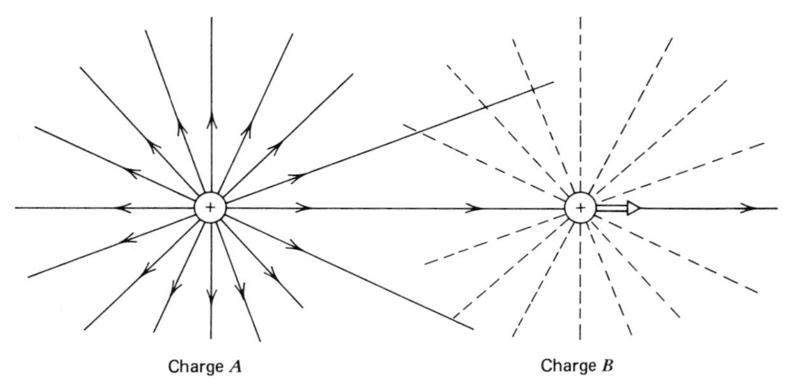

Figure 1.15 Line of force from charge *A* exerts force on charge *B*. Likewise, lines of force from *B* will exert a reciprocal force on *A*.

about the new location of the charge? Or is time required for this rearrangement to take place? The answer to this question was provided by the mathematical investigations of another British physicist, James Clerk Maxwell. Maxwell found that the lines of force, now called field lines, did indeed require time to rearrange themselves. When a charge was moved, all the electric field lines developed kinks that traveled outward away from the charge, rearranging the field lines as they went (Figure 1.16). The kinks, in effect, carried the message that the charge had been moved. Moreover, if a charge was moved periodically up and down, a continuous wave would be produced along the electric field line, and would travel at a speed equal to the *speed of light!* This astounding result led Maxwell to suggest that light was nothing other than an electric field wave (or more properly, an *electromagnetic* wave, since he also found that electric and magnetic waves *always* occur together). In 1888 Heinrich Hertz confirmed Maxwell's suggestion by an ingenious experiment that demonstrated the existence of electromagnetic waves.

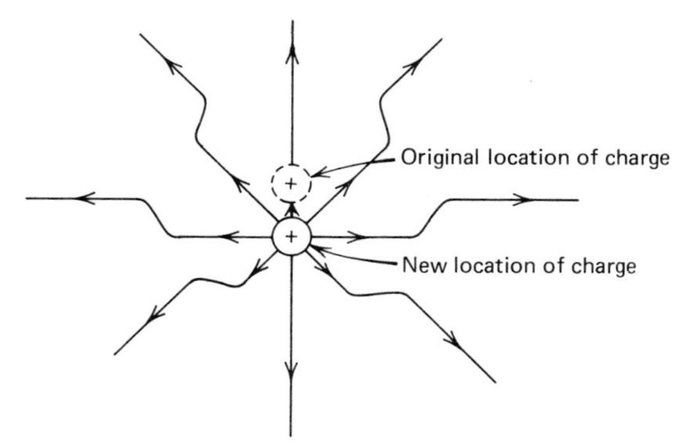

Figure 1.16 Temporary kinks in the lines of force of an electric charge that has just been moved.

Recall that the visible spectrum involves wavelengths in the range of 400 to 700 nm. Maxwell's electromagnetic waves were not limited to this narrow range but could have any wavelength whatever. Thus the visible spectrum apparently formed a very small part of much larger *electromagnetic spectrum* (Figure 1.17). Once it was realized that these other waves existed, they were looked for and soon found. Today we use such electromagnetic waves as radio, microwaves, infrared light, ultraviolet light, x-rays and γ-rays in a variety of scientific, commercial, and medical tasks. Although some of the names suggest otherwise, they are all part of the electromagnetic spectrum, differing only in wavelength.

As Figure 1.17 shows, electromagnetic waves differ from each other in wavelength. In empty space they all travel at the same speed, however, which we found in Section 1.2 was equal to 3×10^8 m/sec. This speed is usually denoted by the symbol c, which is called the speed of light in empty space. Therefore we can write equation (3) as

$$c = \lambda f$$

Thus each wavelength of light also corresponds to a definite frequency. For example, consider light of wavelength 600 nm. Then

$$f = \frac{c}{\lambda}$$

$$c = 3 \times 10^8 \text{ m/sec}, \lambda = 600 \text{ nm} = 6 \times 10^{-7} \text{ m}$$

So

$$f = \frac{3 \times 10^8 \text{ m/sec}}{6 \times 10^{-7} \text{ m}}$$

$$f = 5 \times 10^{14}/\text{sec}$$

$$f = 5 \times 10^{14} \text{ Hz}$$

Either frequency or wavelength can be used to describe a given light wave. Since different wavelengths *in the visible part of the spectrum* correspond to different colors, so also do different frequencies.

To make the triumph of the electromagnetic wave theory of light complete, it was necessary only to discover the details of the "medium" that presumably supported the waves. This question was finally resolved by Einstein in 1905

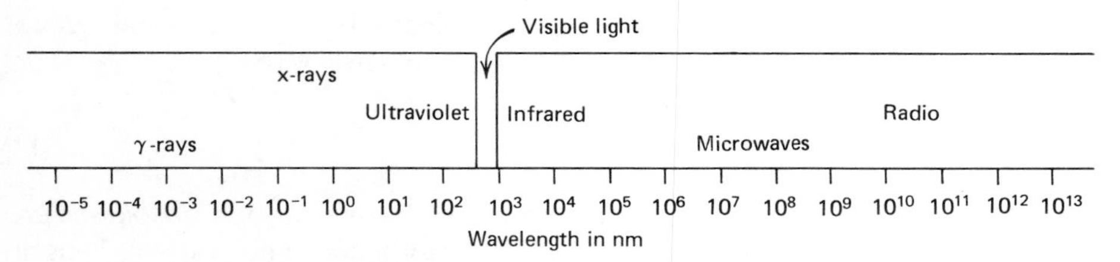

Figure 1.17 The electromagnetic spectrum.

when, in a momentous paper, he showed that *no medium was necessary* for the propagation of light. Electromagnetic waves, although they resemble mechanical waves in many ways, are a unique entity. A slight digression on the mechanical model of the universe should make this idea more understandable.

Until recently, most physicists thought about physical problems in terms of a mechanical model. They might discuss atoms or molecules or the electrical medium, but they visualized these things in mechanical terms. An atom might be seen as a tiny billiard ball, perhaps surrounded by springs, or it might be visualized as a blob of jello, or even a tiny center of force. The electrical medium was pictured as a kind of mechanical medium. It was thought that to explain any phenomenon, a mechanical explanation was required. This was perfectly natural since at the macroscopic level (the size at which we humans can see) things do appear to be mechanical in nature. In the twentieth century, however, it has become quite clear that at the microscopic level things are not mechanical, but rather must be described in their own terms. Thus, although light in many ways behaves like a mechanical wave, it is not a *mechanical* wave. For many purposes light can be described by wave terminology. However, as we shall now see, some aspects of light require a different explanation.

1.8 THE PHOTON THEORY OF LIGHT

In the same 1905 journal in which Einstein demonstrated that light required no medium he also proposed another startling idea. Light might behave like a wave, but it also behaved like a particle! This idea had an interesting origin. One of the more troubling problems in physics at the end of the nineteenth century was the theoretical description of how light radiated from a hot object. The problem revolved around a theoretical object known as a blackbody. Such an object was imagined to be a perfect absorber of light at all wavelengths, thus the name blackbody. Likewise, the object was expected to radiate a particular spectrum of wavelengths, the details of which depended on the temperature of the object. No real object was expected to behave exactly like this theoretical blackbody. However, for a real object at a fairly high temperature, physicists did expect the light emitted to approximate closely that predicted for a black-body at the same temperature. As Figure 1.18 shows, what they found was a significant discrepancy that grew larger at shorter wavelengths. In fact, for real objects at high temperatures there was always some wavelength at which more light was given off than at any other temperature, whereas in the theoretical case no such peak was expected. For real objects this so-called peak wave-length, λ_m, was found to vary with temperature according to

$$\lambda_m \propto \frac{1}{T} \tag{5}$$

where T is the temperature of the object in degrees *above absolute zero* (absolute zero is the temperature below which nothing can be cooled). Thus an object placed in a typical open fire would produce most of its light in the

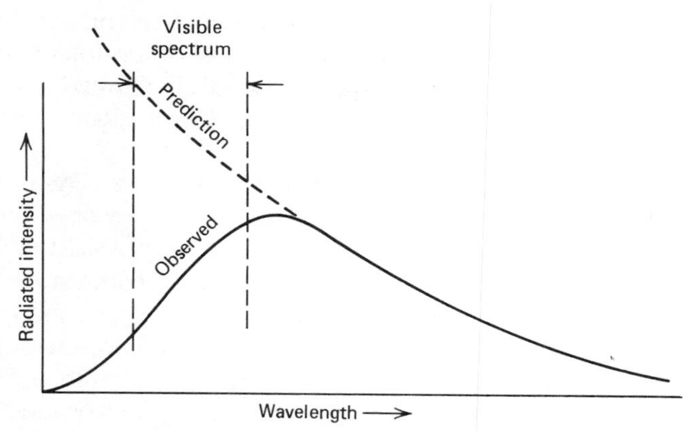

Figure 1.18 Difference between the actual light emitted from a hot object (solid curve) and that predicted for a blackbody at the same temperature (dashed curve). Note the two curves coincide at longer wavelengths but progressively deviate at shorter wavelengths.

infrared portion of the spectrum and would glow dull red. A much hotter object like the sun would have its peak wavelength in the middle of the visible spectrum and would appear white. An extremely hot object could even have its peak wavelength in the ultraviolet, and would take on a bluish color. The total radiant flux emitted by the hot object was found to vary as

$$F \propto T^4 \tag{6}$$

(Radiant flux is the sum of all the light emitted each second, and will be discussed more fully in Chapter 2.) The experimental facts contained in eqs. (5) and (6) plus the *shape* of the curve in Figure 1.18 needed theoretical explanation.

In 1900 the German physicist Max Planck was able to use an idea, which he personally considered an artificial mathematical device, to explain theoretically the observed facts of blackbody radiation. Planck assumed that electromagnetic radiation was produced in discrete units rather than in continuous amounts. It should be recalled (eq. 4) that the amount of energy carried by a mechanical wave (its intensity) is determined by the square of the wave's amplitude. And since classical physics supposed that a wave could have any amplitude whatever, there should be no restrictions on the energy content of a wave. Planck's idea was that when electromagnetic radiation of a given frequency, f, was produced, it was always done so in units each of which had an energy given by

$$\boxed{E = hf} \tag{7}$$

The constant h in this equation is now called Planck's constant, and has the same numerical value for electromagnetic radiation of any frequency. Here is the meaning of eq. (7). When light of frequency f is produced, it must be done so in units of energy *hf*. One unit can be produced, or two, or a million. But it is impossible to produce one and one-half units, or any other noninteger amount. Planck did not really believe in his idea even though it allowed him to provide a theoretical explanation of blackbody radiation. Albert Einstein, on the other hand, took the idea seriously.

In a paper published in 1905 Einstein showed that Planck's idea of light units (or photons as they are now called) could explain a puzzling phenomenon known as the photoelectric effect. In 1902 the German physicist Philipp Lenard noticed that if light of a definite color (frequency) illuminated a metal surface, electrons were ejected at a definite maximum velocity. Increasing the *intensity* of the light caused *more* electrons to be ejected but did not change the maximum velocity of the electrons. On the other hand, equally intense *different colored lights* (lights of different frequencies) produced electrons with *different maximum velocities,* the higher the frequency of light the higher the velocity of the electrons. The wave theory of light could not explain these facts, but Einstein realized that Planck's photon idea could. A beam of light of frequency *f* could be thought of as a stream of photons, each having energy *hf*. When this stream struck the metal surface, some of the photons would penetrate the metal and would be absorbed by individual electrons. The electrons would thus acquire the energy of the photons and, if an electron were close to the metal surface, it could use this energy to break free of the metal. Because the photons carried a definite amount of energy, the electrons would be limited in the speed they could acquire. Increasing the intensity of the light beam would simply mean providing *more* photons, each of which would still have the same energy *hf*. Thus *more* electrons would be ejected, but their speed would not change. But, changing the frequency of the light would clearly involve changing the energy of each photon. For example, *increasing* the frequency would *increase the energy, hf,* of each photon. This would in turn provide more energy to any electron that happened to absorb a photon. The electron would thus be ejected at a higher speed.

Einstein's explanation of the photoelectric effect was so clear and compelling that it rapidly gathered adherents and radically changed our conception of light. As we shall see later, the idea of energy being quantized (coming in discrete units) could also be applied to other areas of physics. In fact, Einstein's 1905 paper was really the beginning of a revolution in physics.

1.9 LIGHT AND ENERGY

The idea that the energy in a light beam consists of discrete amounts or photons simply highlights the fact that light is a form of energy. When we speak of energy, we mean the ability to cause transformations to occur. Thus ice can be changed to water, or an object can be set in motion by the addition of energy. Light is just one form of energy. There are many other kinds of energy as well, such as heat, chemical, electrical, nuclear, and mechanical energy. Whenever a transformation takes place in nature, energy is being changed from one form into another. One particularly important type of transformation from our point of view involves the conversion of light energy into chemical energy on the retina of the eye or in the film of a camera. In both cases, the energy contained in individual photons is used to produce chemical reactions that are necessary for the visual or photographic process. Both the eye and the camera use visible light photons to initiate the appropriate chemical reactions. Photons from the infrared, microwave, and radio parts of the spectrum do not contain sufficient energy to be effective. On the other hand, ultraviolet, x-ray, and γ-ray photons carry too much energy, and would be damaging, especially to the retina of the eye. In fact, x-ray and γ-ray photons carry so much energy that they are capable of damaging cells anywhere in the human body. This is why exposure to these kinds of radiations must be kept to a minimum.

IN CONCLUSION

We return again to the question of the physical nature of light. Is light a wave or is it a particle? This is a semantic trap into which we need not fall. The classical ideas of waves and particles are mechanical in nature. Light is *not* mechanical. Its properties are uniquely its own, and simply resemble in some ways those of waves and those of particles.

Light can be described by a frequency, f, and a wavelength, λ. It always has the same speed in empty space, usually denoted by the symbol c. Photons can interfere with each other like waves, and can diffract around corners like waves. Photons can be polarized like transverse waves. Light has so many properties that resemble mechanical waves that we frequently use the term "light waves" throughout the book. But we must never forget that these waves are really streams of individual photons, each with a definite energy. There will be many occasions when the photon theory of light will provide the only reasonable explanation of the phenomenon under discussion.

One more point should be made. We have developed several concepts in this chapter to give the reader a clear idea of the physical nature of light. Some of these ideas, especially geometrical and wave optics, will be much more fully developed, primarily in Chapters 6, 7, and 8. Our immediate task, however, and the subject of the next four chapters, is to investigate color phenomena.

PROBLEMS AND EXERCISES

1. When Roemer observed the eclipses of the moons of Jupiter, he found that these eclipses became more and more delayed as the earth moved toward its maximum distance from Jupiter (Figure 1.1*b*). This was due to the time required for light to traverse the diameter of the earth's orbit. We now know that the speed of light is 3×10^8 m/sec and the diameter of the earth's orbit is about 3×10^{11} m. Using this information, how long would the time delay observed by Roemer have been?

2. In the seventeenth century controversy between the particle theory and the wave theory of light, there was one area where the two theories made diametrically opposite predictions that could, in principle, be tested. This had to do with the phenomenon of refraction (Figure 1.9). Newton's particle theory of light explained the bending of light as due to a slight impulse exerted by the more dense medium on the particles of light as they crossed the boundary between the two media (Figure 1.19*a*). This would cause the speed of the particles (and thus the speed of light) to *increase* as they entered the more dense medium. On the other hand, the wave theory explained the bending of light as due to a shortening of the wavelength of light as it entered the more dense medium (Figure 1.19*b*). Such a shortening implied a *decrease* in the velocity of light.

(a) Explain, using eq. (3) and the fact that the frequency of the wave stays constant across the boundary (why?), why a reduction in the wavelength implies a decrease in wave speed.

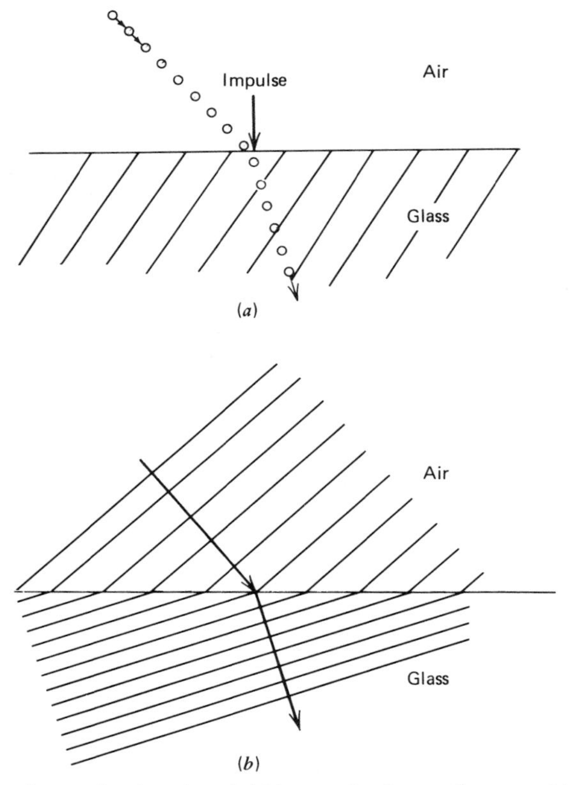

Figure 1.19 Two views of refraction. (*a*) Newton's view – the speed increases in glass. (*b*) Wave theory's view – wavelength (and thus speed) decreases in glass.

(b) Why do you suppose that no direct experiment was done to test whether the speed of light increased or decreased as light went from air to a more dense medium such as glass?

3. If the frequency of a wave is 10 Hz, what is the period of the wave? (The abbreviation Hz stands for a *hertz,* which is one cycle or oscillation per second.)

4. A given wave has a wavelength of 10 cm and a frequency of 200 Hz. What is the speed of the wave?

5. The speed of sound on a particular planet is found to be 800 m/sec. A normal person can hear sound in the frequency range of 20 to 20,000 Hz. What wavelength range is this?

6. A given wave has its amplitude increased until the wave becomes 100 times as intense as it originally was. How much has the amplitude of the wave increased?

7. A typical Am radio frequency is 1000 kHz (one kHz = one kilohertz = 1000 Hz). What wavelength does this correspond to? How about an FM radio frequency of 100 mHz (one mHz = one megahertz = one million Hz).

8. Equation (5) describes the location of the peak of the curve in Figure 1.18. A typical tungsten light operates at nearly 3000°K and produces the greatest amount of light at the infrared wavelength 1100 nm. A human body has a temperature of just over 300°K. What is the peak wavelength emitted by the human body?

9. Show that eq. (7) can be rewritten in the form

$$E = \frac{hc}{\lambda}$$

where c = the speed of light. [*Hint:* Look at eq. (3).]

10. One of the important consequences of eq. (7) is that any type of electromagnetic radiation is made up of discrete units of a definite size. When these photons interact with matter, they can cause significant changes to occur if the energy of the photons is sufficiently high. In the human body, an energy of about 1 eV can cause molecular damage. Assuming that

$$h = 4.1 \times 10^{-15} \text{ eV-sec}$$

calculate the energy of photons of the following wavelengths:

2 m (radio wave)
600 nm (visible light)
5 nm (ultraviolet)
0.1 nm (x-ray)
0.001 nm (γ-ray)

2 THE ORIGIN OF COLOR

In the last chapter we discussed the physical nature of light in detail. However, other than a brief description of Newton's experiments with a prism, we did not investigate the connection between light and color. This is a subject that we must approach with care.

We are able to see because light comes to our eyes from various objects in the field of view. The spectral composition of the light coming from different objects is generally different (by spectral composition we mean the relative amounts of light from different parts of the visible spectrum). It is the differences in the spectral composition of light from various objects that is responsible for the different, perceived colors. Our task would be easier if we could say that a particular spectrum of light always produces the same color sensation. As we shall see in Chapter 4, however, this is simply not true. The same spectrum of light can produce a variety of color sensations under different circumstances. Conversely, a given object will appear to retain its color even when the room illumination is changed in such a way that the object reflects to our eyes a different spectrum of light. Thus, even though there is a strong connection between spectral composition and perceived color, the connection is not exact. With this qualification in mind, let us now proceed to investigate the relationship between light and color.

2.1 THE TERMINOLOGY OF ILLUMINATION

Most of the time, the light coming to our eyes from different objects is reflected light. On the other hand, we occasionally see light coming directly from a light source. In either case we need some way to describe the quantity of light that comes to our eyes from various portions of the field of view. This is usually done in one of two ways. The first approach relies on the fact that light is a form of energy. Thus light coming from a given direction can be regarded as a stream of energy. A description of light in terms of energy utilizes physical units.

The second approach used to describe the quantity of light coming from a particular object concentrates on the visual effect produced by the light. Some wavelengths of light produce greater visual effects than do others. When these effects are taken into account, the resulting system of units is called luminous units.

PHYSICAL UNITS

Since physical units depend on energy for their definition, we must first understand how to measure energy. As we said earlier, energy is the ability to cause transformations. Thus, energy is required to lift a heavy object, to melt ice, or to break a stick. One common unit of energy is the joule (J), which is the energy required to lift a one kilogram object approximately 10.2 cm at sea level. Another very much smaller unit of energy is the electron-volt (eV) which is given by

$$1 \text{ eV} = 1.6 \times 10^{-19} \text{ J}$$

Such a tiny unit is more appropriate for measuring the energy of a single photon of visible light. For example, the energy of a single photon of red light of wavelength 650 nm is about 1.9 eV.

Using energy units, we can speak of quantities that measure the amount of light radiating from an object or striking a surface. The *radiant flux* is the amount of light energy that leaves an object or is received by an object each second. It is thus measured in joules/second (1 J/sec is also called a watt). Thus a 100 watt light bulb radiates a total of 100 joules of light energy each second. We can also speak of the radiant flux emitted or received in a particular wavelength band. For example, a 100 watt light bulb might only radiate 5 watts of radiant flux between the wavelengths 400 and 700 nm (the visible spectrum), the remaining 95 watts being radiated at wavelengths longer than 700 nm. It is also common to refer to the radiant flux emitted in a very narrow wavelength band, say between 549.5 and 550.5 nm. This might be called the radiant flux *per nm* at the wavelength 550 nm.

Radiant flux refers to the *total* light energy emitted or received by an object each second from all directions. It is often necessary to know how much light is being emitted in a given direction. This quantity is measured by the *radiant intensity*. This is the radiant flux emitted by an object per unit solid angle in a given direction. As with the radiant flux and with other physical units as well, it is also possible to refer to the radiant intensity in a beam of light within a specified band of wavelengths.

The radiant intensity measures the total light emitted from an object in a given direction. It is often necessary to know the radiant intensity of light coming from a unit area of an object, for example, from a square centimeter of surface. This quantity, called the *radiance,* is a measure of how much light is available to form an image on the retina of the eye or on the film in a camera. For convenience, throughout the rest of the text, we shall use the more common term *intensity* when referring to radiance.

Finally, it is often desirable to describe the quantity of light that strikes *each unit area* of a surface. The *irradiance* is the radiant flux incident on a unit area of a given surface. Thus it can be measured in watts/cm^2, for example. A surface that was illuminated to have an irradiance of 2 watts/cm^2 would be receiving 2 joules of energy each second on each square centimeter of surface.

LUMINOUS UNITS

The only real difference between luminous units and physical units is that luminous units take into account the sensitivity of the eye to different wavelengths of light. That is, if we shine two different monochromatic (pure color) beams of light of different wavelength but *equal radiant intensity* into the eye, they will not in general look equally bright. Luminous units take this fact into account by scaling the radiant intensity *at each wavelength* up or down according to the sensitivity of the eye to each wavelength. Thus it is possible to have two light sources that produce the same total radiant intensity in a given direction but, because each contains a different distribution of radiant intensities at different wavelengths, do not look equally bright.

In arriving at a specific set of numerical values for luminous units, it is necessary to choose a standard luminous object and to define arbitrarily its brightness. This is done with a device that utilizes platinum heated to the melting point. The apparent brightness of the device is then used to define luminous units.

The resulting unit of *luminous intensity* is the candela. This meausres the apparent brightness of a beam of light heading in a specific direction. The luminous intensity, and all other luminous units as well, can also be specified for a band of wavelengths. Thus a source may have a total luminous intensity of 10 candelas in a particular direction, but between the wavelengths of 500 and 510 nm the luminous intensity may amount to 0.5 candelas.

The luminous intensity, like the radiant intensity, measures the light emitted from an entire object in a given direction. More closely related to the visible brightness of an object is the luminous intensity emitted per unit area from the object's surface in a given direction. This is called the *luminance,* and is measured in candelas/cm^2. It is the luminance that determines how bright an object actually appears.

The total luminous intensity emitted from a source in all directions is the *luminous flux.* This is comparable to the physical unit radiant flux. The luminous flux is measured in lumens. A light source that emits a luminous intensity of one candela uniformly in all directions would have a luminous flux of 4π lumens (about 12.56 lumens). Most light bulbs are rated in lumens since this measures how effective the bulb is in producing *useful* illumination. Two bulbs with equal lumen ratings will produce equal amount of illumination as perceived by most observers. These bulbs might *not* necessarily produce equal amounts of *radiant* flux, however. Conversely, two 100 watt light bulbs (both producing a radiant flux of 100 watts) might have different luminous flux ratings. For example, one might be rated at 1400 lumens while the other might be rated at 1900 lumens. Even though both radiated the same amount of energy each second (100 J/sec), the 1900 lumen bulb would look brighter.

The final unit of interest is the *illuminance,* which is the luminous flux incident on a unit area of surface. The illuminance is measured in lumens/m^2 (the lux) or lumens/ft^2 (the foot-candle), and is a measure of the apparent illumination, as seen by the eye, that falls on a surface.

We can summarize these results as follows:

	Physical Units	Luminous Units
Total light output	Radiant flux (watts)	Luminous flux (lumens)
Light emitted in a specific direction	Radiant intensity (watts per steradian)	Luminous intensity (candelas)
Light emitted from a unit area in a specific direction	Radiance (watts/cm^2 per steradian)	Luminance (candelas/cm^2)
Light striking a unit area	Irradiance (watts/cm^2)	Illuminance (lumens/m^2)

With these units clearly in mind we can now turn our attention to a discussion of various kinds of light sources (which we also refer to as *illuminants* when they are used to illuminate something).

2.2 CONTINUUM SOURCES

Many light sources produce light of every wavelength in the visible spectrum. When light from such a source is analyzed by a prism (Figure 2.1) the full visible spectrum is seen with *no gaps* (Color Plate 1), thus the name "continuum" spectrum or continuum source. Some continuum sources produce far more light in some parts of the spectrum than in others and, thus, appear highly colored. There are, however, common continuum sources such as sunlight, incandescent light, and some fluorescent lights, that produce relatively balanced outputs (Figure 2.2). Such sources are generally referred to as white light sources since they do not appear to distort significantly the color of objects they illuminate. Figure 2.2 shows clearly, however, that the spectra of these sources are quite different, and experience shows that objects do look slightly different under these different illuminations. Thus, although we may loosely refer to sunlight, incandescent light, and fluorescent light as white, it would be nice to have a "per-

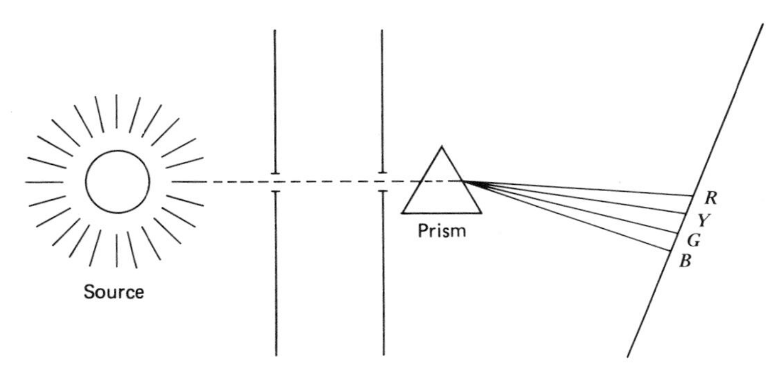

Figure 2.1 Method for separating the light in any source.

fectly'' white source. Such a perfect source would have no tendency to enhance any portion of the spectrum and, thus, would contain equal intensities of all wavelengths. The spectrum of such an ''equal energy'' source is shown in Figure 2.3. Although no such real source exists, it is often convenient to discuss how a scene would look if it were illuminated by an equal energy spectrum.

BLACKBODY RADIATION

A very important class of continuum sources is represented by objects that are heated to incandescence. This subject was discussed briefly in the last chapter, and the sources of light were referred to as blackbody radiation. Although it is not commonly realized by most people, every object radiates some light. Relatively cool objects like a chair in your room or a classmate for that matter

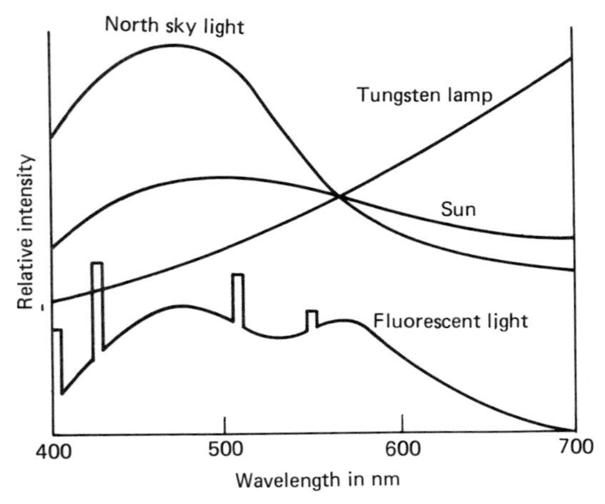

Figure 2.2 Typical spectral distribution curves for some common continuum white light sources. These curves tell us how much light each source produces at every wavelength in the visible spectrum.

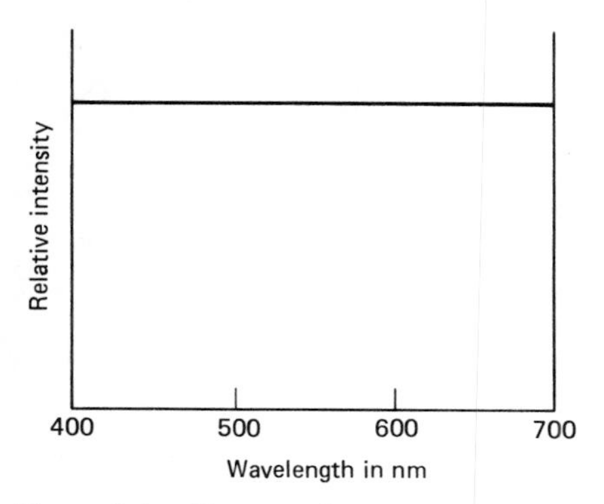

Figure 2.3 The equal energy spectrum.

radiate so little light that you are not aware of it. But consider an iron poker in a fire, the filament of an incandescent bulb, or the sun. Each of these objects is hot enough to radiate substantial amounts of visible light. The emission of light from hot objects follows a simple law, known as Planck's law after the German physicist Max Planck who first successfully described it. Planck's law is stated in terms of the *absolute temperature* of an object, the number of degrees the object is above *absolute zero.* Scientists usually use the Kelvin temperature scale in stating Planck's law since this scale starts at absolute zero and uses the same size degrees as the Celsius scale. On the Kelvin scale 273° is the freezing point of water and 373° is the boiling point of water. Room temperature is about 300°K, the filament of a light bulb about 2700°K, and the surface of the sun about 5800°K. Although a complete statement of Planck's law must be done mathematically, we can get the general idea of it with the help of a few curves on a graph.

Figure 2.4 shows the spectrum of light emitted from identical perfect black-bodies at 2000°K, 4000°K, and 6000°K. The visible spectrum is clearly delineated, but many additional wavelengths of the electromagnetic spectrum are also shown. First, notice that as the temperature is increased, a great deal more light is emitted. Of course, the usefulness of a hot object as a light source depends on the light emitted in the visible portion of the spectrum. Figure 2.4 clearly demonstrates that the visible output of light increases dramatically as the temperature is increased. Second, notice that each curve has a peak occurring at a definite wavelength. This peak moves to the left, toward shorter wavelengths, as the temperature is increased. At 2000°K, for example, the peak is in the infrared portion of the spectrum. This means that in the visible portion of the spectrum a great deal more red light is emitted than anything else. At 6000°K, on the other hand, the peak is right in the middle of the visible spectrum. There is a much better balance of wavelengths. If we consider still higher temperatures, the peak would be in the ultraviolet portion of the spectrum, and the visible spectrum would be dominated by blue. One interesting group of hot

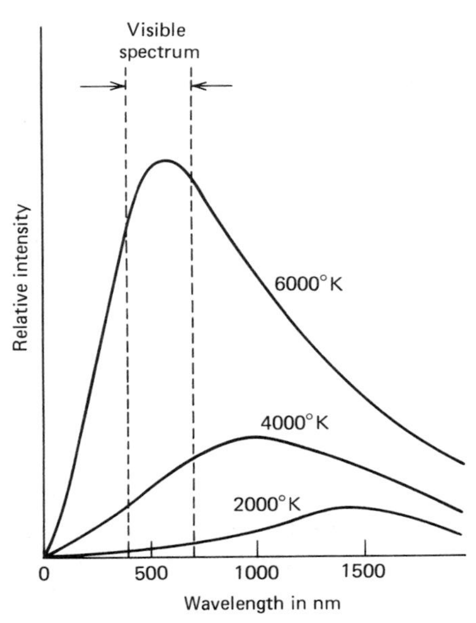

Figure 2.4 Continuous spectrum emitted by three identical blackbodies at three different temperatures.

objects that actually have colors ranging from red through white to blue are stars. In fact, by looking at the color of a star, astronomers can determine its temperature.

Why are we interested in blackbody radiation? Because a large number of lights in everyday use closely approximate Planck's blackbody radiation law. In fact, most commercial incandescent lamps are given a "color temperature" as one way of describing the kind of light they produce. What this means is that a lamp with a 2700°K color temperature produces approximately the same spectrum of light as a blackbody at 2700°K. (Hot objects in reality do not exactly follow Planck's law; only perfectly black objects do.) Anyone who deals with color must pay attention to such things as color temperature because, for example, a low color temperature means the light will have an overabundance of red. This may not be desirable in some cases. The sun, on the other hand, has a color temperature of 5800°K, which is a well-balanced white light. North light from the blue sky has a much higher color temperature than direct sunlight. This is because north light contains an overabundance of blue and an object must be very hot to have a blackbody spectrum dominated by blue.

For many years mankind derived virtually all of its light from hot objects of one kind or another. In the course of the twentieth century, however, light sources of an entirely different kind have been developed. We shall describe the most common of these sources in the next section.

2.3 BRIGHT-LINE SOURCES

Although incandescent lights are the most common type of light in the home, another class of light sources is in wide use in modern society. These lights are known as bright-line sources for reasons that will soon become obvious. Included in this class of illumination are sources such as neon lights and other advertising signs, mercury and sodium vapor lamps, and fluorescent light in certain features of its spectrum. In terms of the visual perception produced, there is no way to distinguish a brightline source from a continuum source. That is, if you look directly at a bright-line source (without the aid of a prism) it will look just like any other light source. Continuum and bright-line sources can only be distinguished with the aid of some device, like a prism, which separates the different wavelengths.

To understand what is different about bright-line sources, it is necessary to examine the light they emit with a spectroscope. The operation of a spectroscope is described in detail in Appendix B. Basically, a spectroscope uses a prism to spread out a beam of light so that the wavelengths present can be seen. In the later nineteenth century investigators studied many different bright-line sources in this way. The light studied could be characterized according to what was seen, namely, a spectrum consisting of a series of sharp bright lines with dark gaps between them. Examples of such spectra are shown in Color Plate 2. In other words, the spectra of light emitted by these sources did not consist of all wavelengths, or even many wavelengths, but rather a definite set of individual wavelengths that was unique to the element or compound being studied.

The typical construction of these sources consisted of a clear glass tube that contained the material to be studied in a gaseous state at low pressure. When a high voltage was established across the ends of the tube, an electric current flowed through the gas causing it to glow. The most well-known source of this type used hydrogen as the gas in the tube.

Physicists in the nineteenth century were at a loss to explain how any material could produce a bright-line rather than continuous spectrum. Since hydrogen is the simplest of elements, it was studied in great detail in the hope that understanding its spectrum would lead to greater insight into the nature of the spectra of more complex elements. J. J. Balmer, a Swiss school teacher, found that the wavelengths of the visible bright lines of the hydrogen spectrum formed a regular mathematical progression. However, he could find no underlying physical explanation for this relationship. Others studied the lines in the infrared and ultraviolet portions of the hydrogen spectrum and found that the wavelengths of these lines formed mathematical progressions similar to what Balmer had found. However, no real progess was made in understanding the reason for the bright lines until early in the twentieth century. Then events moved quickly.

First, Rutherford developed the idea of the planetary or nuclear atom. In this view the atom is mostly empty space, like the solar system. At the center is the nucleus with most of the mass and a definite number of protons, each carrying a single positive charge. In orbit around the nucleus are the tiny electrons, each with a negative charge. The number of protons and electrons are always equal in any neutral atom. This model of the atom was very compelling and, in fact, is essentially the picture we use today. But there were certain problems that arose almost immediately. For example, according to classical electromagnetic theory, when the electrons orbitted the nucleus they should continuously radiate energy and eventually spiral into the nucleus. Since this clearly does not in fact occur, some new idea was needed. A second problem was the inability of the nuclear model to explain the details of the bright-line spectra.

Faced with this dilemma Niels Bohr, a Danish physicist, in 1913 set forth a model of the hydrogen atom that provided an explanation of the origin of bright-line spectra. His model essentially ignored the problem of the electrons spiraling into the nucleus; he postulated simply that this did not occur if the electrons occupied certain well-defined "allowed orbits." The location of these allowed orbits was determined by a set of rules that governed the energy an electron was permitted to have as it orbitted the nucleus. In this model the electron could exist, at least temporarily, in any of the allowed energy levels, but nowhere else. Under normal circumstances, the electron would occupy the lowest level closest to the nucleus (the circle labeled "1st level" in Figure 2.5).

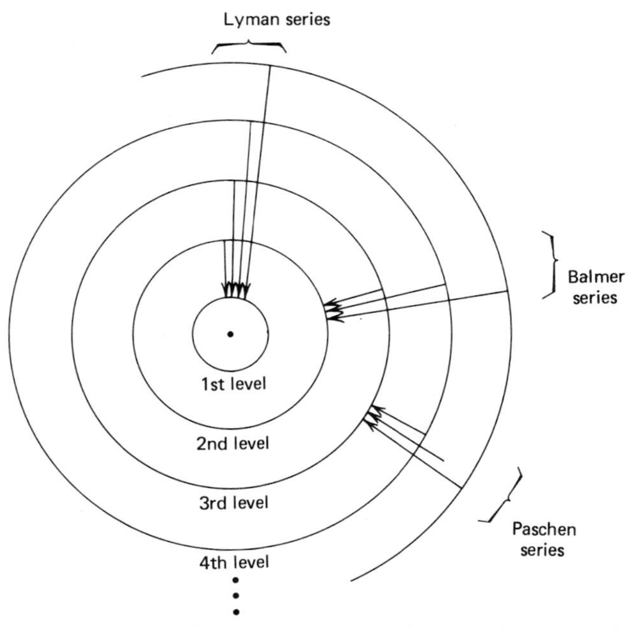

Figure 2.5 The Bohr model of the hydrogen atom showing the energy levels and some of the possible transitions.

However, if given sufficient energy by some means, for example electrical stimulation, the electron could be caused to move to one of the higher levels. This situation would be unstable, however, because the electron would be in what is called an "excited state." Such an excited electron would normally fall back to a lower level, ultimately ending up in the lowest level or ground state. In the process it would lose energy in the form of light each time it fell to a lower level.

Since Bohr's model clearly stated the energy of each level, the energy difference between each level could be calculated. Thus, if an electron made a transition from one level to another and if it was assumed that all of the energy released was emitted as one photon, the frequency (or wavelength) of the emitted light could be calculated by using Planck's quantum law, which we discussed in Chapter 1. For example, Figure 2.5 is a picture of Bohr's model of the hydrogen atom. If an electron falls from the third level back to the first level, the energy radiated as light is given by

$$\text{Energy radiated} = \text{energy of 3rd level} - \text{energy of 1st level}$$

or.

$$E = E_3 - E_1$$

The frequency of the photon emitted can then be found from Planck's law:

$$E = hf$$

or

$$f = \frac{E}{h} = \frac{E_3 - E_1}{h}$$

The corresponding wavelength can then be found using eq. (3) from Chapter 1.

$$\lambda = \frac{c}{f} = \frac{hc}{E} = \frac{hc}{E_3 - E_1}$$

In problem 6 at the end of the chapter the value of hc is given as 1240 eV-nm. Furthermore, problem 6 also contains Bohr's rule for calculating the energy of each hydrogen energy level. From this rule we find that

$$E_3 = 12.09 \text{ eV}$$

and

$$E_1 = 0 \text{ eV}$$

Therefore

$$E_3 - E_1 = 12.09 \text{ eV}$$

Then

$$\lambda = \frac{hc}{E_3 - E_1} = \frac{1240 \text{ eV-nm}}{12.09 \text{ eV}}$$

$$\lambda = 102.6 \text{ nm}$$

Thus this particular transition produces light in the ultraviolet portion of the spectrum.

With Bohr's formula for the energy levels of hydrogen it is possible to calculate all the wavelengths of light that can be produced by a gas of hydrogen. When this is done it is found that the resulting wavelengths do not fill the spectrum but, instead, are rather widely spaced. In fact, in the visible portion of the spectrum only four wavelengths appear (Color Plate 2). A number of other wavelengths are produced in the ultraviolet and the infrared regions of the spectrum. Thus the spectrum consists of individual bright lines rather than a continuum of color.

Bohr's model was quite successful in predicting the numerical values of the energy levels of hydrogen and, thus, the details of the hydrogen spectrum. The numerical success of the model was not good in explaining more complex atoms containing several electrons. Nevertheless, the idea that atoms are organized internally into discrete sets of energy levels has proved to be correct. Each element has its own unique internal structure and thus its own unique bright-line spectrum.

As successful as Bohr's model was, it still left unexplained exactly why special energy levels existed within the atom and why electrons in levels did not radiate. It also failed to predict the relative intensity of the various bright lines. These failures have been remedied in the more complex modern quantum theories of the atom. However, Bohr's model still serves as a useful picture in describing the origin of bright-line spectra.

In viewing a bright-line source, we do not normally observe the bright lines directly because our eyes blend or mix the colors together; hence we see colored or perhaps even white light. As with hydrogen, it frequently occurs that some of the light is produced in the ultraviolet region of the electromagnetic spectrum. This light, while not directly visible, can be used to produce other visible light.

Fluorescent lights make use of this principle to produce "white light." Careful examination of a fluorescent light tube shows that it contains small particles of mercury inside a tube coated with phosphors. Phosphors are chemical substances that have an internal energy structure somewhat similar to that of atoms except that the energy levels have been broadened into bands. Thus phosphors usually produce bands of colors in the spectrum rather than sharp bright lines. In a fluorescent light, a high voltage is used to produce a current that vaporizes the mercury and excites the electrons within the mercury atoms. The excited electrons fall back to lower levels and, in doing so, they release the energy they have captured, thus giving off a bright-line spectrum. Some of these lines are in the visible region of the spectrum but others are ultraviolet. These UV lines are of proper energy to cause the phosphors, which coat the inside of the tube, to become excited. The deexcitation of these phosphors is what produces the continuous spectrum of the fluorescent light. It should be evident that in order for excited phosphors to produce *visible* light, light of *higher* energy must be used. Thus ultraviolet light can cause the production of visible light, but infrared light could not.

A spectral energy distribution curve for a typical fluorescent source is shown in Figure 2.2. As is indicated by the diagram, this type of light has much greater intensity in the blue end of the spectrum than in the red. The visible bright lines of the mercury discharge show up as peaks on the continuous spectrum produced by the phosphors. Different spectral compositions can be achieved by choosing different combinations of phosphors and a close approximation to "daylight white" can be approached if required. The major difficulty in securing such an energy distribution is in strengthening the radiation at the red end of the spectrum. This can be done by adding appropriate phosphors to the tube's interior coating to acquire more balance in the light. Unfortunately, the inclusion of these red phosphors tends to reduce the luminous efficiency of the lamp, which is one of its major advantages over the incandescent lamp. However, if one needs good color rendering this sacrifice must be made. "Balanced fluorescent" bulbs are commercially available from different companies.

Other "vapor-type" lamps are now in widespread use mainly because of their efficiency in producing visible energy from the electrical energy they consume. These lamps usually contain sodium or mercury to produce the ultraviolet radiation that "excites" the phosphors. While their luminous efficiency is high, their poor color rendering means that they are not used where color viewing is important.

2.4 REFLECTANCE

We said earlier that in most viewing situations both the spectral content of the illuminant and the selective reflectivity of objects in the field of view have an effect on the perceived color. With white light illuminants such as sunlight, incandescent light, and fluorescent light, the selective reflectivity of various objects is the primary determinant of color. (Refer to Color Plate 3, which shows the reflectance spectra of several colors.) If the illuminant is sufficiently unbalanced, however, it too can play a major role in determining what is seen. In this section, we discuss the way in which the selective reflectivity of an object is described (the so-called reflectance curve). With this knowledge we can then determine the exact spectrum of light an object will reflect when it is illuminated by a particular source.

When light is incident on any material, three things generally occur. Some of the light will be reflected, some may be transmitted through the material, and some will be absorbed within the material. The degree to which these phenomena take place depends on the nature of the material and the particular wavelength of light being used. Opaque materials will not transmit any light; thus, for these kinds of objects only reflection and absorption need be considered. Translucent objects reflect, transmit, and absorb light. Generally, we describe the optical properties of a material by three graphs that show the *percentage* of incident light reflected, transmitted, and absorbed at each visible wavelength. Such a series of graphs is shown in Figures 2.6*a*, 2.6*b*, and 2.6*c*. These graphs happen to describe the properties of a transparent piece of red glass. Notice that at every wavelength, the percentage of light reflected, transmitted, and absorbed always adds up to 100%. This is just an example of the principle of conservation of energy. That is, we must account for all the light energy that is incident on the red glass as either being reflected, transmitted, or absorbed.

The curves in Figure 2.6 do not tell us the total amount of light that is reflected, transmitted, or absorbed; they tell us only the *percentage* of incident light that is reflected, transmitted, or absorbed. To determine the *absolute* amount of light that is reflected, for example, we must know not only the reflectance curve but also the intensity of incident light at each wavelength. As an illustration, consider the reflectance curve shown in Figure 2.7 for a blue surface. This particular material is very effective at reflecting from the blue end of the spectrum, but is very poor at reflecting long wavelength light. Suppose this sur-

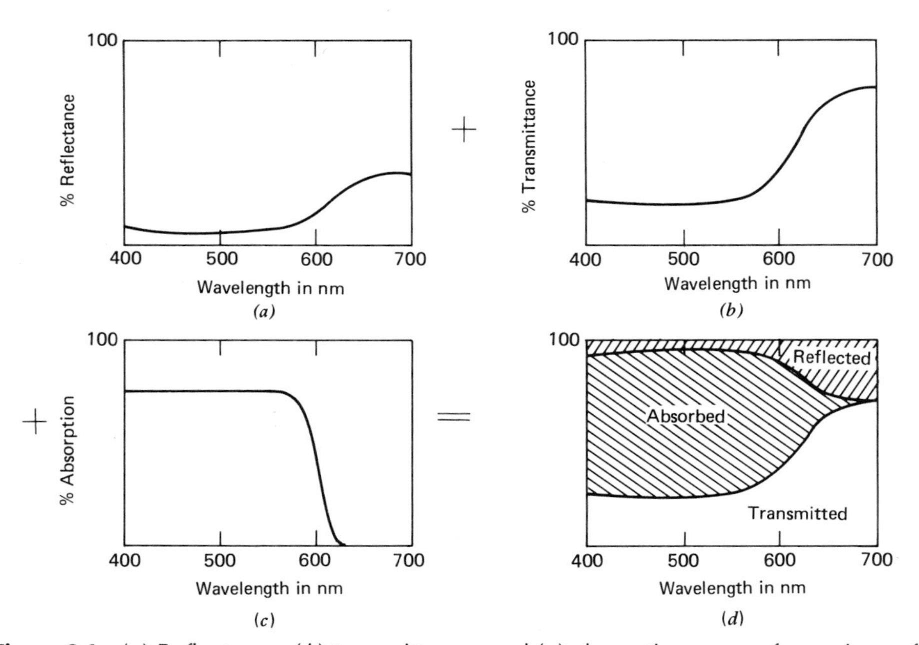

Figure 2.6 (*a*) Reflectance, (*b*) transmittance, and (*c*) absorption curves for a piece of red glass. Note that at each wavelength these three curves add up to 100% (*d*).

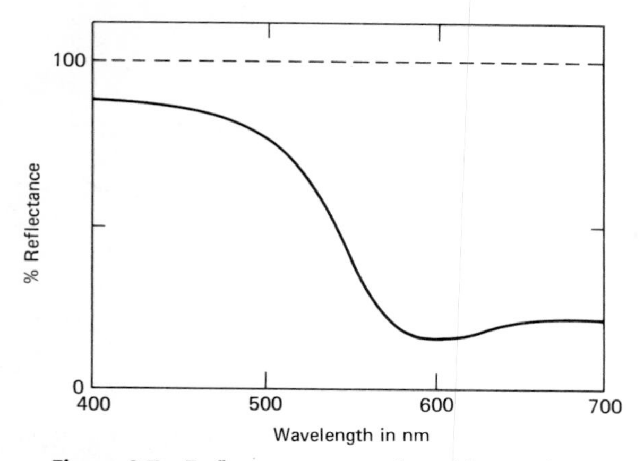

Figure 2.7 Reflectance curve for a blue surface.

face is illuminiated by a source with the spectral content shown in Figure 2.8*a* (which is highly schematic and impossible to actually produce). In the wavelength range from 400 to 500 nm the blue surface reflects about 80% of the light that strikes it. The illuminant in Figure 2.8*a* contains about 5 units of intensity at each wavelength in this range. Thus the absolute intensity of reflected light is (0.8) × (5 units) or 4 units of reflected light at wavelengths between 400 and 500 nm. This is shown in Figure 2.8*b*. In the wavelength range from 500 to 600 nm, the percentage of light reflected is very different at different wavelengths. However, our source contains *no light* in this wavelength range. Thus *no light* is reflected between 500 and 600 nm. Finally, between 600 and 700 nm, approximately 20% of the incident light is reflected at each wavelength. Figure 2.8*a* shows that in this wavelength range the intensity of incident light at each wavelength is about 10 units. Thus the absolute amount of reflected light at these wavelengths is (0.2) × (10 units) or 2 units of intensity. In general, to obtain the intensity of light reflected *at a given wavelength,* you multiply the fraction of

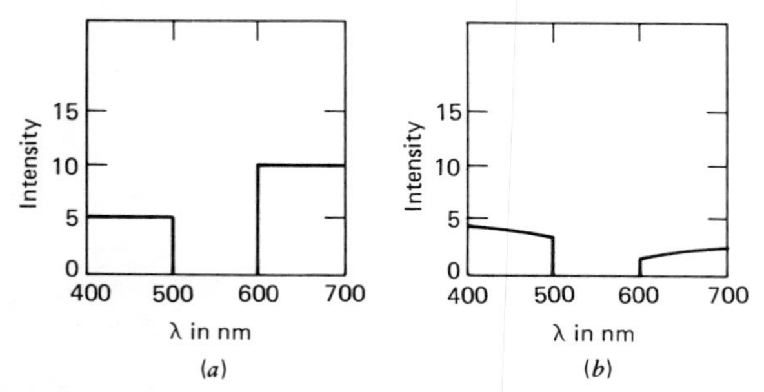

Figure 2.8 (*a*) Spectrum of source used to illuminate blue surface described by Figure 2.7. (*b*) Resulting intensity of reflected light.

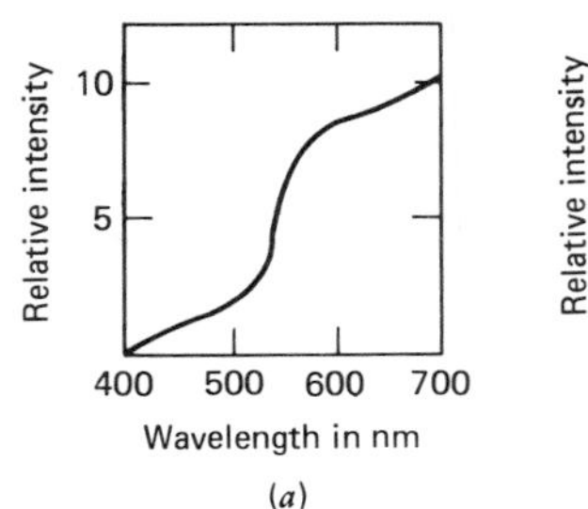

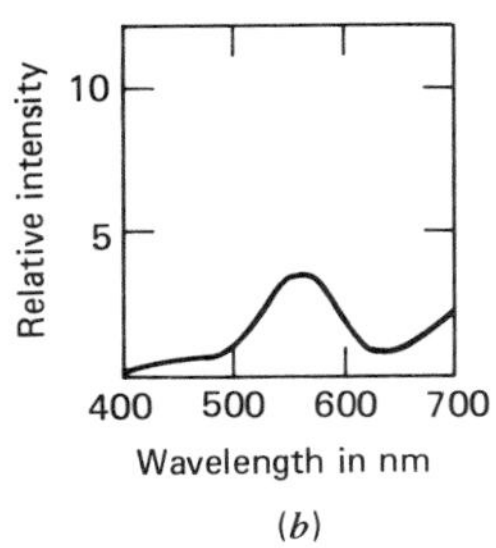

Figure 2.9 (*a*) Spectrum of light incident on the blue surface described by Figure 2.7, (*b*) Spectrum of reflected light from blue surface. The peak in (*b*) occurs where the curves of Figure 2.7 and 2.9*a* have substantial overlap.

light reflected (read from the reflectance curve at the given wavelength) times the intensity of light in the source at that wavelength.

Consider a second, more realistic example. Suppose we illuminate the blue surface described in Figure 2.7 with a source that has the spectral composition shown in Figure 2.9*a*. The intensity of reflected light that results is shown in Figure 2.9*b*. Notice that this reflected light has a very pronounced peak in the green portion of the spectrum. This occurs because the only wavelengths that the illuminant and the reflectance curve have in common are those in the middle part of the spectrum. The illuminant contains very little blue light, so these wavelengths are lacking in the reflected light. Likewise the blue surface is very inefficient at reflecting red light, so these wavelengths are also lacking in the reflected light. Only in that portion of the spectrum in which the illuminant contains a reasonable intensity and also the surface reflects fairly well will there be a significant amount of reflected light. In the example described here, the highly saturated yellow illuminant will make the blue surface actually appear green.

Let us consider one final example. Suppose the illuminant shown in Figure 2.10*a* is incident on our blue surface. The resulting spectrum of reflected light is shown in Figure 2.10*b*. Notice that very little light is reflected at any wave-

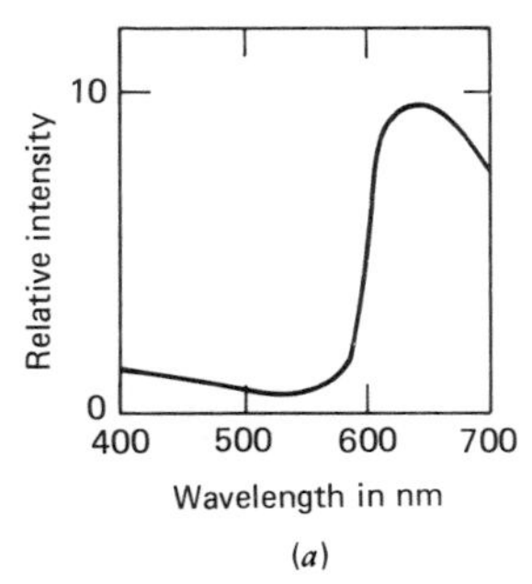

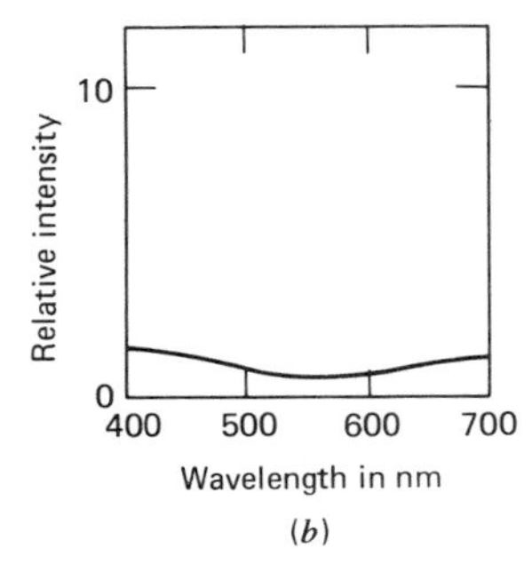

Figure 2.10 (*a*) Spectrum of light incident on the blue surface described by Figure 2.7. (*b*) Spectrum of reflected light from blue surface. Note that very little light is reflected at any wavelength because the curves of Figure 2.7 and 2.10*a* have virtually no overlap.

length. This occurs because the spectral curve of the red illuminant and the reflectance curve of the blue surface have no range of wavelengths where there is substantial incident light and reasonable reflection efficiency. Thus the surface will appear quite dark, perhaps almost black.

2.5 TRANSMISSION

In addition to the possibility of reflection occurring when light is incident on a material, it may also happen that some light is transmitted through the material. Many translucent materials are highly colored and are used as filters, in conjunction with normal white light sources, to produce highly colored illuminants. The spectrum of the transmitted light depends on the spectrum of the incident light striking the filter and on the transmission characteristics of the filter. For example, Figure 2.11*a* shows the visible light coming from an incandescent light bulb. If this light falls on a green filter whose transmission curve is shown in Figure 2.11*b*, the spectrum of the transmitted light will be as shown in Figure 2.11*c*. A detailed examination of these figures shows that 10 units of light energy

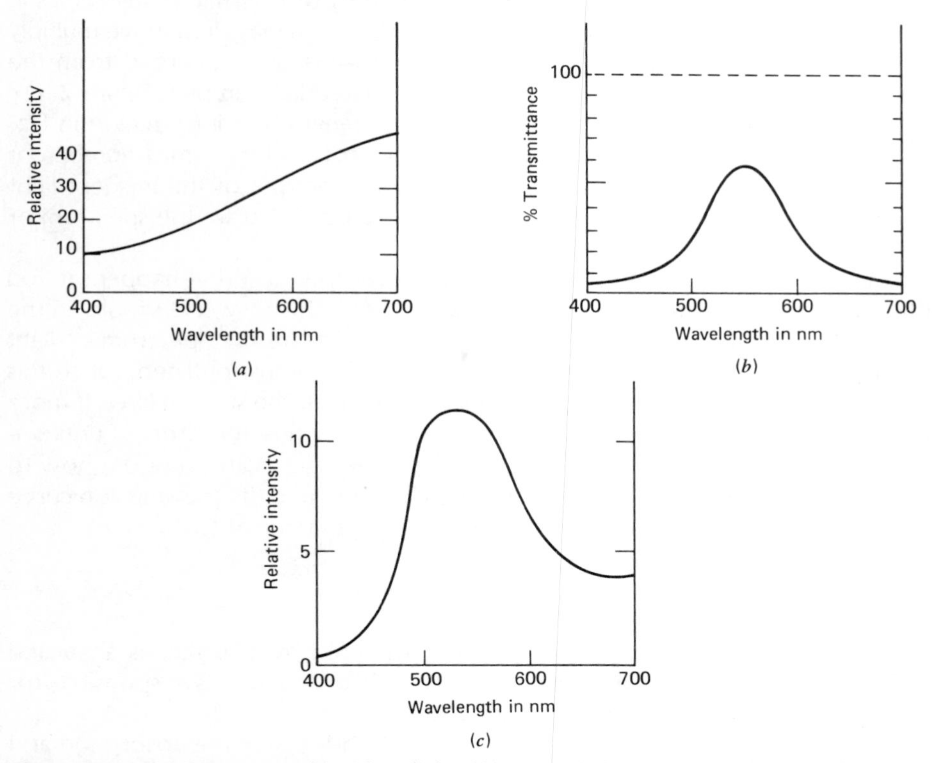

Figure 2.11 (*a*) Spectrum of incident light, (*b*) transmission curve, and (*c*) resulting light intensity for light from a common incandescent bulb passing through a green filter. Note that (*c*) is the product of (*a*) and (*b*).

Wavelength in nm	Incident Intensity	Percent Transmittance	Transmitted Intensity
400	10	5	$10 \times 0.05 = 0.5$
450	13	6	$13 \times 0.06 = 0.7$
500	15	22	$15 \times 0.22 = 3.3$
550	22	60	$22 \times 0.60 = 13.2$
600	28	23	$28 \times 0.23 = 6.4$
650	34	10	$34 \times 0.10 = 3.4$
700	40	5	$40 \times 0.05 = 2.0$

Figure 2.12 Examples of calculations of transmitted intensity drawn from the curves of Figure 2.11.

is available at a wavelength of 400 nm, but that at this wavelength only 5% of the incident light will be transmitted. Thus the actual amount of light transmitted at 400 nm is

$$10 \text{ units} \times 0.05 = 0.5 \text{ units}$$

Notice that this process of determining the quantity of transmitted light is, as in the case of reflectance, one of multiplication. At a given wavelength we multiply the amount of incident light times the fraction transmitted (as read from the transmission curve) to get the amount of transmitted light. In fact, Figure 2.11*c* is obtained by multiplying, wavelength by wavelength, the information in Figures 2.11*a* and 2.11*b*. Figure 2.12 shows this process for a small number of wavelengths. Thus we see that by knowing the spectrum of the incident light and the transmission characteristics of a filter, we can calculate the spectrum of the light that will pass through the filter.

There is one obvious question that might occur to you. What happens if you place two or more filters in series with each other? The answer is simple. In the case of two filters, for example, we can easily calculate the spectrum of light passing through the first filter by the process we have just outlined. Once this is known, it becomes the spectrum of *incident* light for the second filter. If more than two filters were present we would simply continue the process until we ran out of filters. With this discussion as a hint, see if you can figure out how to start with the transmission curve of a given filter, and find the transmission curve if you increase or decrease the filter thickness by a factor of two.

2.6 ABSORPTION

Light energy that is neither reflected nor transmitted by a surface is absorbed and usually converted into heat energy. This absorbed light is responsible for the heating effects that occur when light strikes a surface.

The more light that is reflected or transmitted the smaller the absorption and the less heating that takes place. This is most evident in the heating of common objects by sunlight. Black or dark-colored painted objects or dyed fabrics have extremely low reflectance curves and transmit little or no light. Consequently, they absorb most of the radiant energy that falls on them. This energy is converted to heat, which explains why dark clothing and dark-colored automobiles tend to heat up much more rapidly than light colored ones when in the presence of bright sunlight. On the other hand, white and other colors that have high reflectance curves over large areas of the visible spectrum tend to reflect most of the incident radiation. Because little energy is absorbed by such colors, only a small amount of energy is converted into heat. Thus people wear light-colored clothing in the summer to reflect most of the energy, and dark-colored clothing in the winter to absorb this energy.

Greenhouses also utilize this same effect to capitalize on the energy from the sun. In cool periods of the year visible light enters through the glass and is absorbed by the dark soils and vegetation. This energy is then reemitted as infrared light or heat. The glass of the greenhouse is opaque to the heat waves and thus traps this energy. As the climate warms and less energy is needed, the outside of the glass enclosure is often "white washed" to reflect the majority of the incident light, thus cutting down on the heating effect.

2.7 PRIMARY COLORS

Objects have color because of the way they selectively reflect light. For conceptual purposes, it is convenient to break the visible spectrum into three roughly equal parts consisting of the blue end, the green middle, and the red end. In this conceptual division, we say that objects have color because they selectively reflect red, green, and blue light with different efficiencies. Thus a piece of yellow paper reflects a high percentage of the red and green light that strikes it, but a small percentage of the blue light. A glance at Color Plate 3 will show how different colored surfaces reflect red, green, and blue light with different efficiencies. This division of the spectrum into three separate portions provides the basis for the reproduction of color using three so-called primary colors.

ADDITIVE PRIMARIES

The choice of primary colors is not unique. But for what are called "additive primaries," the three colors used are usually a red, green, and blue, which have spectra with very little overlap. The spectra shown in Figure 2.13 are typical of those used. We must, of course, explain what is meant by "additive primaries."

Consider a white screen – one that reflects essentially all the light that strikes it. Suppose you have three projectors, one each for red, green, and blue light. Now all three projectors are adjusted to project a circular image on the screen and, if all three images exactly coincide, there will be reflected from the screen the combined light from all three projectors. By adjusting the relative intensities of the projectors it is possible to produce any mixture of red, green, and blue that is desired. For example, the projectors can be adjusted so that the circle on the screen appears white. When this has been done you will find that roughly equal amounts of red, green, and blue light are reflected from the screen. Figure 2.13 indicates why this makes sense. Adding equal amounts of the three spectra shown there will produce a spectrum with nearly equal amounts of all wavelengths. Later, when we study colorimetry and color vision, we shall see that it is not *necessary* to have equal amounts of all wavelengths to produce white, but this is certainly one way to do it.

If you continue to adjust the relative intensities of the three projectors, you will find that essentially any color desired can be produced. In every case, what your eye sees is a simple sum of the three spectra coming from the projectors. In other words, your eye is *adding* the *light* from the three projectors to produce the different color sensations. This is why we call these three colors (red, green, and blue) the *additive* primaries.

Now suppose we take the three projectors with their intensities set to produce white and tilt them slightly so that none of the images exactly coincide. The result is something like Color Plate 4. Where the light from all three projectors overlaps we get white. Where light from only one projector strikes the screen, we see red, green, or blue. The other three colors produced result from adding together the light from various combinations of two projectors. For example, where red and green overlap, a spectrum such as the one shown in Figure 2.14*a* will result, producing the sensation of yellow. Likewise, red and blue (spectrum in Figure 2.14*b*) produce magenta, while blue and green (spectrum in Figure 2.14*c*) produce cyan. These three colors, yellow, magenta, and cyan, are very special. Each of them has a spectrum with essentially one-third of the visible spectrum *missing*. For example, yellow can be thought of as white with the blue end of the spectrum removed. Another way of looking at this would be to say that blue light and yellow light together make white light. Pairs of colors like this are said to be complementary. It should be apparent why red and cyan, and green and magenta are also pairs of complementary colors.

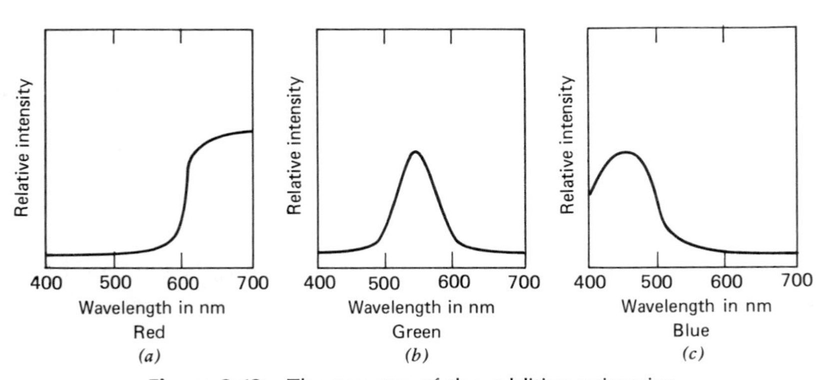

Figure 2.13 The spectra of the additive primaries.

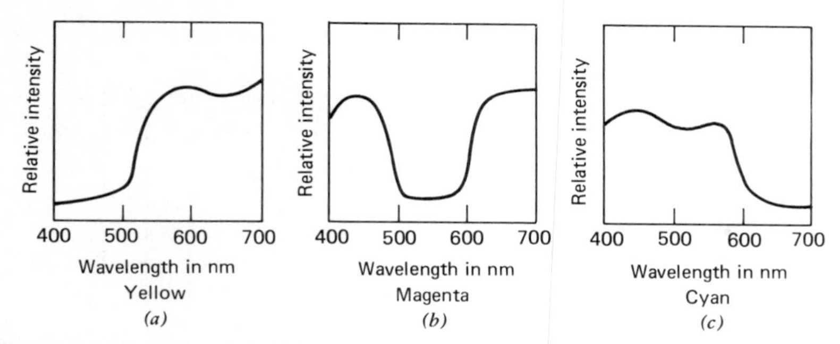

Figure 2.14 (*a*) The effect of adding red light and green light, (*b*) blue light plus red light, and (*c*) blue light plus green light. These same curves could also represent the transmission characteristics of a yellow, magenta, and cyan filter, respectively.

SUBTRACTIVE PRIMARIES

We said that magenta, yellow, and cyan were somewhat special because they each have a spectrum that is made up of essentially two-thirds of the visible spectrum (one-third missing). Why is this important? If you think a minute you will realize that using additive primaries to produce color requires essentially three separate sources of light, one each for red, green and blue. But most color reproduction involves only one source of light. When you look at a color photograph there is only a single source of illumination. Likewise, with a slide projector there is only one light bulb. If we are to have color reproduction under these circumstances, a method is needed that uses only one light source and yet can reproduce the sensation of essentially any color. The method is a subtractive process.

Suppose we have three kinds of filters, magenta, yellow, and cyan. That is, each type of filter has a transmission curve like Figures 2.14*a*, *b*, *c* respectively. One way of thinking about each of these filters is to realize that each has the ability to *subtract* away light from a specific third of the visible spectrum. That is, the magenta filter primarily removes green light, the yellow filter removes blue light, and the cyan filter removes red light. The transmission curves in Figure 2.14 are only examples of typical transmission curves for specific filters of definite thickness. If the filters were made somewhat thinner, then more light would be transmitted, especially where the curves are quite low. For example, a thinner yellow filter would not subtract out as much blue light, and so on. If the filters were available in a wide variety of thicknesses, then it should be possible to stack them together in such a fashion that any amount of red, green, or blue light is removed from the original incident light. For example, Figure 2.15 shows white light projected through a pair of filters, magenta and yellow. The magenta filter *removes* most of the green light while the yellow filter *removes* most of the blue light. Thus primarily red light finally gets through. By stacking various combinations of the basic magenta, yellow, and cyan filters in varying thicknesses, nearly any color can be produced. In effect we can control the amount

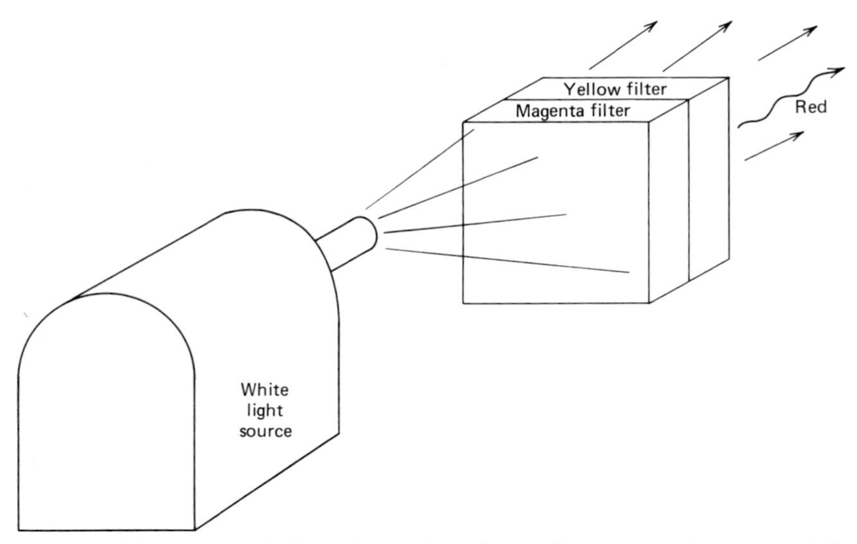

Figure 2.15 Light projected through a series of two filters to produce essentially red light.

of red, green, and blue light that is transmitted by using a cyan, magenta, and yellow filter, respectively. Examples of how various colors could be produced are shown in Figure 2.16. It is of course possible, by the method discussed in Section 2.5 under *transmission,* to calculate exactly how much light from a given source will pass through any combination of filters. This is illustrated in Figure 2.17 for a white light incident on a pair of cyan and yellow filters. We know qualitatively that this must produce green since both red and blue will be substantially removed. But Figure 2.17 shows us how to calculate exactly what finally is transmitted. In the industrial and commercial world, it is often necessary to make such exact calculations.

Since most paints, dyes, and inks are essentially filters, the subtractive primaries are also called pigment primaries. That is, for example, a mixture of yellow and cyan inks behaves like a combination of a yellow filter and a cyan filter. The

Color Desired	Filters Required
Red	Yellow and magenta
Green	Yellow and cyan
Blue	Cyan and magenta
White	None
Black	Yellow, cyan, and magenta
Light red	Thin yellow and magenta
Orange	Thick yellow and thin magenta

Figure 2.16 Some combinations of filters required to produce a selection of colors.

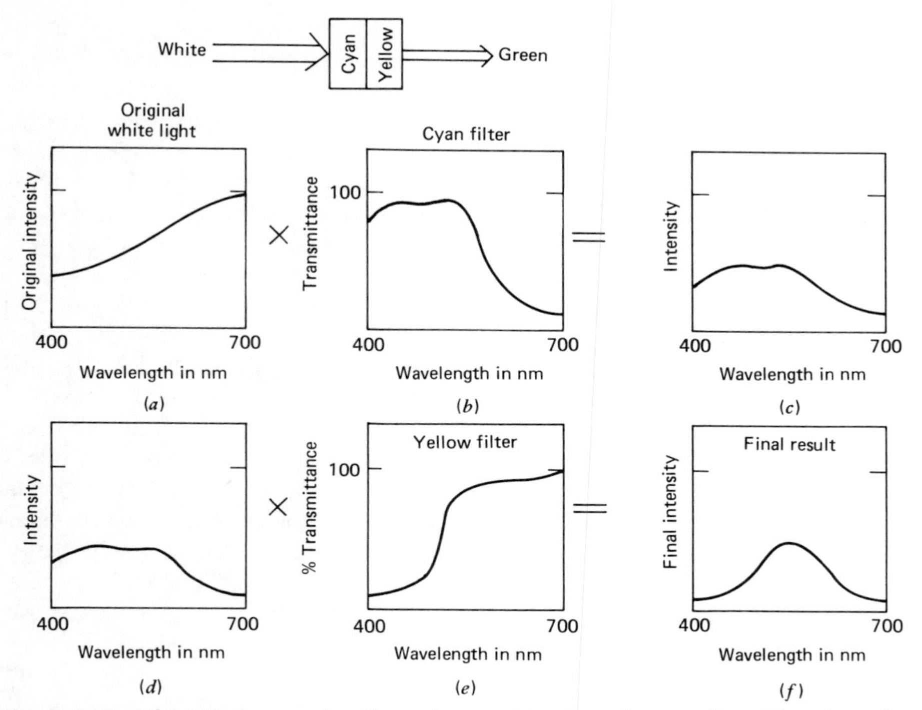

Figure 2.17 A detailed example of how the combination of a cyan filter (*b*) and a yellow filter (*e*) produce a green spectrum (*f*). Note that (*c*) and (*d*) are identical curves shown twice for clarity. Curve (*c*) shows how much of the incident light, (*a*), gets through the cyan filter (*b*). Curve (*c*) [or (*d*)] then becomes the incident light on the yellow filter (*e*). The light that finally gets through is shown in curve (*f*).

resulting color, which depends on the relative proportions of yellow and cyan, and the thickness with which the ink is laid down, will be some shade of green.

All color reproduction processes make use of either additive or subtractive primaries. In the next section we shall discuss some of these processes in greater detail.

2.8 APPLICATION OF COLOR CONCEPTS

COLOR TELEVISION

If you were to observe the painting La Grande Jatte by Georges Seurat (Color Plate 5), you would be experiencing an example of additive color mixing by the eye. This nineteenth century technique of painting thousands of tiny dots close together so they blend into a variety of subtle colors when viewed at a distance is known as pointillism. The array of colors being emitted by television tubes is produced essentially in this same manner.

The screen of a color television consists of many tiny sets of phosphors or fluorescent dots. Each set of dots consists of one phosphor that will produce each of the additive primary colors (red, green, and blue). Depending on the color that is to be reproduced by the television, these dots are stimulated or excited to different degrees.

To produce a single picture on the television screen, 525 rows of dots must be stimulated by the electron beams. To deceive your eyes into believing motion is taking place 30 complete pictures are flashed on the screen each second. The amount of energy provided for the stimulation of the dots is determined by the television camera, which views the original scene dot by dot and line by line.

The color television camera scans the original scene – left to right and top to bottom – 30 times each second. The light that enters from any spot on the original scene is passed through a series of filters and mirrors, which analyze and separate the incoming light into its component red, green, and blue parts. This information is then converted into three electrical signals and transmitted via electromagnetic waves to the television receiver in the home. In the receiver these electrical signals activate "electron guns" (one for each color), which give off controlled beams of electrons, the intensity of which depends on the strength of the incoming signals. These electrons strike the phosphors and cause them to glow with the primary colors of light. The dots are so small that they are unresolved by the eye and the colors they produce are therefore mixed additively by the eye.

For example, let us consider how the image of a yellow flower might reach us from its origin. Figure 2.18 shows the light as it leaves the flower and enters the television camera. On entering the camera, the yellow light is resolved into red and green light by the filters. The red portion of the light produces a signal to the red amplifier while the green amplifier receives a different signal. Since the yellow light contains no wavelengths from the blue end of the visible spectrum, the blue amplifier does not receive any energy from this light.

The television station now broadcasts these signals, which are received by the sets in our homes. The incoming red signal activates the "red electron gun" in the picture tube. This causes a stream of electrons to be carefully directed to the red phosphors, which are then excited and glow red. Likewise the green signal activates the green phosphor dots. The light from these tiny dots glowing red and green is unresolved by our eyes and is thus additively mixed and perceived as yellow light.

This single spot on the television screen represents one single spot on the flower. To produce the image of the entire flower requires that thousands of dots be activated by this same technique. When the hue of the colored flower changes, the phosphors on the picture tube will be activated by different amounts that, when mixed by our eyes, give rise to the perception of various hues or colors. With 30 complete pictures being flashed on the screen each

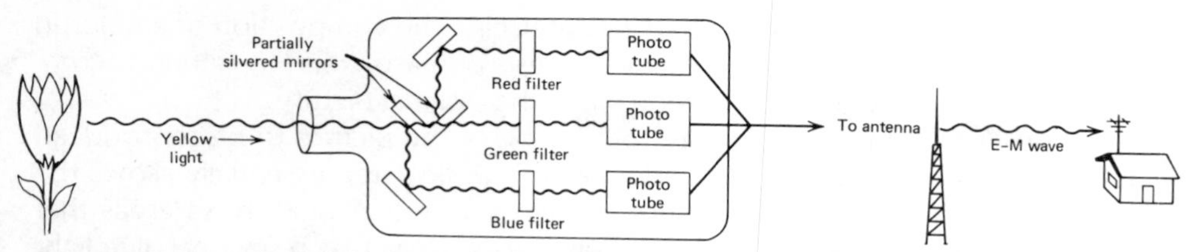

Figure 2.18*a* Yellow light from the flower enters the television camera and is broken into red and green light. The red and green light activates phototubes that send electrical signals to the transmitter. Electromagnetic waves carry the coded information to the receiver in the home.

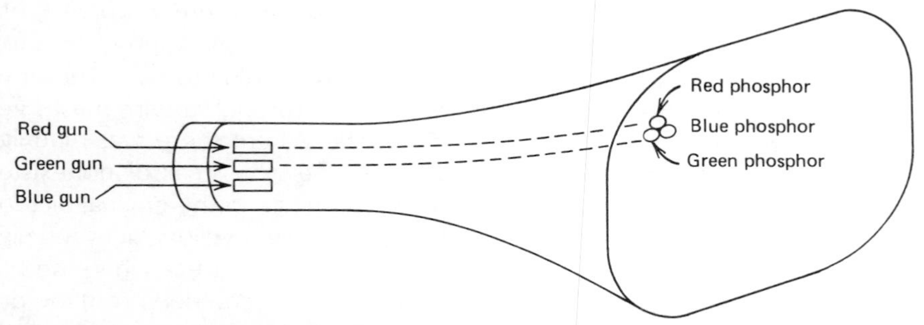

Figure 2.18*b* The incoming electrical signals stimulate the red and green guns causing them to activate the red and green phosphors. The light from these phosphors is additively mixed by the eye-brain system, giving a yellow sensation.

second, this additive color mixing process goes on continually, unnoticed by our eyes unless one of the electron color guns becomes weak or maladjusted.

We become aware of color distortion in the television when the electronics controlling the stimulation of the phosphor dots become unbalanced, producing colors unlike those in the original scene. Adjustments are then required to strengthen or weaken one or more of the devices used to stimulate the color phosphors.

COLOR PRINTING

Color television is neither the first nor the most familar example of mixing colored dots to give a full spectrum of colors. An examination with a magnifying glass of the color plates in this book, color magazine illustrations, or the Sunday comics will show that the same technique is used to produce color in printing.

To reproduce a color picture an engraver works in an analogous fashion to a chemist doing analysis and synthesis of a compound. By electrolytic analysis the chemist finds that water contains two parts of hydrogen and one part oxygen. By synthesis these elements can be recombined to produce water. In a similar way, the engraver uses filters to determine the composition of a colored picture in terms of the primary colors. These primary colors are then recombined in the printing process to reproduce the original picture.

The process begins with the photographing of the picture to be reproduced through a fine mesh screen and a blue filter. This first step essentially allows the photographer to look at the picture in such a way that only those areas that reflect blue light are seen. Or, conversely, those areas that reflect no blue light can also be clearly identified since they will look black through the blue filter. The printing plate that is made from this step must be used to insure that blue light is reflected from the same areas in the print as it was in the original picture, and no other areas. When the plate is made, any area that is struck by blue light will be exposed and etched away. Areas on the plate where *no blue light* is present will not be etched, and it is on these areas that ink will be laid down. The ink that must be used will allow every color except blue to pass. Thus it will be yellow ink. Since the plate photograph is taken through a wire mesh, generally the plate consists of a pattern of fine dots. Where there was blue light present in the original photograph, the dots will be very small or nonexistent (areas that were blue, purple, cyan, magenta, and white in the original picture). In areas where less blue light was present (e.g., gray, light yellow, light red, light green) the dots will be larger but still easily identifiable. In areas where no blue light was present (e.g., black, deep green, deep yellow, and deep red) the dots will be so large that they will form a solid area. Notice that in each case, the presence or absence of yellow ink controls the quantity of blue light that will be reflected in the final print.

Now a second photograph is taken through a red filter with the original screen rotated approximately 30 degrees (this rotation is necessary to eliminate unwanted geometric dot patterns in the final plate). The plate produced from this step is used to control the presence of red in the print. Therefore, cyan ink (which blocks only red) must be used. Any area that reflects red light in the original picture (e.g., red, magenta, yellow, white) will be an etched area on the plate and will not print. Likewise, areas that do not reflect red light (e.g., blue, green, cyan, black) will print cyan ink. The presence or absence of cyan ink controls the *quantity of red light* that will be reflected in the final print.

The screen is now rotated again and another photograph is taken through a green filter. This will be used to produce a plate that will print magenta ink (which blocks only green). The presence or absence of magenta ink controls the *quantity of green light* that will be reflected in the final print. The analysis of the picture is now essentially complete.

To reconstruct the original picture, the three primary colored plates will now be printed on the same sheet. The printer will begin by producing a yellow copy of the original scene. Great care must be taken to insure that the yellow ink is very close to the complement of the blue filter that was used in producing the plate. This is necessary to have the best color rendition. The cyan plate is now printed on top of the yellow print, care being taken to keep the plates in register (positioning one exactly on top of the other). Again, the cyan ink should be very close to the complement of the red filter. Finally, the magenta print is made on top of the first two prints. In some cases a fourth print using black ink is made in order to enhance the quality of contrast in the final reproduction. Now let us examine a small area of the final result (without the final black overprint), such as that shown schematically in Figure 2.19, and in Color Plate 6. There are three different colors of dots present, magenta, yellow, and cyan. Some of these dots will overlap while others will not, but whether or not this overlap occurs is irrelevant. Each colored dot controls a different part of the spectrum and, thus, the dots are basically transparent to each other. What is important in determining the final color perceived in the little area of Figure 2.19 is the percentage of this area covered by each type of colored dot. For example, if the little area had no dots at all, then a very high percentage of the incident light would be reflected (nearly 90% if the paper was white) and the little area would look white. On the other hand, if 80% of the area was covered by yellow dots, this would reduce the amount of reflected blue light by 80%. Figure 2.20 illustrates how the amount of red, green, and blue reflected light has been reduced by the dots in Figure 2.19. The complete printing process is shown in Color Plate 7.

The quality of the final reproduction is determined by the number of inks used, the degree to which the inks and their corresponding filters are complementary, and the number of dots per inch.

Newspapers usually use relatively coarse screens that produce 60 dots per linear inch. Higher quality prints such as magazines or books employ a finer screen having 120 or 240 separate dots per linear inch. The more closely spaced

Figure 2.19 Closeup view of a small section of a color print. Different shapes have been used in the interest of clarity to illustrate the magenta (□), yellow (△), and cyan (○) dots.

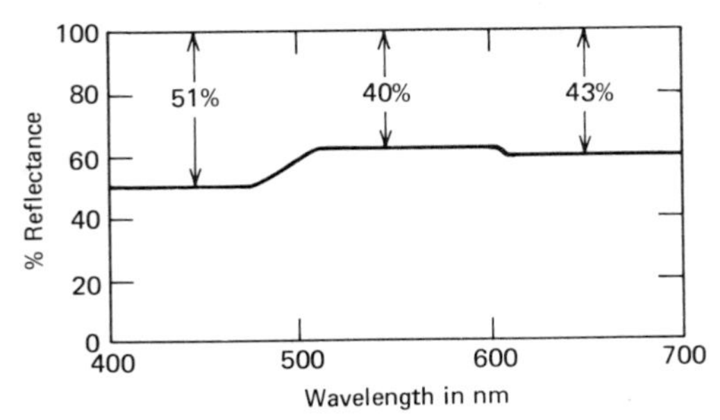

Figure 2.20 Approximate net reflectance spectrum from the area shown in Figure 2.19. This area would probably appear as an unsaturated yellow green.

the dots are, the more effective will be the blending of the colors by the eye and the better the color rendition of the original scene.

THE MIXING OF PIGMENTS

Oils

A rather interesting application of the principles of light and color occurs when an artist mixes a small number of basic pigments to produce a virtually unlimited number of colors. A typical oil-based paint consists of a transparent medium in which is suspended a very fine absorbing material called the pigment. It is the pigment that produces the color by selectively absorbing certain wavelengths of light while reflecting others.

Figure 2.21 shows schematically how a layer of oil paint would look on a canvas. Notice that the particles of a pigment are very numerous and are randomly scattered throughout the oil medium. As light of various colors (wavelengths) reaches the surface of the paint it will usually encounter the transparent medium first. The light will pass through the medium until it encounters a particle of pigment. In the example of Figure 2.21, the pigment is red. Thus any light that is not from the red end of the spectrum will essentially be absorbed after one or two encounters with a particle of red pigment. Red light, on the other hand, will reflect from one particle to another until it finally escapes from the paint and returns to our eye. Thus the portion of canvas painted with this particular pigment looks red.

Now, what happens when different pigments are mixed together? The key to understanding this lies in realizing that a given light wave (or photon) will usually reflect from several particles of pigment before it leaves the painting. Thus, if two different pigments are present in roughly equal quantities, then in order for a particular color to be reflected from the painting, it must be able to reflect from *both kinds of pigments.* Consider the two reflectance curves shown in Figure 2.22, which describes a cyan and a yellow pigment. Now, for the sake of argument let us suppose that, on the average, any light that escapes from the

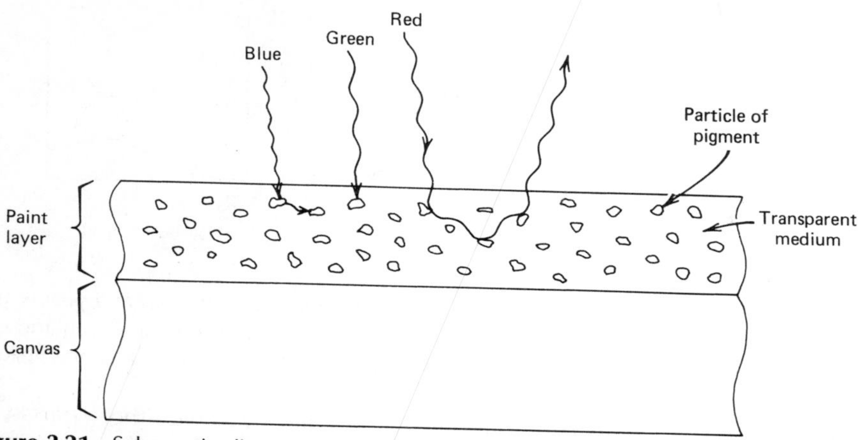

Figure 2.21 Schematic diagram of a layer of oil paint on a canvas. Particles of pigment in this example are red. Thus the particles will absorb most of the blue and green parts of the spectrum but will reflect the red portion of the spectrum.

paint must make four reflections from individual particles of pigment. If we mix the cyan and yellow pigment in roughly equal amounts, we can expect that light that reflects from the paint will reflect from two particles of cyan pigment and two particles of yellow pigment. To find out what the spectrum of this light will be, we start with an equal energy white light. Then we let this light be reflected twice from the cyan pigment (this is done by taking the upper curve in Figure 2.22*a* and multiplying it times itself). The resulting spectrum is then reflected twice from the yellow pigment. The result, which is shown in Figure 2.23*a*, is clearly a green.

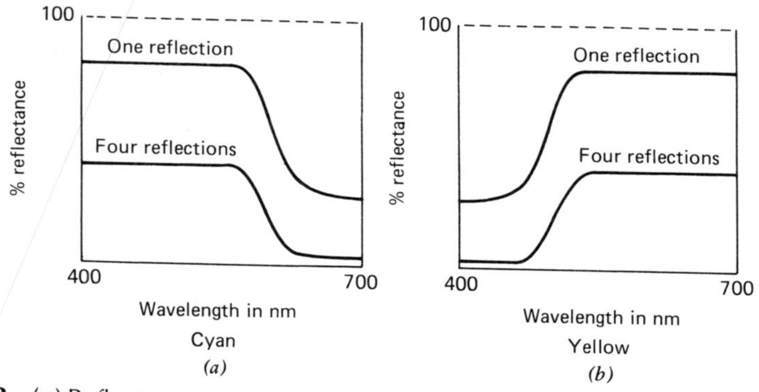

Figure 2.22 (*a*) Reflectance curve for cyan pigment, and (*b*) reflectance curve for yellow pigment. In each case, the upper curve represents the reflectance curve for a single reflection from one particle of pigment. The lower curve represents the net effect of four successive reflections from four different particles of pigment.

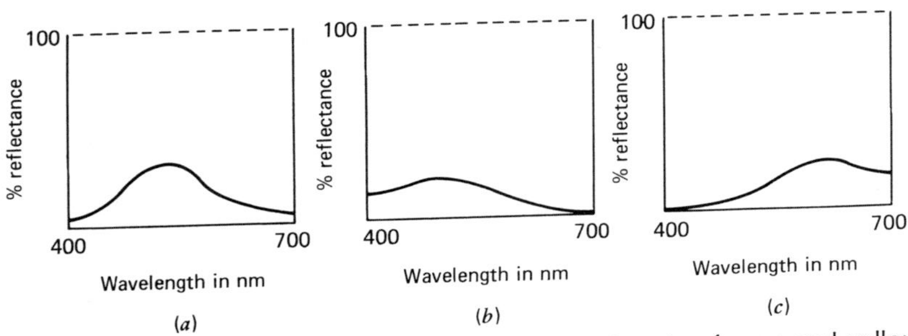

Figure 2.23 (*a*) Reflectance curve for a mixture of equal parts of cyan and yellow pigment (from Figure 2.23); (*b*) reflectance curve for a mixture of three parts cyan and one part yellow; (*c*) reflectance curve for a mixture of one part cyan and three parts yellow.

Suppose, on the other hand, that we mix one part cyan with three parts yellow. We would then expect, on the average, to have the light reflect once from a particle of cyan pigment and three times from particles of yellow pigment. The resulting spectrum is shown in Figure 2.23*c*, and is a yellow-green. If we mix one part yellow with three parts cyan, similar reasoning results in the spectrum of Figure 2.23*b*, which is a blue-green. Thus the two pigments, cyan and yellow, can be combined to produce a large range of colors.

It should be apparent that although the microscopic process involved here is one of *reflection,* the net effect is similar to combining transparent inks or filters.

Watercolors

In watercolors we have a situation that is truly one of transmission. That is, watercolors are essentially filters laid down on a white surface. Light passes through the watercolor, reflects from the white surface, and passes back through the watercolor. The light is thus filtered twice. The color is determined by the transmission curve of the watercolor. Painting a second color over the first is equivalent to placing two filters together. The resulting color depends on the transmission curves of both layers of paint.

The saturation of the color can be controlled by the dilution of the paint. Adding water has the effect of making the paint layer thinner. This reduces the effective filter thickness and makes the color lighter. Obviously, painting over an area more than once will have just the opposite effect.

INTERIOR DECORATING AND FASHION DESIGN

The basic principles of interior decorating and fashion design deal mostly with the subjective preferences of human beings and are therefore outside the scope of this text. However, in selecting materials that are compatible with various illuminants, we shall find that the concepts we have developed so far are very useful. On the other hand, failure to appreciate the principles of light and color can often lead to some embarassing mismatches.

In Section 2.3 it was pointed out that the spectrum of light reflected from a surface depends on both the spectral content of the illuminant and the reflectance curve of the surface. A change in either of them will usually result in some change in color also. Thus both must be constantly kept in mind. For example, a room that has been carefully designed to produce a warm soft mood under incandescent illumation might produce a jarring effect if fluorescent light were used. On the other hand, careful selection of materials might allow the same room to assume two entirely different appearances under two different illuminants. The important point is that the choice of materials must be made with the illuminant in mind. Anyone who has tried to pick out some item, say upholstery or clothing, of a particular color knows the difficulties. The fabric appears to be one shade in the store under fluorescent lighting. Outside in direct sunlight it may look entirely different. Finally at home under incandescent illumination it undergoes yet another subtle change in color. A proper choice can only be made if the material is viewed under illumination that is the same as that which exists where the material will ultimately be used.

A similar problem arises in fashion design. Complete coordination in fashion often requires that several diverse materials be compatible with each other. Under one illuminant they may be, but under another illuminant they may be incompatible. In Figure 2.2 you can see how very different our three most common light sources (daylight, incandescent ight, and fluorescent light) really are. One is well-balanced, one is rich in red, and one is rich in blue. When we change from one of these illuminants to another there are rarely any gross changes in colors. But there can be some rather disturbingly subtle changes that may create problems if the original colors are very closely matched. For example, a variety of pinks that appeared well-coordinated under sunlight might produce a jarring effect under fluorescent lighting. It is unreasonable to require that a particular outfit look perfectly coordinated under any possible illumination. This is just not possible. But proper design should anticipate the kinds of illuminants likely to be encountered and choose materials accordingly.

STAGE LIGHTING

Another area that draws heavily for its effects on the principles presented in this chapter is the production of color in stage lighting. Light can be used for setting moods, differentiating the time of day or the seasons, for shock value, distortion, or for other reasons. To produce these effects requires that the lighting designer have access to a variety of light sources that can be manipulated by adding or subtracting the colors needed. Filters or gels are used to alter the spectrum of the light being emitted from the light sources. These filtered light sources then undergo selective reflection from the surfaces they strike.

If the reflecting surface and the incident light are carefully chosen, it is possible to have an array of colors reflected from a backdrop without the necessity of changing that structure. For example, let us suppose a viewing area is to represent an area of sky that at different times is to be a light blue of midday, a red

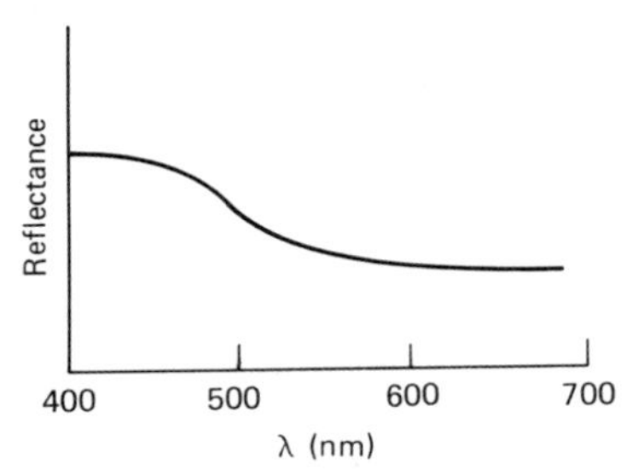

Figure 2.24*a* Spectrum of blue backdrop to be illuminated with different sources.

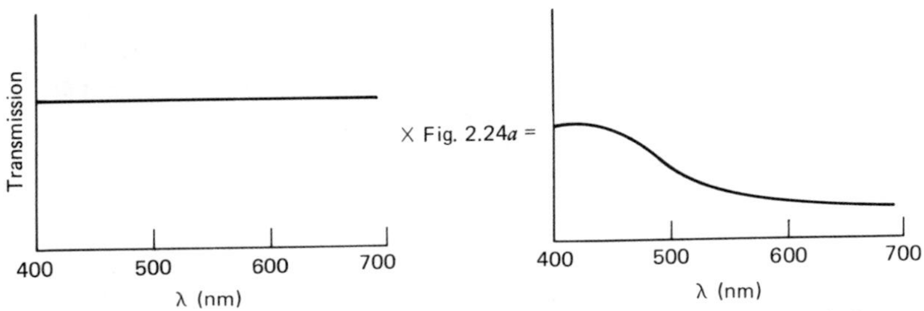

Figure 2.24*b* White light used to illuminate surface of Figure 2.24*a* and the resultant reflected light.

for a sunset, and a dark blue of an oncoming storm. This might be accomplished by using the following scheme.

The backdrop could be painted the light blue color whose spectrum is shown in Figure 2.24*a*. When this surface was then illuminated with a "white" light for the normal daylight scenes, the desired blue sky would be produced. To produce the red background for sunset a red filter could be used on the light, cutting out all the blue and green wavelengths. Since the light blue pigment reflects a large amount of red, the surface will now appear that color. Finally to produce a dark blue or grayish color the total illumination on the stage could be greatly reduced and the backdrop illuminated with faint blue light. Figure 2.24 illustrates the light that would be reflected from the backdrop in each of these cases.

IN CONCLUSION

The concepts discussed in this chapter allow us to understand what controls the light that comes to our eyes from various objects. We have also begun to appreciate the connection between the spectral composition of light and the perceived color. The concept of primary colors has been introduced, and we have seen how these colors can be used to reproduce a wide range of hues. Finally, we have looked at a few applications of basic principles in order to illustrate their use. We have not yet developed, however, a way of accurately

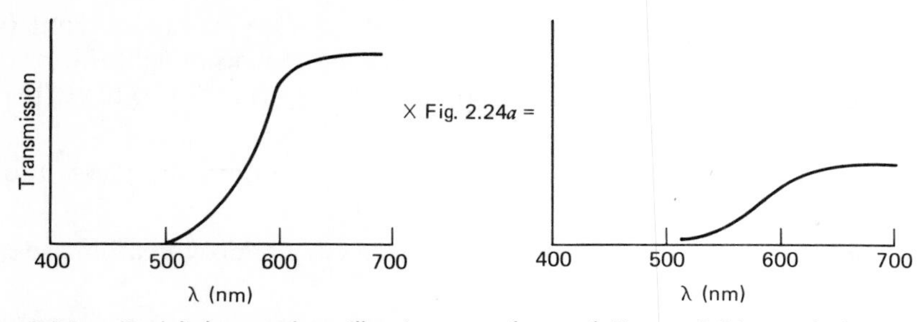

Figure 2.24*c* Red light used to illuminate surface of Figure 2.24*a* and the resultant reflected light.

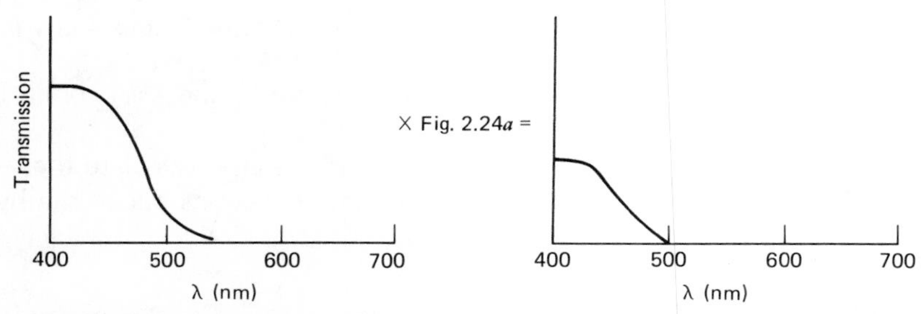

Figure 2.24*d* Blue light used to illuminate surface of Figure 2.24*a* and the resultant reflected light.

describing color. The science of color description is called colorimetry, and is the subject of the next chapter.

PROBLEMS AND EXERCISES

1. Which of the following would you classify as white light sources? (a) Standard flashlight, (b) mercury vapor street light, (c) campfire, (d) sunlight, (e) neon advertising sign, (f) standard slide projector bulb, (g) sodium vapor street light.
2. By examining Color Plate 1, list the colors associated with each of the following wavelengths: (a) 400 nm, (b) 450 nm (c) 520 nm, (d) 550 nm, (e) 600 nm, (f) 700 nm.
3. By examining Color Plate 1, list the *range* of wavelengths appropriate for each of the following colors: (a) violet, (b) green, (c) yellow, (d) red.
4. One feature of blackbody radiation is that the wavelength emitted at maximum intensity (the peak of the radiation curve) depends on temperature according to

$$\text{Peak wavelength (in nm)} = \frac{2.89 \times 10^6}{T(°\text{K})}$$

(a) Calculate the peak wavelength at which you radiate light (your body temperature is about 310°K). What kind of light is this?
(b) How hot would a blackbody need to be in order to have its peak wavelength at 550 nm?

5. Examine Figure 2.2. Place the sources whose spectra are given in order of increasing color temperature.

6. The energies of the various allowed levels in a hydrogen atom are given by

$$E_n = 13.1\left(1 - \frac{1}{n^2}\right) \text{eV}$$

where n is the number of the level (1, 2, 3, 4, etc). The symbol eV stands for electron volts and is a small unit of energy. Now Planck's law is

$$\lambda = \frac{hc}{E} = \frac{1240 \text{ eV-nm}}{E} \qquad E \text{ is in units of eV, and } \lambda \text{ is in nm.}$$

Using the energy equation above and Planck's law, calculate the wavelength of light emitted when the following transitions occur in a hydrogen atom:
(a) Level 6 to level 1
(b) Level 4 to level 2
(c) Level 7 to level 3
In what part of the spectrum is the light produced in each case?

7. Using the wavelengths from Color Plate 2 and Planck's law from problem 6, calculate the energy associated with each line in the sodium spectrum.

8. It was explained in the text that the ultraviolet light produced by mercury is used in a fluorescent light to excite phosphors, which then emit visible light. Why can't infrared light be used for this purpose also?

9. On the basis of the spectra shown in Color Plate 3, make a sketch of the approximate spectra of each of the following colors: (a) light green (b) pink (c) light orange (d) yellow-green (e) very deep red.

10. The two curves below give the light incident on a filter and the transmission curve of the filter, respectively. Calculate the transmitted spectrum.

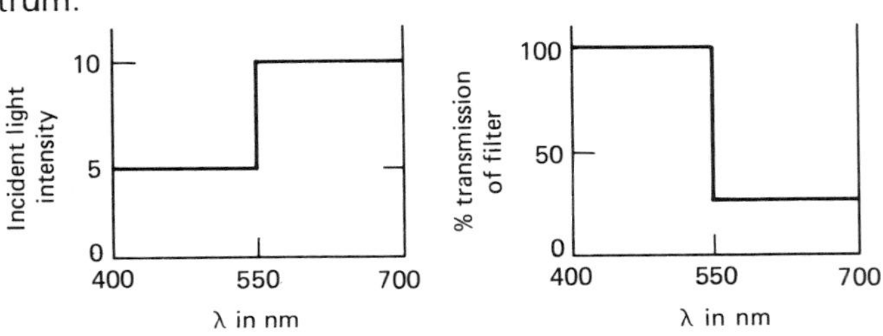

Figure 2.25

11. The transmission curve of a green filter is given below. Calculate the transmission curve of a filter (a) twice as thick, (b) half as thick. How would you describe the change in the shade of green in each case?

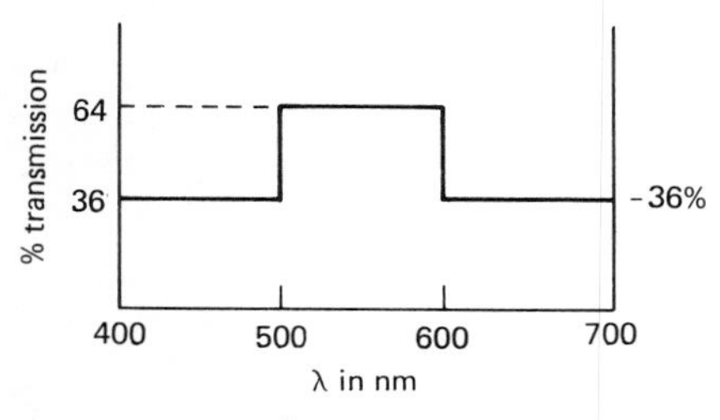

Figure 2.26

12. Examine the reflectance spectrum shown below. Sketch an example of an incident spectrum of light that would cause this surface to reflect light that is mostly: (a) red, (b) yellow, (c) green, (d) blue.

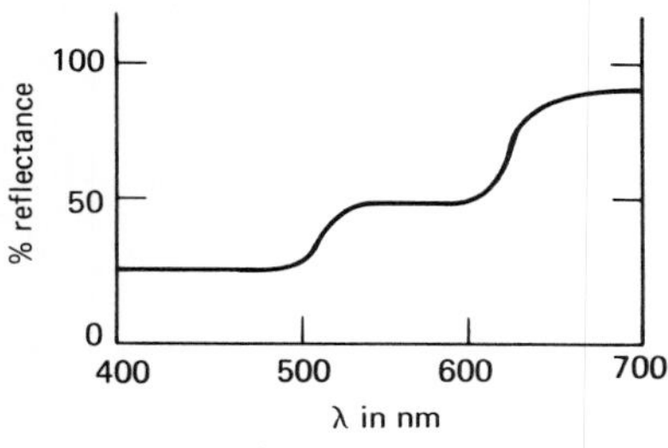

Figure 2.27

13. Explain how you would obtain the following colors by combining various intensities of the additive primaries. (a) yellow, (b) pink, (c) white, (d) orange, (e) purple, (f) very light cyan.

14. Explain how you would obtain the following colors by combining subtractive primary filters of various thicknesses. (a) red, (b) green, (c) blue, (d) black, (e) white, (f) pink, (g) yellow, (h) orange.

15. A color television normally uses three separate electron "guns" to activate the three basic phosphors: red, green, and blue. Thus when a television camera is pointed at a white object, the portion of the picture corresponding to that object has all three guns equally activated. But suppose, because of faulty electronics, that the green gun is much weaker than it should be. Then, when the television camera is pointed at objects of the following colors, what colors will be seen on the TV screen? Red, blue, yellow, magenta, white, black, green.

16. The three basic inks used in color printing are yellow, cyan, and magenta. What combinations of these must be printed on top of each other on white paper to produce the following colors: red, blue, black, white, green, orange, pink, brown.
17. Discuss why bluing is often added to white laundry when it is washed.
18. Refer to Color Plate 3 and assume that the spectra shown are transmission curves. Using the given spectra for magenta and yellow, determine what light will pass through if these filters are placed back to back and illuminated with an equal energy source.
19. Yellow and blue, magenta and green, and red and cyan are usually given as complementary colors. Name some other combinations of colors that could be mixed visually to give white.
20. Suppose a performer is wearing a light blue costume on stage. What color will this appear if illuminated by yellow light? By violet light?

3
COLORIMETRY — DESCRIBING AND MEASURING COLOR

The judgment of color and its faithful reproduction are of the utmost importance in the commercial and industrial world. The quality of numerous products are judged by the colors they display or fail to display. In all phases of life we are continually making judgments based on the color of the materials that we encounter.

The grocery shopper looks for the familiar color of the package of cereal that he wishes to purchase. If the color of this box varies from what it should be, the customer will more carefully inspect the box. If the color is lighter than usual, or faded looking, the cereal may be rejected because the faded color indicates that the box has been on the shelf too long and, thus, the cereal has lost its freshness. In the meat department the shopper's eye examines the color of meat to ascertain if it looks fresh. If the color of the beef does not meet the expected standard of redness, the cut will be rejected for being too old. As the shopper continues through the store, more and more products undergo the scrutiny of color memory. Are the vegetables green enough? Does the cooking oil have too much color? The bread looks too brown — maybe it has been overbaked.

While colors in the grocery store are judged by one's memory, at other times it is the consistency of colors among varving types of products that are judged. Consider the importance of color control to the car manufacturer and buyers. First the basic color must be appealing. Then this color must be faithfully reproduced on a variety of materials. The color of the paint on the metal should match as closely as possible the color of the plastic on the dashboard. Is the color of the seat covering the same as that of the plastic dash? If the car should need body work and repainting, can the color be reproduced?

These and other judgments are constantly being evaluated by consumers in their daily lives. The manufacturers and producers must have some standards and techniques to ensure that the customer's color concerns are being met. In this chapter we shall investigate a number of widely used systems by which color can be specified and judged.

There are two widely used approaches to the precise description of color. One kind of system analyzes the spectrum of light reflected from a surface and assigns a set of tristimulus values that specify the color. With this kind of system, any two surfaces that reflect spectra having the same tristimulus values will also appear the same color. Thus specification of the tristimulus values uniquely describes the color appearance of a surface.

The second basic approach to colorimetry involves the use of standard colored samples against which other materials can be compared. These colored samples can be arranged in a variety of systematic ways; thus there are several such colorimetry systems. In this chapter we shall primarily be interested in the Munsell and Ostwald Systems, which use colored samples, and in the CIE system, which analyzes the reflected light.

3.1 NEWTON'S COLORIMETRIC SYSTEM

In Chapter 1 we described Newton's classic demonstration that white light contains a spectrum of colors. Newton also experimented with the mixing of colors and devised a simple colorimetric system. In order to mix colors, Newton used two prisms so that he had two separate spectra that could be combined in various ways. By using two opaque cards with slots, Newton was able to select the parts of the spectra from each prism that he wished to mix (the arrangement is shown in Figure 3.1). In this simple way, Newton was able to observe the effect of combining light from various parts of the spectrum. The results of his observations led Newton to propose a simple "center of gravity" or barocentric colorimetric system. First, Newton arranged all the colors of the spectrum around the periphery of a circle as shown in Figure 3.2*a*. There were seven unequal segements around this circle corresponding to the seven basic hues that Newton saw in the spectrum. Points interior to the circle represented hues that were less saturated than the pure spectrum colors (saturation is a color's purity or freedom from dilution). At the very center of the circle was pure white. This system allowed predictions to be made about the result of mixing any combination of colors.

For example, consider a mixture of 2 units of red light and 1 unit of yellow light. We begin by drawing a line from the center of the red segment of the circle to the center of the yellow segment of the circle as shown in Figure 3.2. The mixture is represented by a point along this line at the "center of gravity" of the two colors. That is, the line is divided into two segments, which are in the same proportion as the amounts of the two colors being mixed. The short line

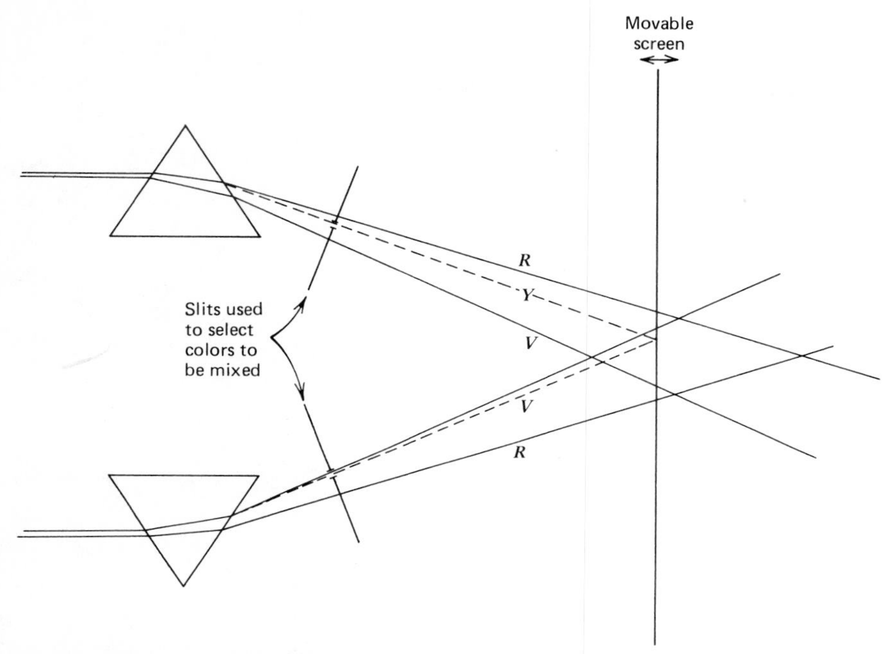

Figure 3.1 Newton's color mixing apparatus. In this diagram yellow and violet light are being mixed.

segment goes on the end toward the color present in the greater amount, red in this case. Figure 3.2*a* shows that the mixture should be a light described as a 3 units of red-orange (slightly desaturated compared to the pure spectrum).

Another example is provided by the mixing of 2 units of white light with 1 unit of green light. This mixture is found in the same way as the previous example, and is shown in Figure 3.2*a*. Any number of colors can be added in this way. For example, when we added 2 units of white light to 1 unit of green light we got 3 units of pale green light. Suppose to this mixture we add 6 units of blue light. This will yield 9 units of a slightly desaturated blue as shown in Figure 3.2*a*. Another color could be added to this, and so on.

Newton's system gives fairly good qualitative results in most cases, but it is quantitatively inaccurate. In addition, there is a whole range of hues, the purples or magentas, that are missing entirely (hue refers to the basic color, that is, red, yellow, green, etc.). These hues should fit between the reds and the violets around the periphery of Newton's diagram. Since these hues are not represented by any pure light in the visible spectrum but, instead, result from mixtures of reds and violets, this segment of the diagram would have to be a straight line rather than an arc of a circle (See Figure 3.2*b*). This modified Newtonian system would represent a definite improvement over Newton's original system, but it

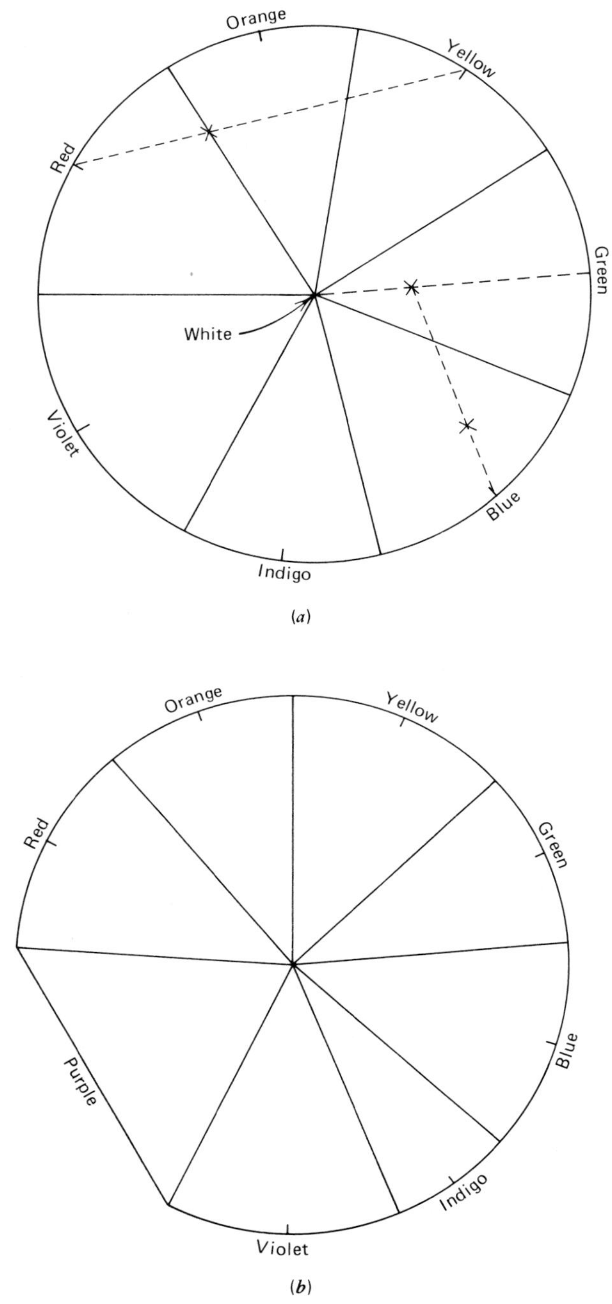

Figure 3.2 (*a*) Newton's barocentric color system illustrating color mixing. (*b*) Modified color system that includes the purples.

would still suffer from quantitative inaccuracies. To construct a truly accurate system of this kind, it is necessary to determine the facts of color mixing with greater precision than Newton was capable of.

3.2 THE CIE SYSTEM

In 1931 the Commission International de l'Eclairage (the CIE-International Commission on Illumination) reviewed the best available experimental data on color mixing, and constructed a barocentric colorimetric system. This system is really just an accurate refinement of the modified Newtonian system discussed in the previous section. To help us to understand how the CIE system was developed, we will first instroduce the concept of metamerism.

METAMERISM

We have already discussed in Section 2.7 how three primary lights can be added to produce a wide range of colors. The primary lights in this case are not monochromatic, but are instead produced by a standard white light coupled with a red, a green, and a blue filter, respectively. In order to clarify exactly how three primary lights can be used to match various colors, let us envision the following situation. An observer with normal color vision is placed in front of a perfectly white screen. On the left half of the screen is projected some arbitrary spectrum of light. On the right half of the screen is projected a combination of the three additive primaries. Our observer is asked to adjust the intensities of the three primaries until the left and right halves of the screen match exactly, in color and brightness. With the kind of primaries produced with filters it will be possible to produce this match provided the color to be matched is not too highly saturated. Now, even though the two halves of the screen *look* the same (are *equivalent stimuli*), they do *not* have the same spectral composition. Pairs of equivalent stimuli that have different spectral compositions are said to form a *metameric* pair. The phenomenon is called *metamerism* and is the basis of all color reproduction techniques. The CIE system is also based on a form of metamerism – the laws of color matching.

THE LAWS OF COLOR MATCHING

The laws of color matching are derived from careful experiments that use a set of three monochromatic lights as additive primaries. An attempt is made to match some combination of these three primaries with a standard intensity of each wavelength of the pure spectrum. It should be pointed out that the actual choice of the wavelengths used for the three primaries is completely arbitrary. The final CIE-XYZ system, which is the system we are working toward, is identical regardless of the choice of primaries.

Suppose we choose three monochromatic primaries of wavelength λ_R, λ_G, and λ_B (R, G, and B refer loosely to red, green, and blue) and try to use these primaries to match some other monochromatic light of wavelength λ. For convenience let us define the following quantities:

$$I_R = \text{intensity of red primary } (\lambda_R)$$

$$I_G = \text{intensity of green primary } (\lambda_G)$$

$$I_B = \text{intensity of blue primary } (\lambda_B)$$

$$U_\lambda = \text{unit intensity of monochromatic light of wavelength } \lambda$$

Now let us try to match a monochromatic light of wavelength λ and unit intensity with some combination of our three primaries. After diligently trying to produce such a direct match we must conclude that it cannot be achieved. That is, for monochromatic light of *any* wavelength in the visible spectrum (except for λ_R, λ_G, and λ_B themselves) it will not prove possible to find any direct combination of intensities, $I_R + I_G + I_B$ that will exactly match U_λ in hue, saturation, and brightness. However, further experimentation shows us that it is possible to achieve a condition where a match occurs. Depending on the value of λ, one of the following three matches will always prove possible.

$$U_\lambda + I_R \equiv I_G + I_B \qquad (1)$$

or

$$U_\lambda + I_G \equiv I_R + I_B \qquad (2)$$

or

$$U_\lambda + I_B \equiv I_R + I_G \qquad (3)$$

These three equations are metameric equations and should be read as follows. Equation (1) states that:

A unit intensity of light of wavelength λ plus light of wavelength λ_R and intensity I_R	exactly matches	Light of wavelength λ_G and intensity I_G plus Light of wavelength λ_B and intensity I_B

Equations (2) and (3) should be read similarly. All the quantities in equations (1), (2), and (3) are intrinsically positive. That is, they all represent the direct physical addition of real lights. This is further emphasized by the use of the $\equiv$ sign, which signifies a metameric match. As a practical matter we can conveniently summarize the information contained in equations (1), (2), and (3) by the single equation

$$U_\lambda = I_R + I_G + I_B \qquad (4)$$

Notice that we have dropped the ≡ sign and replaced it with an = sign. This is intentional. Equation (4) is no longer a true metameric equation in the sense that no direct combination of I_R, I_G, and I_B can match U_λ. In equation (4), *one* of the quantities I_R, I_G, I_B is in fact *negative* and should be on the other side of the equation. Exactly which one depends on the value of λ. For every wavelength, λ, in the visible spectrum the triplet of numbers I_R, I_G, I_B will take on a unique set of values. These values can be displayed graphically by plotting the relative intensity of each primary necessary to match a unit intensity of the pure colors of the visible spectrum. Such a plot is shown in Figure 3.3. The actual shape of these curves depends on the choice of wavelengths λ_R, λ_G, and λ_B. However, when the information contained in such curves is further analyzed to produce the CIE-XYZ system, the same result is obtained regardless of the original choice of λ_R, λ_G, and λ_B.

The data of Figure 3.3 tell us how to match a unit intensity of any monochromatic light in the visible spectrum. To match something other than a unit intensity of λ, we simply scale the values from Figure 3.3 accordingly. For example,

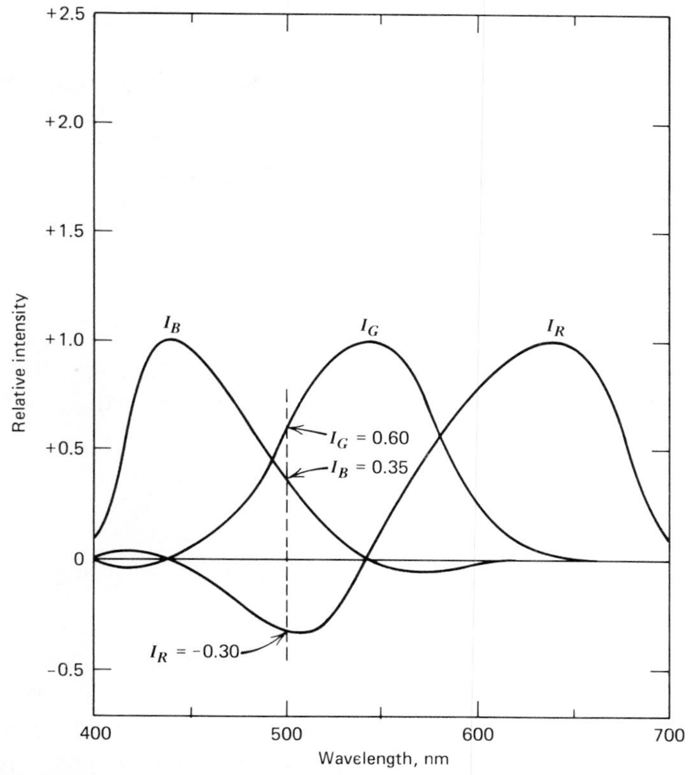

Figure 3.3 Relative intensity of three monochromatic primaries necessary to match a unit intensity of the colors of the visible spectrum.

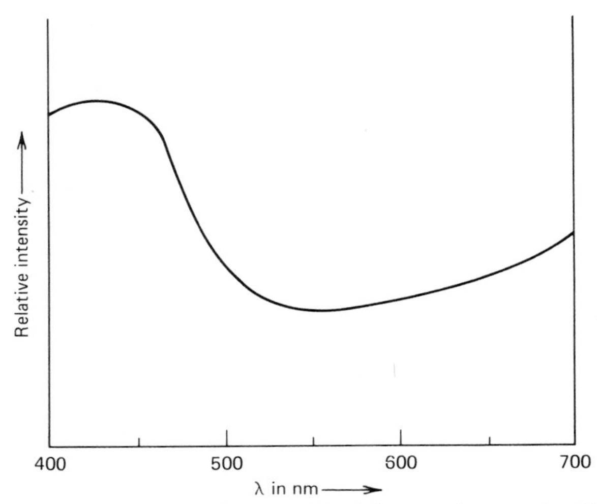

Figure 3.4 Example of spectrum (in this case blue) that can be matched by a combination of three primary colors.

let us first use Figure 3.3 to determine the intensities I_R, I_G, and I_B necessary to match a unit intensity of 500 nm light. The dashed vertical line on Figure 3.3 is drawn at 500 nm. Where this line crosses the three curves on the figure indicates the intensities of I_R, I_G, and I_B needed to match a unit intensity of 500 nm light. From the figure we see that the required values of I_R, I_G, and I_B are

$$I_R = -0.30 \text{ units}$$

$$I_G = 0.60 \text{ units}$$

$$I_B = 0.35 \text{ units}$$

We note in passing that in this case it happens to be I_R that is negative; therefore the true metameric match would be (in terms of some unit of intensity)

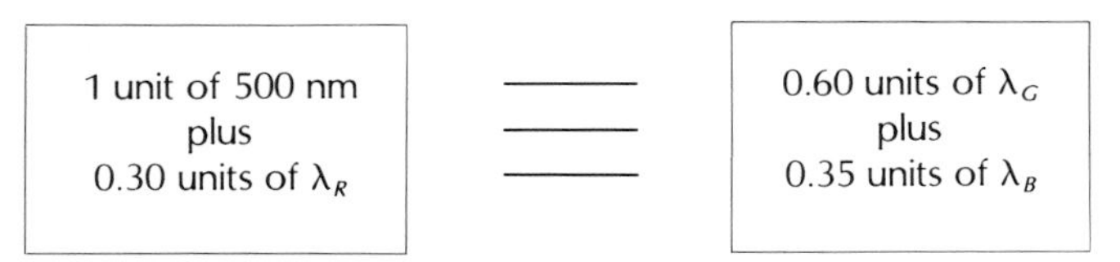

Algebraically the match is simply written

$$\begin{matrix} 1 \text{ unit} \\ \text{of } 500 \text{ nm} \end{matrix} = -\begin{matrix} 0.30 \text{ units} \\ \text{of } \lambda_R \end{matrix} + \begin{matrix} 0.60 \text{ units} \\ \text{of } \lambda_G \end{matrix} + \begin{matrix} 0.35 \text{ units} \\ \text{of } \lambda_B \end{matrix}$$

In order to match 2 units of 500 nm light we simply double the values of I_R, I_G, and I_B. Clearly any amount of 500 nm light can be matched in this way. Light of other wavelengths is matched in a similar fashion.

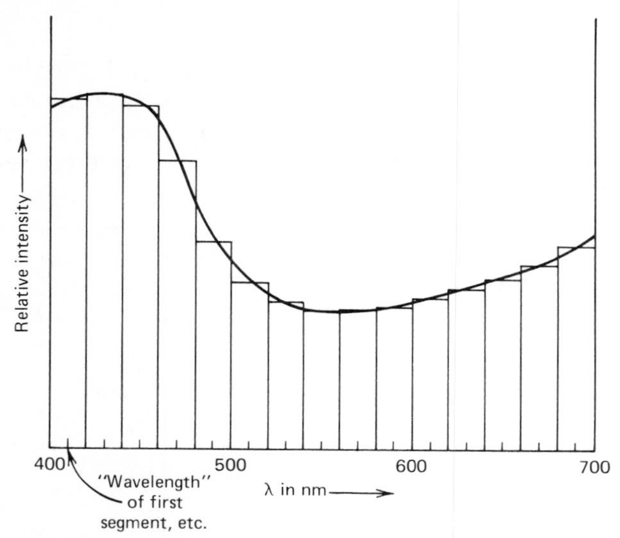

Figure 3.5 The spectrum of Figure 3.4 broken into small segments of approximately one color each. The light of each segment can be reduced to a combination of λ_R, λ_G, and λ_B. For the entire spectrum, all the combinations of λ_R, λ_G, and λ_B are added to give I_R, I_G, and I_B.

The data in Figure 3.3 can be used to match any monochromatic light. However, this information can also be used to match any complex spectrum as well. That is, any complex spectrum can be reduced to an equivalent combination of some set of values I_R, I_G, and I_B. Conceptually, such a reduction is easy to understand, but the actual computation is somewhat time-consuming. For example, suppose we want to know how to match the spectrum shown in Figure 3.4. We begin by breaking the spectrum into small segments as shown in Figure 3.5. Then, to compute how much of the red primary is needed, for example, we multiply the amount of light in the first segment of Figure 3.5 times the corresponding value of the red primary curve in Figure 3.3 at this wavelength. This tells us how much red primary is needed to match the light in the first segment of Figure 3.5. We do the same thing for the second, third, etc. segments until we have completed the entire spectrum. All these numbers are then added up. This sum gives us the amount of the red primary, I_R, needed to match the entire spectrum in Figure 3.5. In effect we have multiplied the curve in Figure 3.5 by the red primary curve in Figure 3.3, point by point. If we do the same thing for the green and blue primary curves, we get I_G and I_B, respectively. The point is that each wavelength in any complex spectrum can be reduced to a certain amount of λ_R, λ_G, and λ_B, and then all these amounts for all the wavelengths in the spectrum can be added up to give I_R, I_G, and I_B for the entire complex spectrum in question.

THE XYZ SYSTEM

Systems such as that of the previous section are not usually used for colorimetric calculations for two reasons: (1) for many spectra, one of the intensities I_R, I_G, I_B is often algebraically negative; (2) no choice of primaries is really any more basic than any other set that might have been chosen. For these reasons the CIE

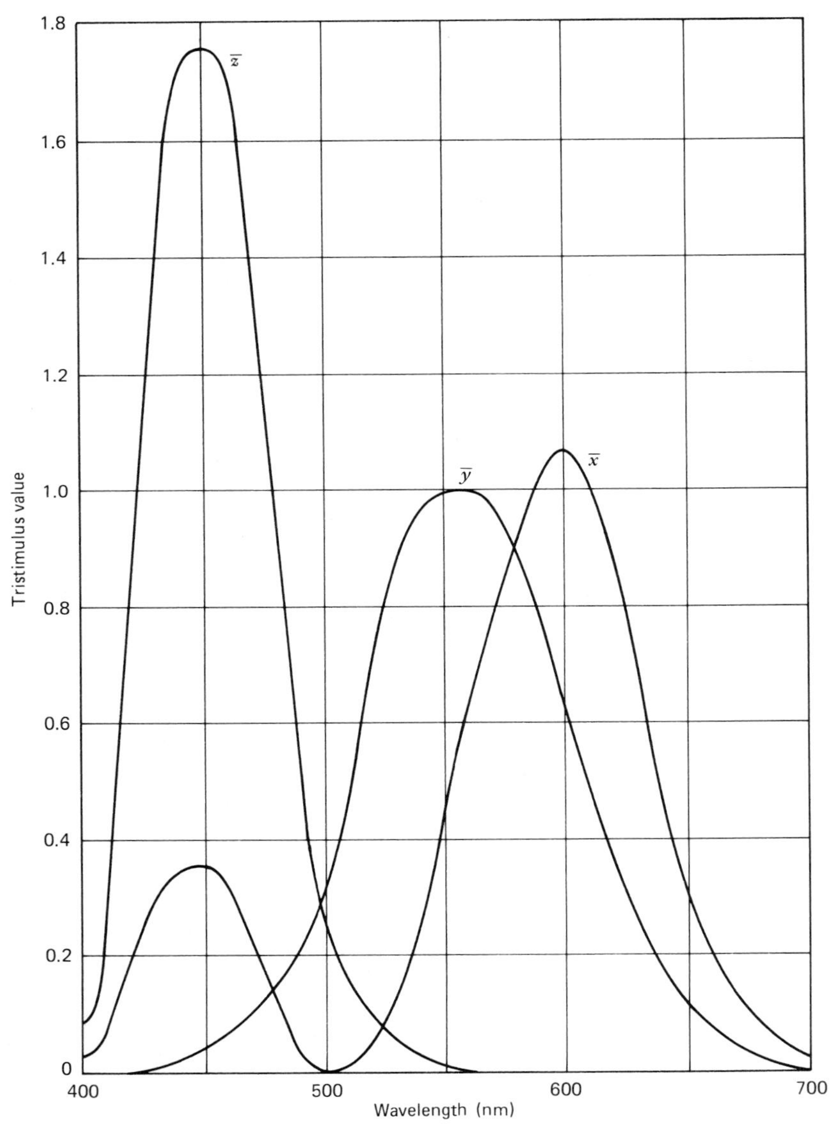

Figure 3.6 1931 CIE color matching functions for the XYZ system.

created the XYZ system that can be derived from any system of real primaries, but that has some distinct advantages.

1. The XYZ primaries are "imaginary" primaries that can be algebraically derived from the color matching data representing any system of real primaries (such as the data of Figure 3.3). But these new primaries are derived in such a way that the color matches they express never require negative numbers.
2. The "green" primary is arbitrarily given a matching curve that is identical to the visual sensitivity curve of the human eye. That is, this curve represents how sensitive the human eye is to light of different wavelengths. Multiplying the green curve times the spectrum to be analyzed automatically gives the apparent brightness of the spectrum.

The color matching curves for the primaries of the XYZ system are shown in Figure 3.6 where $\bar{x}$ corresponds to the "red" primary, $\bar{y}$ to the "green" primary, and $\bar{z}$ to the "blue" primary. To use the XYZ system to analyze a particular spectrum we proceed just as we did in the previous section with the λ_R, λ_G, λ_B primaries. To find I_X (or just X in the usual notation) we multiply the $\bar{x}$ curve times the spectrum in question, and similarly for Y and Z. Appendix A contains a detailed description and table on how to do this. The result of this process for all three primaries gives

X = the amount of "red" primary needed to match the spectrum
Y = the amount of "green" primary necessary—also the total brightness of the light
Z = the amount of "blue" primary needed to match the spectrum

The X, Y, and Z values are called the *tristimulus values* of the spectrum. They tell us the color and brightness of the spectrum in question. This analysis reduces any spectrum of light to the three tristimulus values. Any two spectra that have identical tristimulus values, even though the spectra might differ in detail, will produce the same visual sensation provided they are observed under the same circumstances. Figure 3.7 shows an example of two different spectra with identical tristimulus values. Such spectra would be metamers, that is, they form a metameric match.

CHROMATICITY COORDINATES

The tristimulus values X, Y, and Z do indeed specify the color of a given spectrum. On the other hand, it is difficult for most people, when presented with a particular set of tristimulus values, to form a mental image of what the color in question actually looks like. To overcome this problem a new set of numbers, which are easily calculated from the tristimulus values, are produced. These numbers, called the *color coordinates,* are defined as follows:

$$x = \frac{X}{X + Y + Z} \tag{5a}$$

$$y = \frac{Y}{X + Y + Z} \tag{5b}$$

$$z = \frac{Z}{X + Y + Z} \tag{5c}$$

In effect, x, y, and z represent the relative amounts of the three imaginary primaries required to match a particular spectrum. The real usefulness of these color coorindates will become apparent in a moment. But first, let us examine them a little more closely. From the definitions of x, y, and z it is apparent that the following equation is always true.

$$x + y + z = 1$$

Thus once two of the three numbers x, y, z are given, the third is automatically determined. This means that the triplet of numbers (x, y, z) really only contains two *independent* numbers. As we shall see shortly, the chromaticity coordinates specify only the hue and saturation of a color, but not its brightness. In order to specify the brightness, we continue to use the value of Y, the *green* tristimulus value. As a matter of convention, the two independent color coordinates that are used to describe the color of a spectrum are x and y. Thus the complete description of the color of a spectrum is given by the values of x, y, and Y. These three numbers provide an alternative to using the values of X, Y, and Z.

The advantage of using the color coordinates x and y to represent a color is that values of x and y can be displayed graphically. Let us make a graph where the vertical axis is marked with values of y from zero to one, and where the horizontal axis is marked with values of x also from zero to one. Such a graph is shown in Color Plate 8, and is called a chromaticity diagram. On a diagram such as this every point (x, y) represents a unique color or chromaticity. It turns

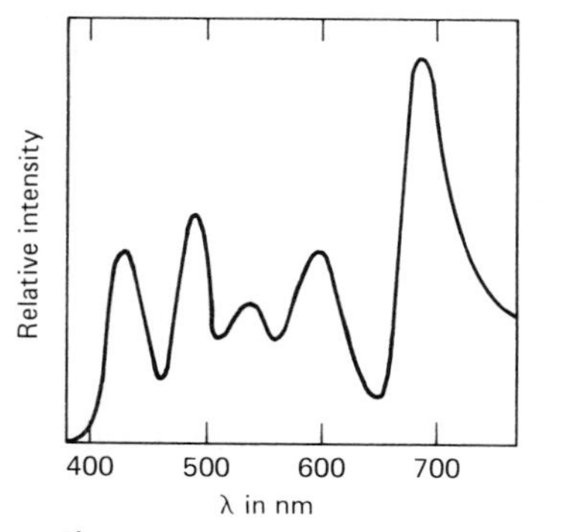

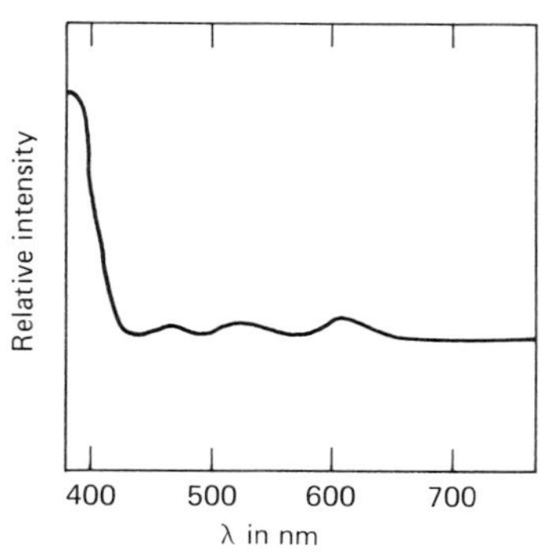

Figure 3.7 Two spectra with identical tristimulus values.

out that the chromaticities (values of x and y) for all physically possible spectra are confined to a single region of the chromaticity diagram, that shown in color on Color Plate 8. This region is called the *color locus.* Chromaticities outside the color locus are impossible to achieve with any spectrum of light from the visible spectrum. The horseshoe shaped outer boundary of the color locus represents the chromaticities of all the pure colors of the visible spectrum. The corresponding wavelengths are labeled around the periphery of this curve, which is known as the *spectrum locus.* The lower portion of the color locus is bounded by a straight line that connects the blue and red ends of the spectrum locus. This straight line is known as the *purple line.* Hues along this line are not produced by any single wavelength of light, but rather, result from the mixture of red and blue light.

Points along the periphery of the color locus (along the spectrum locus and purple line) represent colors of the maximum possible saturation or purity. As we move toward the center of the color locus the saturation diminishes until, at the point $x = 0.33$, $y = 0.33$ the saturation or purity becomes zero. This central point, labeled *E* on Color Plate 8, represents equal energy white.

USE OF THE CIE CHROMATICITY DIAGRAM

The CIE chromaticity diagram represents a barocentric colorimetry system just as did Newton's color wheel system. Thus we can use the CIE chromaticity diagram to analyze color mixtures and other features of color in the same way we used Newton's system. To see how this is done let us focus our attention on Figure 3.8. Points *M* and *N* represent two colors on the spectrum locus. All the chromaticities along the line joining *M* and *N* can be achieved by mixing various combinations of the pure color *M* with the pure color *N*. For example, point *L* represents an equal mixture of *M* and *N*. Now consider points *J* and *K*, which represent two spectra that are not monochromatic. *J* and *K* might represent the color of light coming through two filters, for example. Again, any chromaticity along the line joining *J* and *K* can be achieved by mixing lights *J* and *K*.

Of particular interest in Figure 3.8 are points *P* and *Q*. Notice that these points both represent pure spectrum colors, and that the line joining them passes directly through the equal energy white point (*E*). Pure colors such as *P* and *Q*, which lie at opposite ends of a straight line passing through the white point, are called *complementary colors.* That is, these two colors can be combined in the proper proportion to produce pure white.

Finally, on Figure 3.8, let us examine the small triangular region labeled *RST*. The three spectra represented by *R*, *S*, and *T* can be combined to produce any chromaticity inside this triangle.

Now let us turn our attention to Figure 3.9. On this diagram there are two triangles labeled *RGB* and *rgb*, respectively. The larger triangle (*RGB*) is formed by three monochromatic lights. All the chromaticities inside the larger triangle can be matched by a combination of the lights *R*, *G*, and *B*. However, all the

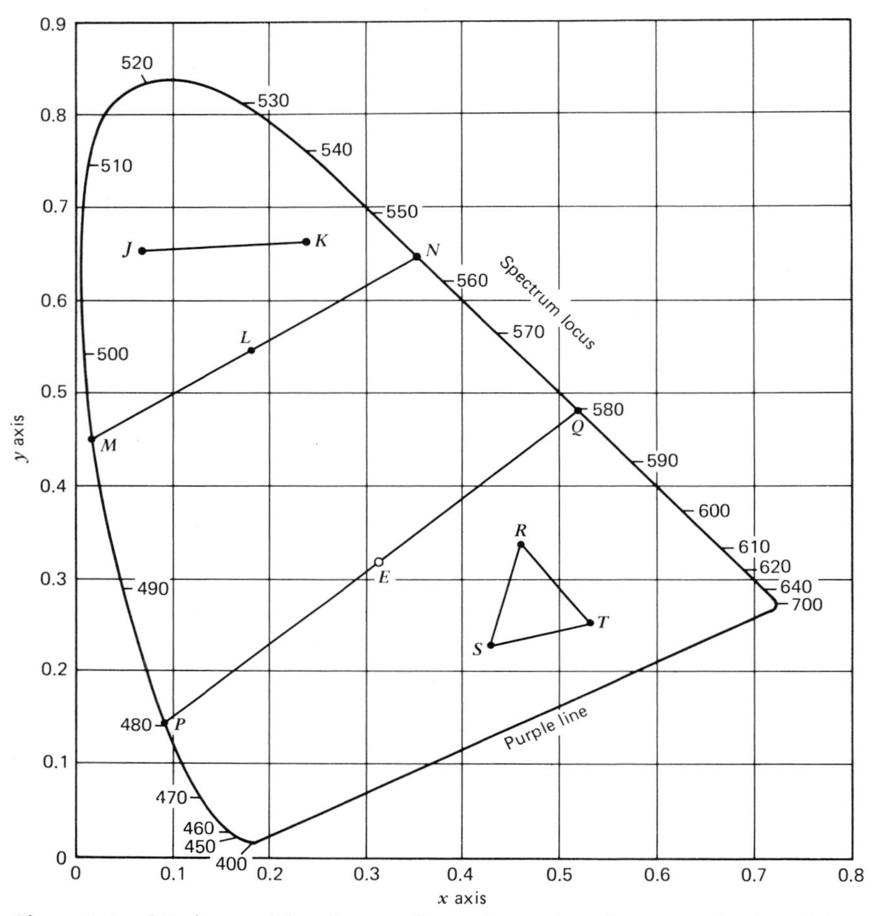

Figure 3.8 CIE chromaticity diagram illustrating various features of color mixing.

points outside the triangle (which includes the entire spectrum locus) cannot be matched. Now we see why three pure primaries cannot match any other pure color, such as *K*, directly. However, the diagram also shows why a particular combination of *R* and *K* (represented by point *M*) will match a certain combination of *B* and *G* (also represented by *M*). That is, there is some combination of intensities for which

$$I_k + I_R \equiv I_B + I_G \tag{6}$$

This is the origin of equations (1), (2), and (3).

The smaller triangle (*rgb*) on Figure 3.9 represents the chromaticities that can be produced by combining three nonmonochromatic primaries produced by standard translucent filters. It is clear that a full range of hues can be achieved

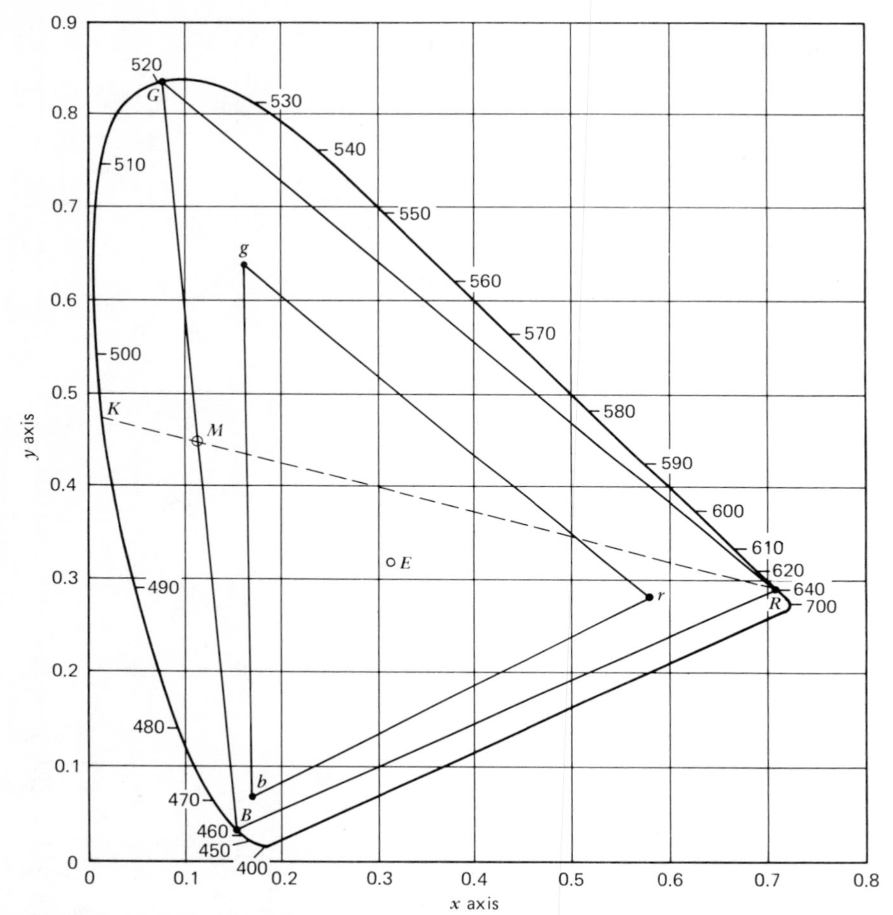

Figure 3.9 Chromaticity diagram showing the area covered by three monochromatic primaries (R,G,B) and three nonmonochromatic primaries (r,g,b).

(the white point is completely surrounded), but the maximum attainable saturation is less than with monochromatic primaries.

DOMINANT WAVELENGTH AND PURITY

The idea of using chromaticity coordinates and the resulting diagram provides a very powerful tool for analyzing color. In fact, with the aid of a chromaticity diagram it is possible to compute a new pair of numbers for a given chromaticity that permits a clearer visualization of the color actually involved. To see how this is done let us turn to Figure 3.10. On this figure point S represents the chromaticity of some spectrum of light. Let us draw a line from point E through point

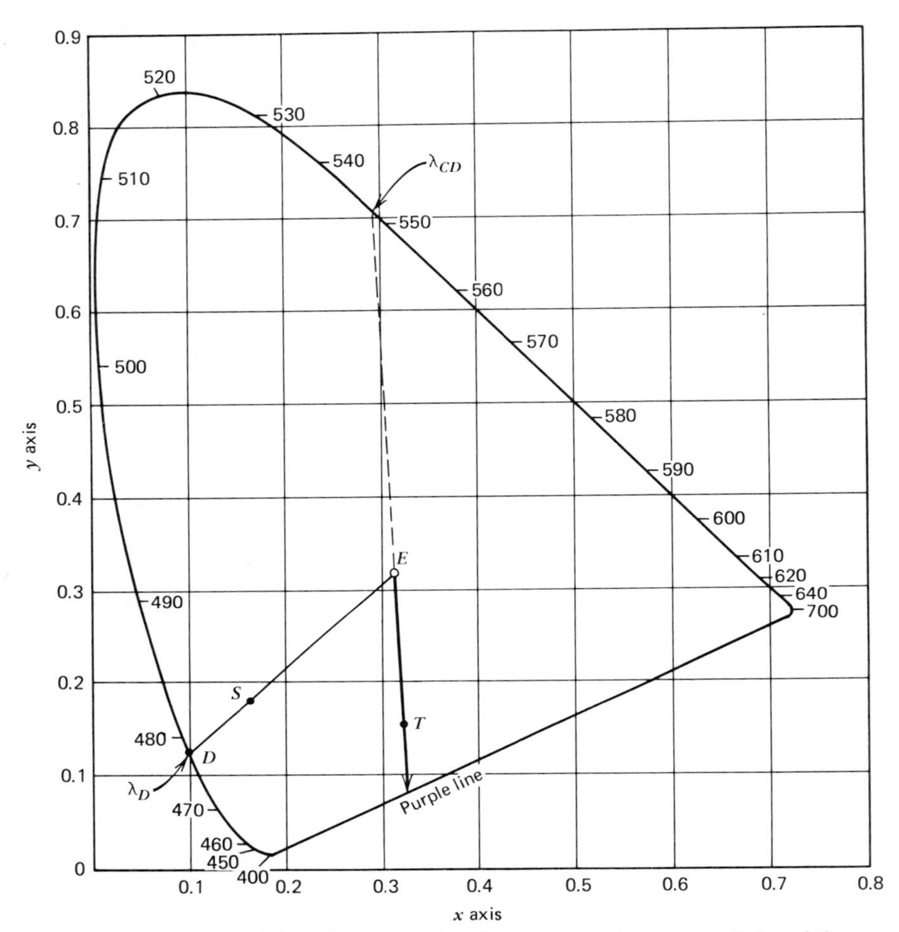

Figure 3.10 CIE Chromaticity Diagram. Point E represents the chromaticity of the equal energy spectrum. Point S represents the chromaticity of some blue light. λ_D is the dominant wavelength of the blue light. The purity of the blue light is $p = \frac{SE}{DE} \times 100\%$.

S until the line strikes the spectrum locus (at point D). Point D will be located at a particular wavelength on the spectrum locus. This wavelength is called the *dominant wavelength,* λ_D, of the spectrum represented by point S. The basic hue of this dominant wavelength is the hue of color S. Thus knowing λ_D gives us an immediate idea of what kind of color S really is. Now we can also define a *purity* for the color. White (point E) is considered to have zero purity, while a pure color such as D is considered to be 100% pure. Thus the purity of point S, expressed as a percent, is given by

$$p = \frac{\text{length of } SE}{\text{length of } DE} \times 100\% \tag{7}$$

Purity is a measure of the saturation of the color, or how far it is removed from white toward the full color of the spectrum locus. The dominant wavelength and purity corresponds basically to the hue and saturation of a color. The brightness must be specified separately, and this is usually done by giving Y. Thus the color of a particular spectrum of light can be completely specified by giving λ_D, p, and Y.

We must add one note of caution. For some spectra, the chromaticity may lie at a point such as T in Figure 3.10. If we try to draw a line from E through T, we find that this line strikes the purple line rather than the spectrum locus. We can still calculate the purity as before, but what about the dominant wavelength? The answer is to extend the line *backwards* (the dash line in Figure 3.10) until the line hits the spectrum locus. The wavelength at which this occurs is called the *complementary dominant wavelength,* λ_{CD}, of the color T.

THE ANALYSIS OF SURFACES

As a practical matter, the CIE method of analysis is quite frequently used to describe the color of surfaces. Two facts now combine to pose a potential difficulty. The first is that the CIE system analyzes the spectrum of light coming from a given object. The second is that the spectrum of light coming from a given surface depends on the spectrum of light that illuminates the surface. Thus it would seem that a particular surface could have a whole variety of chromaticities under different circumstances. For example, in Color Plate 3 the chromaticities of the six colors shown have been computed assuming illumination by an equal energy spectrum. A different type of illumination would produce a different set of chromaticities. In order to reduce this problem to manageable proportions, the CIE has adopted three standard light sources, which are used in most practical colorimetry. The spectra of these sources, which are described below, are shown in Figure 3.11.

Source A. Representative of a gas-filled incandescent lamp. This source is an incandescant lamp operated at a color temperature of 2854°K. It may be purchased from independent testing laboratories or from the National Bureau of Standards.

Source B. Representative of noon sunlight. This source is produced by using source *A* in combination with a special liquid filter giving a color temperature of about 4870°K.

Source C. This source is similar to an overcast sky or average daylight, and has a color temperature of about 6700°K. Like Source *B*, it can be produced by the use of Source *A* with proper liquid filters.

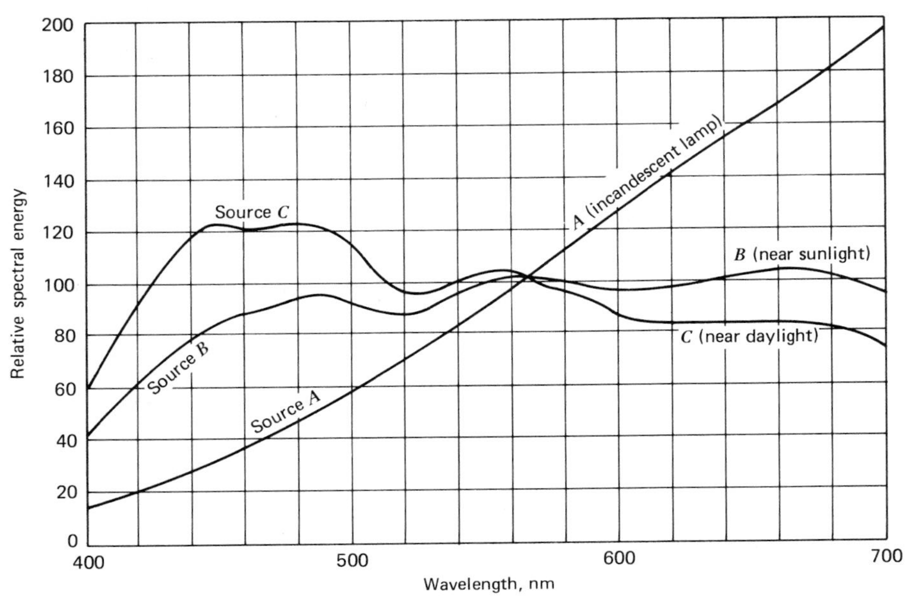

Figure 3.11 Relative spectral intensity of CIE standard sources *A*, *B*, and *C*. Source *A* is typical of the gas-filled incandescent lamp. Source *B* represents noon sunlight. Source *C* represents average daylight.

As an example of how these sources are used in the analysis of color let us begin with Source *B*. When this source is analyzed, it is found to have chromaticity coordinates of $x = 0.35$ and $y = 0.35$. Notice that these coordinates are slightly different from those of the equal energy spectrum ($x = 0.33$, $y = 0.33$). Figure 3.12 shows the chromaticity of illuminant *B* plotted on a chromaticity diagram. Suppose illuminant *B* is used to illuminate a surface whose reflectance curve is known. Since the spectral content of illuminant *B* is known as well as the reflectance curve of the surface, the methods discussed in Chapter 2 can be used to find the spectrum of light reflected from the surface. Once this spectrum is known its chromaticity coordinates can be found by the methods described in this section. Suppose this analysis gives the chromaticity coordinates labeled point *S* on Figure 3.12. Point *S* thus represents the chromaticity of the surface when illuminated by source *B*. If a different source were used, the location of point *S* would probably shift somewhat.

In order to compute the dominant wavelength and purity of *S* we *must use point B as the white point,* not point *E*. The reason for this is simple. If source *B* illuminates a perfectly white surface, the reflected light will have the same spectral content as source *B* (since white reflects all wavelengths equally). The chromaticity of this reflected light will thus be given by point *B*. That is, when illuminated by source *B*, a perfectly white surface will have a chromaticity given

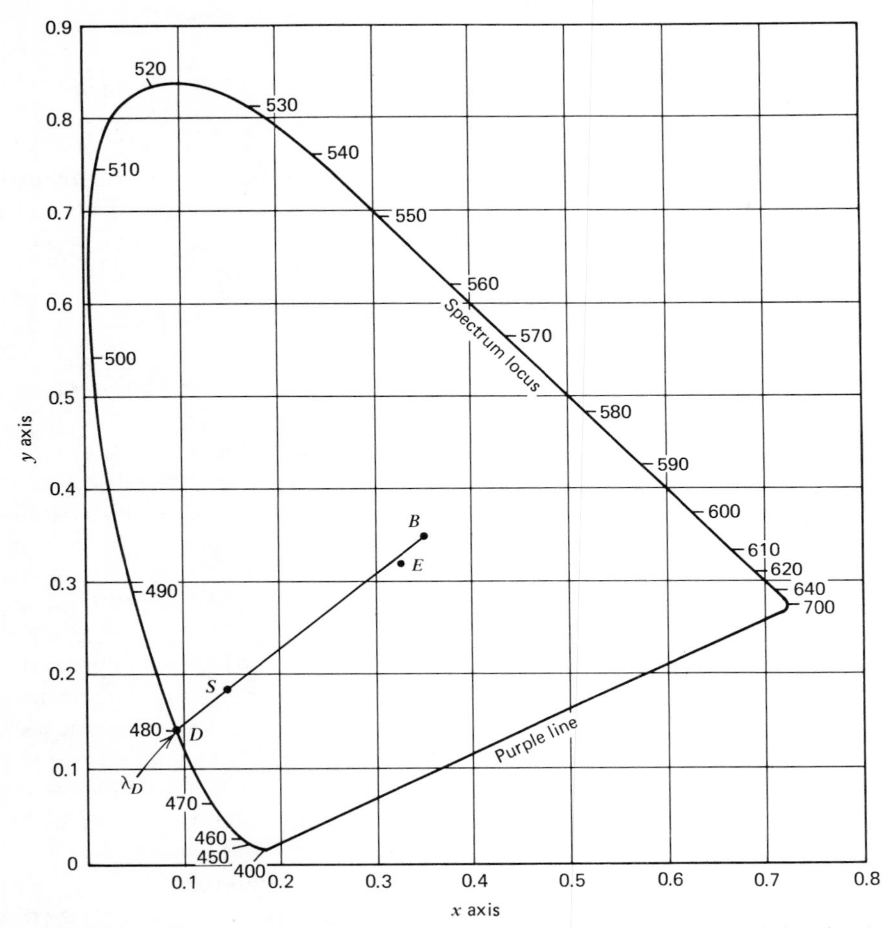

Figure 3.12 A CIE chromaticity diagram illustrating the determination of the dominant wavelength (λ_D) and purity of a spectrum (S) reflected from a surface illuminated by standard Source B. The purity is given by $p = \frac{BS}{BD} \times 100\%$.

by point B. Thus point B is taken as the white point under these circumstances. We now proceed as before to compute the dominant wavelength and purity of point S. This is illustrated in Figure 3.12.

In analyzing a surface we must also consider the value of Y. Frequently this is done in terms of a percent. We begin by imagining that our illuminant (for example, source B) illuminates a perfectly white surface. We compute the value of Y for this situation and call this value Y_0. Next we compute the value of Y for the actual surface of interest, and call this value Y_R. We then say that the value of Y expressed as a percent is given by

$$Y = \frac{Y_R}{Y_0} \times 100\% \tag{8}$$

This means that a perfectly white surface would have a value of Y of 100%, while a perfectly black surface would be characterized by 0%. It is entirely possible for two different surfaces to have the same chromaticity coordinates (and therefore the same λ_D and p), but have different values of Y. An example of two such colors might be a pink and a dark unsaturated red.

A WORD OF CAUTION

The whole point of colorimetry is to provide an accurate way of describing a color so that it can be faithfully reproduced anywhere at anytime. The CIE system provides a way of reducing the spectrum of light coming from any object to a set of three numbers (X, Y, Z or x, y, Y, or λ_D, p, Y). But we must be very careful about what these three numbers mean. A spectrum that is described by a particular chromaticity and brightness *will not always produce the same visual sensation.* The chromaticity of a given spectrum does not tell us what color we shall actually see unless we also know all the conditions under which the spectrum is being observed (e.g., what illumination is being used, what surrounds the object in question, and the state of our eyes' adaption). What the tristimulus values do tell us is how to match the visual stimulus presented by the spectrum in question. That is, under *a given set of conditions,* two entirely different spectra which have identical tristimulus values will appear identical. If the viewing conditions are changed, the actual color produced by the two spectra may change, but the match will remain intact provided the tristimulus values are still equal.

The use of the CIE system in practice is carried out in the following way. Suppose we have a particular color that we want to produce in a fabric. The color to be matched is usually a real standard of some sort with an already measured reflectance curve. Suppose we want to dye our fabric so that it matches our color standard under CIE illuminant *A*. Since the spectrum of illuminant *A* and the reflectance curve of our standard are known, we can compute the tristimulus values of our standard with respect to illuminant *A*. We now experiment with various dyes until our fabric, when illuminated by illuminant *A*, has the same tristimulus values as our standard. We then know that a match will exist, and our fabric will be the desired color.

3.3 THE MUNSELL COLOR NOTATION SYSTEM

The obvious needs that science and industry have for color standards led an American teacher and portraitist, Albert Munsell, to develop a color notation system in about 1915. Munsell's system is a three-dimensional system in which he classified colors according to three perceivable qualities: hue, value, and chroma.

Hue is that quality by which we distinquish one color from another, as a red from a yellow, a green, a blue or a purple. In the Munsell system colors are arranged from red to yellow to green, blue and purple and back to red in a series of steps that psychologically are seen as equally spaced.

Differences of hues are represented in a Munsell solid or tree by different planes around a vertical axis (Color Plate 9). In actual practice a series of pages of color chips are provided for each separate hue that is represented. Usually, the whole solid is divided into 10 vertical *segments* or hues. The five principle hues, red *R*, yellow *Y*, green *G*, blue, *B*, and purple *P* occupy the planes of alternate segments, while the intermediate hues *YR*, *GY*, *BG*, *PB*, and *RP* are situated in the planes of the five remaining segments. To provide for closer grading of hues, each segment is then subdivided into 10 *sections*, numbered 1 to 10, and arranged so that the main hue in each segment is numbered 5 as shown in Figure 3.13. The hue of a sample is therefore designated by a section number and a segment letter. Thus the designation 4Y indicates that the color is to be found in the fourth *section* of the yellow *segment*.

The second variable used in specifying a sample is value. The value of a color is its lightness and darkness. Synonyms sometimes used for value are brightness, brilliance, or luminosity. It may also be thought of as the measure of the portion of the incident light that is reflected. The value scale range is divided into 11

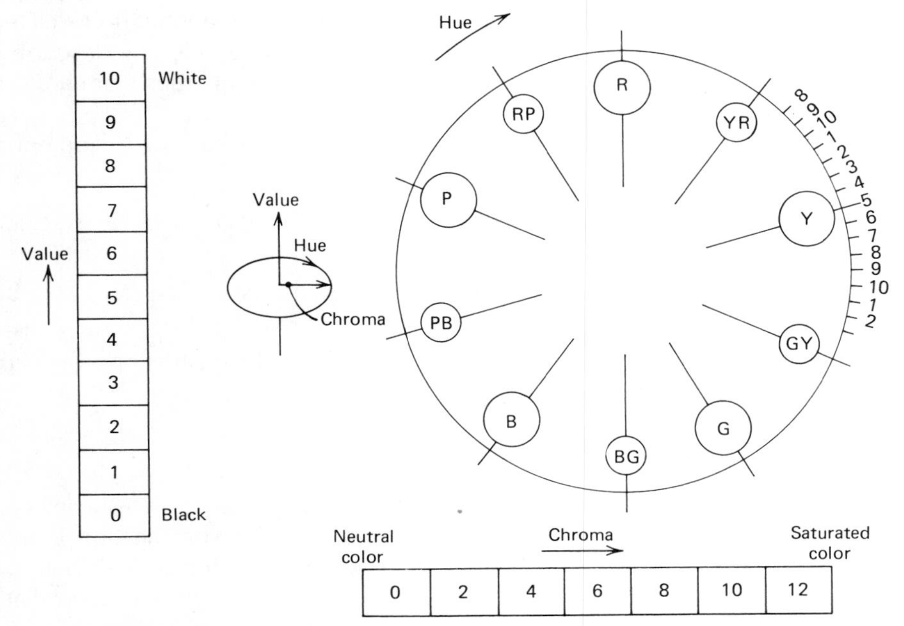

Figure 3.13 Hue, value, and chroma coordinates of the Munsell system.

subjectively equal steps, with perfect black as 0 and a perfect white having a top value of 10. In practice most colors have values somewhere between 1 and 9. On the color solid or tree the lowest values are found at the bottom and lighter values nearer the top. A rough estimate of the value of a color can be found by considering the portion of the incident light that the color reflects. The *square root* of the percentage of the light reflected is a fair estimate of the value of the color. Thus a surface reflecting 50% of the incoming light would have a value of approximately 7. Similarly a surface of value 9 would reflect about 80% of the light.*

The third term necessary for the complete description of a color in this system is chroma. Chroma is a measure of the intensity of the color. It may also be thought of as the purity or saturation of the color, or conversely, its freedom from dilution. The color's chroma rises as it becomes stronger, that is, as the color moves away from a neutral black, white, or gray. The chroma scale begins with a zero for a neutral color and moves to high-numbered values as the color becomes more saturated. Thus we find black, grays, and white, which have no color, at the central axis of our color tree or solid. Colors with increasing chroma are found progressively farther from the central axis.

To completely specify a color in this system it is necessary to give values for each of the variables – hue, value, and chroma. This is done in written notation by first designating the hue of the sample such as 5P followed by its value, such as 4, and then a slanted line / after which its chroma, 8 for example, is found. Thus the complete specification would be 5P4/8, indicating a medium purple hue of value 4 with high chroma of 8. Likewise, if a fabric manufacturer desired to order dye for red stripes in the American flag he could properly describe it as 5R3/14 – an intermediate red hue of value 3 that was 14 steps from gray of the same value.

Besides being convenient and easily used this system has these additional advantages:

A. Loosely stated and unrelated color terms are replaced by a specified notation.

B. If and when new colors are developed by chemists, they can easily be fitted into the system without disturbing it.

C. Color grading and specification of agricultural and industrial products are easily accomplished by direct perceptual comparison.

3.4 THE OSTWALD COLOR SYSTEM

Another material system using color samples similar to those of the Munsell system is the Ostwald Color System. This scheme of color specification is also based on a three-dimensional solid in which the variables for a given hue are the amounts of colored pigment, white pigment, and black pigment (Color Plate

*For exact values see Table of Luminous Reflectances for Munsell Value, page 64, *A Color Notation;* A. H. Munsell; Munsell Color Company, Baltimore, 1975.

9). As shown in Figure 3.14 the black, white, and full color pigments are located at the corners of an equilateral triangle. The colors within the triangle are regarded as an additive mixture of black, white, and color with the sum of the black content B, the white content W, and the color content C, being equal to one; that is,

$$B + W + C = 1$$

As can be seen from Figure 3.15, colors that are located along lines that are parallel to line WC are colors with fixed amounts of black content. These colors are called isotones. The amount of white content along one of these lines is varied to provide uniform spacing of the colors. Lines parallel to BC represent colors that have a constant proportion of white content and are known as isotints. The third type of grouping of colors available within the triangle are those along lines parallel to BW and correspond to colors of equal full-color content and are known as isochromes. These isochromes have the same proportion of color but varying amounts of black and white content. An entire color solid is produced by rotating a triangle about the WB axis. Within this solid colors may then have constant black, white, and color contents but differing hues. Such colors are known as isovalent colors.

This system is particularly appealing and useful to those who are involved in the mixing of chromatic pigments with black and white pigments as in the formulation of a paint or the mixing of paints on a palette. People involved in such activities often find that a collection of Ostwald chips are extremely helpful in

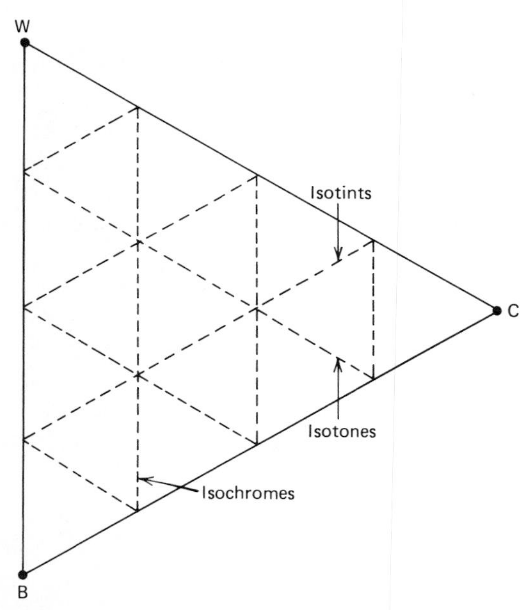

Figure 3.14 Coordinates of the Ostwald system for a given hue plane.

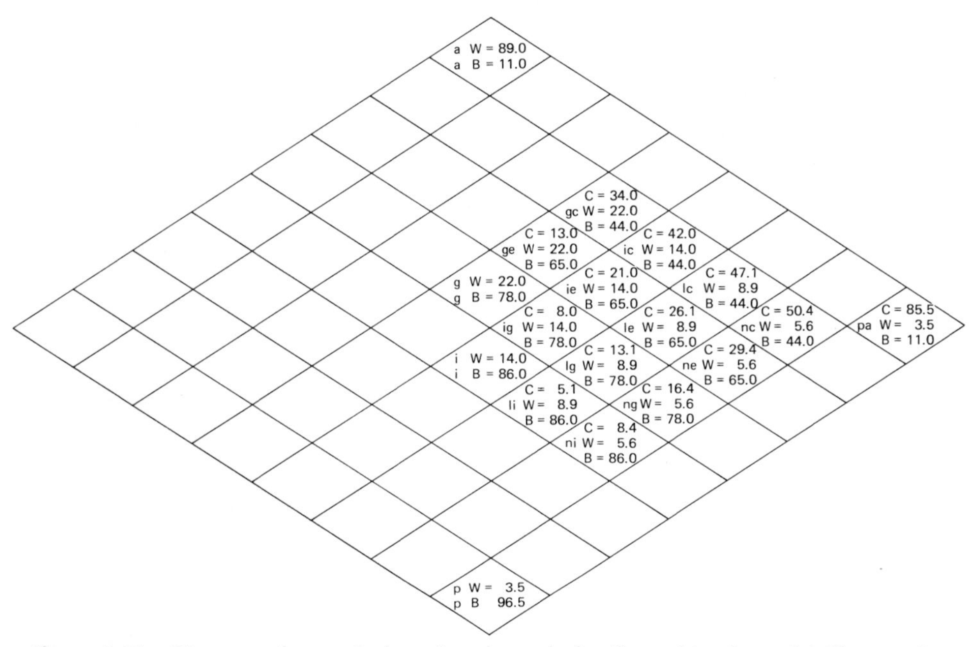

Figure 3.15 Diagram of a vertical section through the Ostwald color solid. The number given after the letter W in each small section is the percentage of white, after letter B is the percentage of black, and after letter C is the percentage of "full color." The left half of the diagram corresponds to the right half except that the "full color" is the complement of that of the right half.

indicating whether black, white, or color must be added to their existing mixture to move the color in the direction they desire the color to go. The existence of the chips and a knowledge of their make-up give a clear indication of what to add to what they presently have. Those who work with printing by screen plates, involving combinations of colored pigments and black pigment with a white paper background showing between the dots, also find this to be a very useful system when determining which direction to take to arrive at a final color.

While this system is convenient for the reasons indicated above, it does have the disadvantage of being set up for a limited number of basic hues. Another problem that could be encountered involves the development of a pigment of one of the hues that is brighter than the existing full color pigment. If account were to be taken of such a new pigment, then this pigment would have to be placed at C and a whole new distribution in that hue would have to be developed.

IN CONCLUSION

This chapter has described three important colorimetric systems that are commonly used to describe color. The *Munsell* and *Ostwald* systems rely on *psy-*

chological attributes of color such as hue, color content, and lightness or darkness. In each case, the basic hues of the system were chosen so that they appeared equally spaced to the normal color observer. New hues can then be described by fitting them in the appropriate slot between the standard hues of the system.

The Munsell system uses *hue, value,* and *chroma* to describe the color, lightness or darkness, and color saturation of a sample. The Ostwald system, on the other hand, uses *hue* combined with *white content* and *black content* to describe the color of a sample.

In contrast to the Munsell and Ostwald systems, the CIE system relies on a detailed numerical analysis of the spectrum of light coming from a colored sample. The spectrum is reduced to a set of three numbers, X, Y, Z the *tristimulus values.* These numbers roughly represent the amount of red, green, and blue light, respectively, present in the spectrum. The value of Y also tells us the *brightness* of the light. The numbers X, Y, Z, can be converted, through equations (5), to the *chromaticity coordinates* x, y, z. A convenient way of graphically presenting color information is to plot the chromaticity (values of x and y) of various colors on a *chromaticity diagram* (a plot of x versus y). In such a diagram the color of a spectrum can equally well be described by its color coordinates and brightness (x, y, Y) or by its dominant wavelength, purity, and brightness (λ_D, p, Y).

As a matter of convenience, it is possible to draw a rough analogy between the Munsell system and the CIE system. Generally speaking we may say that:

Munsell		*CIE*
hue	corresponds to	dominant wavelength (λ_D)
chroma	corresponds to	purity (p)
value	corresponds to	brightness (Y)

PROBLEMS AND EXERCISES

1. Using Newton's barocentric color system shown in Figure 3.2*a*, predict the results of the following mixtures of colored light.
   (a) Equal amounts of red and green
   (b) Equal amounts of red, yellow, and indigo
   (c) Equal amounts of yellow and blue

2. Using Figure 3.3, estimate the relative amounts of the primary colors used in the CIE system (700 nm, 546.1 nm, 435.8 nm) needed to match the following pure wavelengths.
   (a) 450 nm **(c)** 600 nm
   (b) 500 nm **(d)** 575 nm

3. The tristimulus values of a given spectrum are found to be

$$X = 14, \qquad Y = 12, \qquad Z = 4$$

(a) Find the chromaticity coordinates of this spectrum.
(b) Using Color Plate 6, give an approximate description of the color of this spectrum.

4. When a particular surface is illuminated by CIE source *B*, its color coordinates are found to be

$$x = 0.2, \qquad y = 0.6$$

With the aid of Figure 3.9, estimate the dominant wavelength and purity of this color.

5. Using Figure 3.10, estimate the complementary wavelength of each of the following:
(a) 460 nm **(c)** 600 nm
(b) 480 nm **(d)** 640 nm
How would you describe the complement to a wavelength of 520 nm?

6. When the CIE system is used with colored surfaces, the value of Y is often given as a percent. The way this is done is to first calculate the value of Y assuming the surface is perfectly white. That is, calculate the value of Y for the illuminant alone. Call this value Y_0. Next, calculate the value of Y for the light actually reflected from the colored surface. Call this Y_R. Then the value of *Y* expressed as a percent is

$$Y = \frac{Y_R}{Y_0} \times 100$$

Problem: Source *B* illuminates a perfectly white surface. The tristimulus values are found to be

$$X_0 = 35, \qquad Y_0 = 35, \qquad Z_0 = 30$$

When the white surface is replaced by a colored surface, the tristimulus values became

$$X_R = 30, \qquad Y_R = 28, \qquad Z_R = 10$$

(a) Calculate the color coordinates of the colored surface.
(b) With the aid of Figure 3.12, estimate the dominant wavelength and purity of the color of the surface.
(c) Calculate the reflected brightness of the surface (the value of *Y* expressed as a percent).

7. Describe the color from the previous problem in terms of Munsell Notation. (*Note:* The value of the color of a surface is given approximately by $\sqrt{Y}$ where Y is expressed as a percent.)

8. In terms of the Munsell system, how would you describe the difference between a very pale pink and a deep rich red?

4 COLOR VISION

In the previous chapter we discussed the science of colorimetry, the measurement and description of color. We did not, however, attempt to explain the actual mechanisms of color vision. That is the subject of the present chapter.

It will be helpful to begin by breaking the process of color vision into a sequence of stages, each somewhat distinct from the others. First, we have a physical stimulus, light. Under normal circumstances, light from all portions of the field of view enters the eye through the pupil and is focused on the retina (Figure 4.1). The spectral composition of light from different parts of the field of view is different, and presumably it is these differences that ultimately result in different perceived colors. Before any perception can occur, however, the light striking the retina must be absorbed and converted into some kind of physiological message that is transmitted from the retina through a complex network of nerves to the brain (Figure 4.2). In the brain the message must be interpreted, perhaps in concert with other messages from elsewhere in the body and with memory, to result ultimately in our perception of the world. Perception involves much more than just color; it includes size, shape, distance, speed, depth, shadow, and much more. Since we are particularly interested in color, however, we shall concentrate our efforts on theories of color vision. Let us first take a closer look at the observable facts of color vision.

4.1 BASIC FEATURES OF COLOR VISION

TRICHROMACY

Perhaps the most obvious feature of color vision is trichromacy. That is, as we saw in the last chapter, color can be specified by three numbers or terms (such as hue, value, and chroma, or λ_D, p and Y). This apparent trichromatic nature of color vision has formed the starting point for most theories of color vision.

COLOR CONSTANCY

One striking feature of color vision that is not appreciated by most casual observers is the phenomenon known as *color constancy*. In the course of

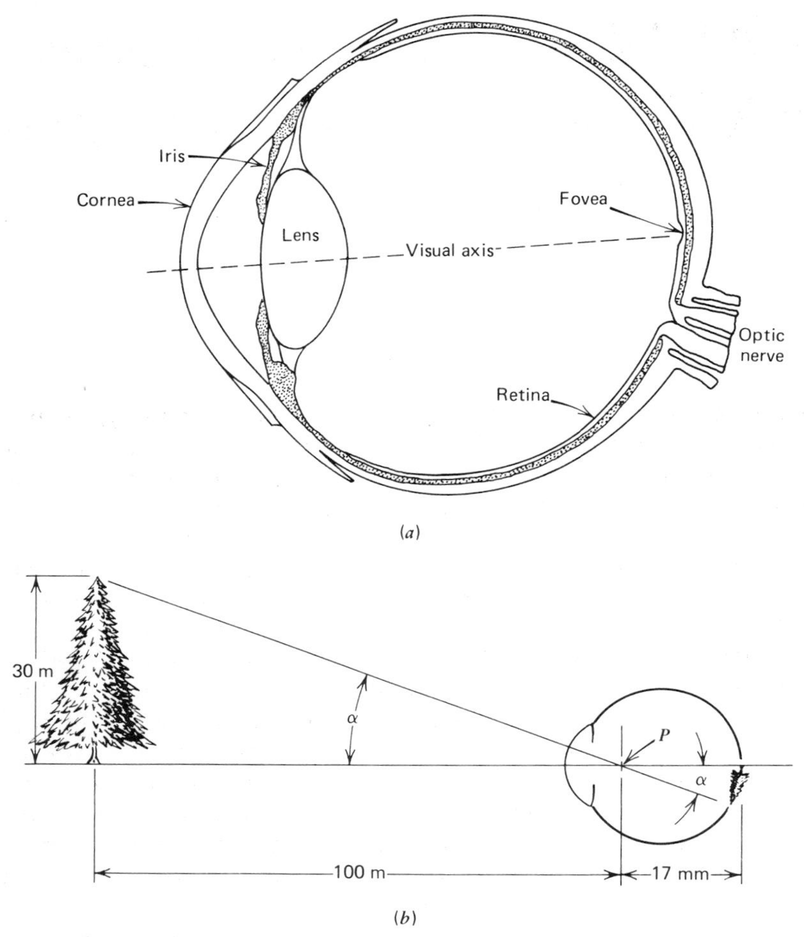

Figure 4.1 (*a*) Simplified diagram of the eye. (*b*) Image location on the retina. (Adapted from Cornsweet, *Visual Perception,* Academic Press, 1970)

everyday life we view the world under a wide variety of different illuminations. The light from the blue sky is radically different from the light of a fire. Yet the same object seen under these very different illuminations appears to have basically the same color. The illumination may change greatly but our perception of the colors of objects remains fairly constant. Except for extreme forms of illumination, it is as if the "true" color of objects somehow is correctly perceived regardless of the illumination.

CONTRAST EFFECTS

Although color constancy is a central feature of color vision, contrast between areas of different color viewed simultaneously or one after the other produces

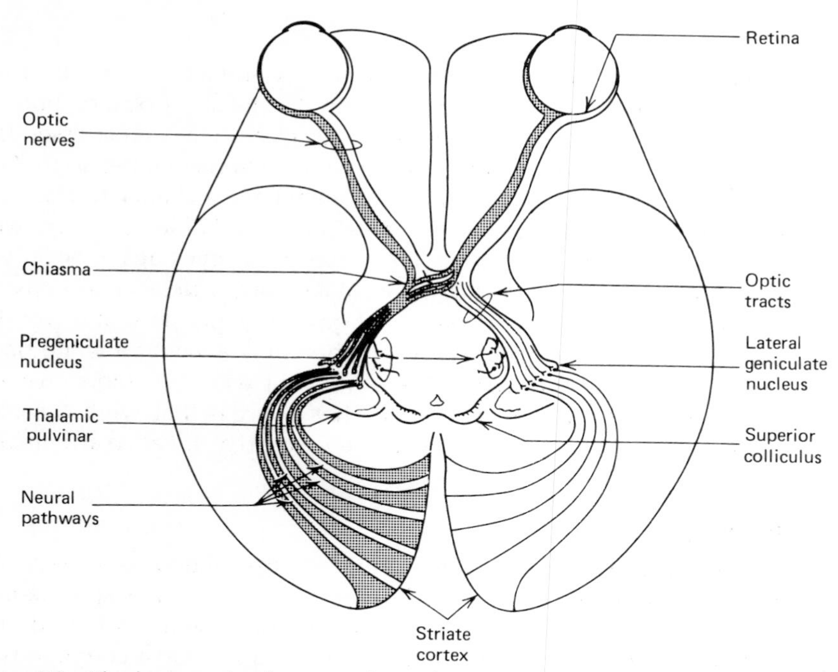

Figure 4.2 The human visual system drawn into an outline of the brain. (Adapted from Polyak, *The Verbebrate Visual System,* University of Chicago, 1957)

some interesting effects. First, let us consider *simultaneous color contrast.* This can be most readily observed in the following way. Shine a circle of white light on a screen. Around the white light shine a region of pale blue light. The effect is that the white circle will take on a yellow appearance. If the pale blue surround is removed and replaced by a pink surround, the white circle will appear cyan. It is the same white spot in both cases, reflecting exactly the same light to our eyes. But it *looks* different. There is also a black-white simultaneous contrast. Consider a circle of faint white light projected on a wall. Surrounding the circle is a slightly brighter area of white light. Under these circumstances the surround will look off-white and the central circular area will look slightly gray. If the intensity of the surround is increased, the surround will look whiter and the circle darker. If the surround is made very bright, the circle will look almost black even though the light coming from the circle has not changed at all.

In addition to simultaneous contrast, there is also a successive contrast effect. If you look for a period of time at an area of one color, and then look quickly at an area of a different color, the color of the second area will be modified. For example, if you stare at an orange piece of paper for a while and then look at a green piece of paper, the green paper will take on a definite blue-green appearance for a period of time.

AFTERIMAGES

Another interesting phenomenon is afterimages. Generally speaking, afterimages can be positive or negative depending on the details of observation. To observe a positive afterimage close your eyes and cover them so that essentially no light falls on the retina. Keep your eyes closed for several minutes so that the residual effects of light that previously entered them have a chance to dissipate. Now open your eyes for about *one-half of a second* and look at some well-illuminated colored object – then close and cover your eyes again. Soon you will see an afterimage of the object in its original colors, a *positive* afterimage. Observing a negative afterimage is somewhat easier. Simply stare at a well-illuminated colored object for a minute or two. Then look away at a white background. You will see an afterimage of the object but with the colors changed into the complement of the original colors. That is, areas that were originally yellow will look blue, and so on. What you see is called a *negative* afterimage.

COLOR BLINDNESS

Anyone with normal color vision can experience the phenomena discussed above. However, important information can be discovered by examining individuals who do not experience normal color vision – so-called color-blind people. Although there are several kinds of color blindness, the most common type is the inability to distinguish red from green. There are two basic forms of red-green color blindness. To understand the difference between them let us envision the following experiment. An observer is presented with a split field, one half of which consists of a yellow light, and the other half of which consists of a mixture of a yellowish-red (*YR*) light and a yellowish-green (*YG*) light. The observer is allowed to adjust the intensity of the *YR* and *YG* lights independently in order to get the mixture to match the yellow light perfectly. A person with normal color vision will always adjust the *YR* and *YG* lights to the same intensity settings in repeated trials of this experiment. On the other hand, a red-green color-blind person will find that many different combinations of the *YR* and *YG* light will match the yellow. In fact, the color-blind observer will be able to match the yellow light with *either* the *YR* light by itself, *or* the *YG* by itself. A color-blind person is said to be protanopic if he requires a very large amount of *YR* light to match the yellow, and deuteranopic if he requires a very large amount of *YG* light to match the yellow. In other words, in addition to confusing red and green, a protanopic person is abnormally insensitive to the long wavelength portion of the spectrum while a deuteranopic person is abnormally insensitive to the middle wavelength portion of the spectrum.

LIGHTNESS CONSTANCY

A final phenomenon to be considered is lightness constancy. This is most easily observed by viewing a mosaic made up of areas of white, black, and various degrees of gray. Such a mosaic can be highly illuminated or dimly illuminated without causing any significant changes in the whiteness of the white areas or the blackness of the black areas. In other words, the absolute level of illumination does not seem to affect our judgment of which areas are white, gray, or black. Even if the illumination is spatially very nonuniform, still our judgment of the lightness or darkness of the various areas of the mosaic remains unimpaired. For example, under highly nonuniform illumination it is possible that a dark gray area on one side of the mosaic might actually reflect to our eyes more light than a white area on the other side of the mosaic. Yet the dark gray will still look dark gray and the white will still look white.

With these features of color vision clearly in mind, let us now turn to the historical development of theories of color vision.

4.2 EARLY THEORIES

NEWTON

In Chapter 1 we mentioned that Newton was the first to make a major breakthrough in understanding color. With the aid of a glass prism he was able to display the colors that make up white light. For Newton, these different colors represented particles or corpuscles of light of different sizes. He was careful to point out that light itself had no color; rather, the different particles of light were capable of producing the *sensation* of various colors. Newton believed that particles of light excited vibrations on the retina that were propagated along the optic nerve to the brain causing the sense of sight. Furthermore, he speculated that particles of light of different sizes would also excite vibrations of different sizes. These vibrations, according to their size, would produce different sensations of color.

White light was supposed to consist of a mixture of corpuscles of all possible sizes. This mixture would excite on the retina a complete range of vibrations simultaneously, leading to the perception of white. If a prism were used to isolate a certain part of the spectrum, the more or less "pure" light that was transmitted would consist of particles of nearly equal size. The resulting sensation produced by such light would be that of a pure color. Newton recognized seven pure colors: red, orange, yellow, green, blue, indigo, and violet. A close examination of the visible spectrum reveals that its colors or hues actually undergo a smooth transition from red through violet. Thus in principle it would be possible to identify hundreds of different hues. It is not clear whether Newton regarded his seven basic hues as fundamental or whether these were simply convenient names he gave to major segments of the spectrum. He did recognize that two pure colors (light from two different parts of the spectrum) could be mixed to produce a completely different color. For example, green light and red light could be mixed to produce yellow. Likewise, red light and violet light would produce magenta, a hue that cannot be seen in the visible spectrum. Newton also observed that as more hues were added, the resulting color became progressively less saturated and began to resemble white. Presumably, if only one size of vibration were excited on the retina a pure color was perceived. As the number of kinds of vibrations increased, the perceived hue became progressively whiter, or less chromatic.

Newton did not speculate on the exact nature of the structure of the retina, nor on what actually was set to vibrating by the light particles. This question was addressed nearly 100 years later by Thomas Young.

YOUNG

Thomas Young, it will be recalled, was an advocate of the wave theory of light. For him light of different colors meant light of different wavelengths or frequencies. Thus he had a somewhat different view of what was occurring when light struck the retina. From the science of mechanics it was well known that flexible structures could be made to vibrate or oscillate, usually at a definite frequency called the resonant frequency of the structure. The visible spectrum consists of a continuous range of frequencies that correspond to the spectral colors. For each of these frequencies to be capable of stimulating an oscillator *of the same frequency* on the retina, it would be necessary to assume that the retina was covered by an enormously large number of oscillators covering all the frequencies of the visible spectrum. Moreover, it would be necessary that *every point* on the retina contain this full range of oscillators so that any color could be perceived at any point. Presumably if, at a point on the retina, a particular oscillator were stimuated by the corresponding frequency of light, a particular hue would be seen. If the *intensity* of light were increased, the *magnitude* of the vibration would be increased, meaning that the *same hue* would be perceived as *brighter.* Young recognized that although this scheme was possible, it was cumbersome and unnecessary.

Young's proposal was that there were only *three* basic kinds of oscillators on the retina, covering it more or less uniformly. All three oscillators would be somewhat sensitive to all frequencies of light, but each type of oscillator would be most sensitive to light from a particular part of the spectrum. One type of oscillator would respond best to red light, another to green, and the third to blue. Figure 4.3 shows a hypothetical graph of the relative sensitivity of each kind of oscillator to all the frequencies (wavelengths) of the visible spectrum. When light of some arbitrary mixture of wavelengths struck the retina, all three types of oscillators (or receptors as we shall call them) would be stimulated to some degree. The relative amount of stimulation produced in the three kinds of receptors would then determine the perceived color. Notice how this theory explains some of the color mixing effects noticed by Newton. For example, a pure yellow light of a single wavelength would stimulate the red receptors and green receptors about equally, but *so would a mixture of a red and green light.* Thus in both these cases the *perception* should be the same *(and it is – yellow).* Young put forward his trichromatic (three-color) theory in a somewhat casual

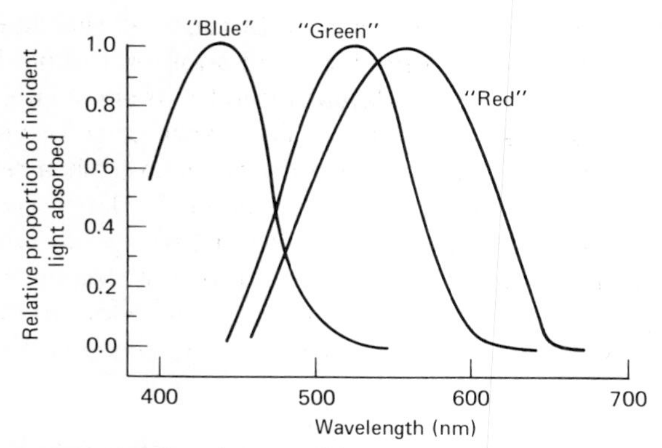

Figure 4.3 Curves representing the possible sensitivity of each of the three kinds of retinal receptors on the human retina.

manner in the course of his lectures on light. The full development and exploitation of the idea was delayed several decades until the subject was taken up by Hermann von Helmholt, a German physicist.

4.3 THEORIES OF THE LATE NINETEENTH CENTURY

HELMHOLTZ

Helmholtz was a physiologist, a physicist, and an experimenter par excellence. He was the first to subject the facts of color vision to detailed quantitative scrutiny. One of the first questions he attempted to answer was how many pure colors from the visible spectrum are needed to match all the hues of the spectrum. In other words, how many primary colors are needed to match any other color? Helmholtz found that he needed at least five and possibly even more. Thus he initially concluded that Young's trichromatic theory must be wrong. But Helmholtz had made a conceptual error. He was using physically pure lights as his primaries, but if Young's theory was correct, physically pure lights would still lead to *all three receptors* being stimulated. That is, a *physically* pure light would not produce a *perceptually* pure response. Such a perceptually pure response would require that *only one* kind of receptor be stimulated. James Clerk Maxwell, who we met in the first chapter, pointed out that according to Young's theory, no real light could possibly do this. Thus there could not exist three pure primary *lights* that could be added to each other in varying amounts to match all possible colors. We have already seen this to be true in our discussion of the CIE colorimetry system.

Maxwell realized that each of Young's receptors would produce some unique kind of perception if it could be stimulated individually. These individual stimulations could be considered the true primaries of a trichromatic system (much like the imaginary x,y,z primaries of the CIE system). Of course we would never actually perceive a pure primary in this case because all real lights stimulate all three primaries simultaneously to one degree of another. Helmholtz realized the wisdom of Maxwell's suggestion, and readopted the trichromatic theory, which is now usually known as the Young-Helmholtz theory of color vision.

In its simplest form the Young-Hemholtz theory is basically a retinal approach to color vision. That is, the color perceived is uniquely determined by the relative stimulation of the three basic kinds of retinal receptors which is, in turn, directly dependent on the spectral makeup of the light striking the retina.

Helmholtz proposed explanations for most of the phenomena discussed at the beginning of the chapter. Let us begin with afterimages. Helmholtz made two assumptions:

1. Even after the light stimulus is removed from the retina, the receptors continue for a short time to be active and send a signal to the brain.
2. After a period of stimulation, the receptors become less sensitive to further stimulation and are thus fatigued.

The first assumption explains the existence of positive afterimages. The second assumption can be used to explain negative afterimages in the following way. Suppose you stare at a red object for a minute or two. Your "red" receptors will then become somewhat fatigued relative to the "green" and "blue" receptors. If you now look away at a white background, the "green" and "blue" receptors will be fully stimulated, but on the area of the retina where the image of the red object had been, the "red" receptors will produce a diminished response. Thus you will see a blue-green afterimage. This same mechanism would explain the successive contrast phenomenon as well. Helmholtz also considered this to be the explanation for simultaneous contrast. He reasoned along the following lines. Under normal circumstances your eyes do not stare fixedly at a single spot but rather roam the visual field continuously. Thus you are constantly creating weak overlapping afterimages of which you are totally unaware. If you are presented simultaneously with two adjacent areas of different color, your eyes will flick back and forth between the two areas involuntarily. The color you see for each area will be a combination of the "true" color of the area and the afterimage of the adjacent area. Consider a white circle surrounded by blue. As your vision swings back and forth between the white and blue areas, the "blue" receptors will become somewhat fatigued. Thus the "white" area will begin to look yellow because its "blue" receptors are not sending a full signal. We shall say more about this effect later.

Next, let us consider red-green color blindness. Helmholtz proposed that what we have called protanopia and deuteranopia were due to lack of "red" receptors and "green" receptors, respectively. Since he took the function of the "red," "green," and "blue" receptors quite literally, he felt that a protanope (lacking "red") would actually see only green (which the protanope would mistakenly call yellow), blue, and various combinations of these colors. We now know that a protanope in fact sees yellow, blue, white, and various combinations of these. This is known because of the existence of individuals known as unilateral protanopes. These are people with one normal eye and one protanopic eye. They can therefore compare what their normal eye sees with what their protanopic eye sees. Helmholtz likewise felt that a deuteranope (lacking "green") would actually see only red and blue. Again, today we know that a deuteranope, like a protanope, sees yellow and blue. The idea that protanopia and deuteranipoa are caused by the lack of "red" receptors or "green" receptors, respectively, may be correct. But clearly Helmholtz's theory must be modified somehow to explain what these individuals actually see.

Color constancy poses a somewhat more difficult problem. At times Helmholtz seems to have felt that memory played an important role in this effect. That is, since we know what the "true" colors of most common objects are, we tend to see them as they are supposed to look regardless of the illumination, and then take this into account in determining the color of other unfamiliar objects in the same field of view. Since this whole thought process is envisioned as unconscious and very rapid, we are normally unaware of it. Helmholtz also proposed a somewhat different explanation of color constancy that would supplement his first explanation. He suggested that what we take as white is the average of all the colors that are in the visual field at any given time. Other colors are then referenced to this average white. Thus if the illumination were to be reddened, for example, this effect would simply be averaged out and go unnoticed. That is, a white object would reflect more red light, but the average color of the field of view would also be more red. Thus white would still look white, and so on. The really interesting question is, exactly how is this averaging process carried out. We will consider this again later.

The case of lightness constancy was not addressed by Helmholtz in quite the way we presented it earlier. Helmholtz felt that the relative brightness of two areas was determined by the *ratio* of their luminosities. By luminosity we mean the physical quantity of light coming from a surface, not how bright it *appears.* Thus, if two areas maintained a constant ratio of illumination, they would also maintain a constant *relative* apparent brightness. In the example discussed earlier, if a black, white, and gray mosaic were illuminated by progressively more intense light, the ratio of reflected light between any two areas would remain constant and thus the apparent *relative* brightness of the areas would remain constant. This does not really explain the fact that white areas continue to look *white* rather than various shades of gray. The case of nonuniform illumination is very difficult to explain according to Helmholtz's theory without additional assumptions.

The theories we have been discussing so far can all be classified as *component* theories. That is, color is supposed to result from the *relative* stimulation of three different types of receptors or components. The total brightness of a particular stimulus is supposed to depend on the *sum* of the outputs from the three components. This type of theory does not really try to explain *why* we see certain hues, but rather attempts to *describe* the connection between perceived hue and the relative stimulation of the three components. For example, no explanation is given for why there are *four* subjectively pure hues, red, yellow, green, and blue, nor is any explanation given for the subjectively pure achromatic colors white and black. The inability of component theories to explain these subjective perceptions has led some investigators to propose an alternative approach.

HERING

One such champion of an alternate theory of color vision was Ewald Hering. In order to fully appreciate Hering's approach to color vision, we need to look more closely at the facts of color perception. Psychologically there seem to be four unique chromatic hues, red, green, yellow, and blue. There are parts of the visible spectrum that appear to be pure green, yellow, and blue. But no single wavelength produces the sensation of pure red. At the long wavelength end of the spectrum the red color appears to have some yellow mixed in with it, while at the short wavelength end of the spectrum the perception is one of red mixed with blue. In order to produce a psychologically pure red it is necessary to mix some long wavelength and short wavelength light together. Now if this psychologically pure red light is combined with psychologically pure green light, the resulting mixture will always appear either red, green, or white, never yellow. Thus "pure" red and green are in fact complementary colors. Likewise "pure" yellow and blue are also complementary colors. These perceptual pairs, red and green, and yellow and blue, seem to form *opponent* pairs. That is, pure red and pure green seem to oppose each other, so that if they are present in "equal" amounts the perception is hueless, or white. The same applies to yellow and blue.

Another illustration of this coupling between yellow and blue, and red and green is the so-called achromatic response. Suppose you take two small pieces of brightly colored paper, one pure blue and the other pure yellow, and paste them a few inches apart on a black background. Now begin to slowly back away from the papers. There will come a time when, although the shapes of the papers are still clearly visible, the colors will simply disappear simultaneously. That is, the blue paper will look dark gray and the yellow paper will look light gray. Apparently, when the stimulus is below a certain angular size (about $\frac{1}{4}^{\circ}$ in diameter) the eye can see neither blue nor yellow hues, but sees only gray. The same kind of thing happens for red and green, but these colors vanish at a smaller stimulus size. What is important is that the colors always disappear *in pairs.* This phenomenon is the reason that our eyes are not bothered by *chromatic aberration* (see Chapter 6), which is the spreading out of colors (like with a prism) that occurs to some extent when light passes through a lens. The small bands of color that surround the images formed on the retina are not apparent because of the achromatic response.

The hues red, green, yellow, and blue are not the only pure psychological colors. White and black are also quite unique experiences. But these achromatic colors do not appear to be in opposition to each other. That is, in the color gray we appear to see both white and black. All in all, then, there are six unique color perceptions: red, green, yellow, blue, white, and black. Any color can be described psychologically as some combination of these six colors. For example, a light orange would appear as a combination of red, yellow, and white. A dark purple would be composed of red, blue, and black. The various browns would include yellow, black, and perhaps red or green in addition. Notice, however, that red and green are never perceived simultaneously, nor are yellow and blue.

These psychological aspects of color led Hering to propose that color vision operated on an *opponent* basis. He proposed that there were three basic kinds of receptors on the retina, red-green, yellow-blue, and black-white, each containing a different biochemical visual substance. When the eye was unstimulated, it was assumed that the quantity of visual substance contained in each receptor was constant at some equilibrium value. When light entered the eye, the result was that in each type of receptor the quantity of visual substance would begin either to increase (assimilate) or decrease (dissimilate) depending on the spectral nature of the light. For example, with respect to the red-green receptors, he assumed that long wavelength and short wavelength light caused dissimilation and produced the sensation of red, while middle wavelength light caused assimilation and produced the sensation of green. If the incoming light contained as much intensity in the long and short wavelengths as in the middle wavelengths, assimilation would equal dissimilation, and the receptors would in effect be neutralized. Likewise, for the yellow-blue receptors the long wavelength light that comprises two thirds of the spectrum caused assimilation and produced the sensation of yellow, while the short wavelength portion of the spectrum caused dissimilation and produced the sensation of blue. Clearly, white light would produce equal amounts of assimilation and dissimilation in both the *R-G* and *Y-B* receptors so that no hue at all would be perceived.

Black-white receptors were supposed to work a little differently. In these receptors, light of any wavelength was supposed to cause dissimilation to a certain extent. A low intensity light would produce a low rate of dissimilation and thus a slight perception of white. However, higher light intensities would produce higher rates of dissimilation and thus a more intense perception of white. Total lack of light would produce neither assimilation nor dissimilation, but Hering postulated that this would lead to the perception of a mean gray rather than black. True black would require assimilation in the black-white receptors and could happen in the following way. If the *B-W* receptors had been stimulated for a period of time, the visual substance would have undergone significant depletion because of prolonged dissimilation. If the stimulus were now removed, the natural metabolic processes in the receptors would begin to regenerate (i.e., assimilate) the visual substance. This would produce the sensation of black, basically a successive contrast effect.

Hering's theory contained another interesting supposition, that of lateral inhibition. His idea was that assimilation (or dissimilation) on one part of the retina for a particular type of receptor would have the effect of inducing the opposite effect in these same kinds of receptors on adjacent parts of the retina. This would explain the phenomenon of simultaneous contrast quite nicely. Consider, for example, a white area surrounded by yellow. In the yellow area the *Y-B* receptors would be undergoing assimilation. In the adjacent white area the *Y-B* receptors would normally be in equilibrium with assimilation equalling dissimilation. Because of the lateral inhibition, however, the *Y-B* receptors in the white area will have the dissimilative process enhanced. Thus the dissimilative process will dominate and produce the sensation of blue. The white area will thus take on a blue appearance. This same effect will tend to produce the sensation of black in contrast to white. That is, a dark area next to a white area will tend to look black because the *B-W* receptors in the dark area will have their dissimilative processes enhanced.

As attractive as Hering's theory is in explaining the psychological aspects of color, and other phenomena such as simultaneous contrast, successive contrast, and afterimages, it has some major weaknesses in the form proposed by Hering. For example, Hering could not really explain the two forms of red-green color blindness. His theory would tend to predict that, in the absence of the *R-G* receptors, only a single type of red-green color blindness should occur. Another major weakness is the lack of any evidence for a single visual substance that can produce two separate effects by assimilation and dissimilation. In spite of its weaknesses, Hering's theory does contain several ideas that have proved to be fruitful.

4.4 CURRENT THEORIES OF COLOR VISION

DETAILS OF THE VISUAL SYSTEM

The theories we have discussed so far assume that the retina is the light-sensitive portion of the eye, and that covering the retina are three varieties of photoreceptors. A detailed examination of the retina reveals that this is partly true. It turns out that there are two basic kinds of receptors on the retina, rods and cones. The rods all contain a single form of biochemical visual substance that is very sensitive to light. Because this substance is present in only one form in the rods, vision that depended on the rods alone would be achromatic — that is, the rods have no way of distinguishing one part of the spectrum from another. The cones, on the other hand, are found to contain three forms of a less sensitive visual substance, although any given cone contains only one form of this material. Thus the cones appear to correspond to the three kinds of receptors of the Young-Helmholtz theory. The distribution of rods and cones on the retina is not entirely uniform. The cones are most closely spaced on the central part of the retina, which corresponds to the center of our field of view. This small part of the retina is called the fovea (Figure 4.4), and covers only about the central 2°

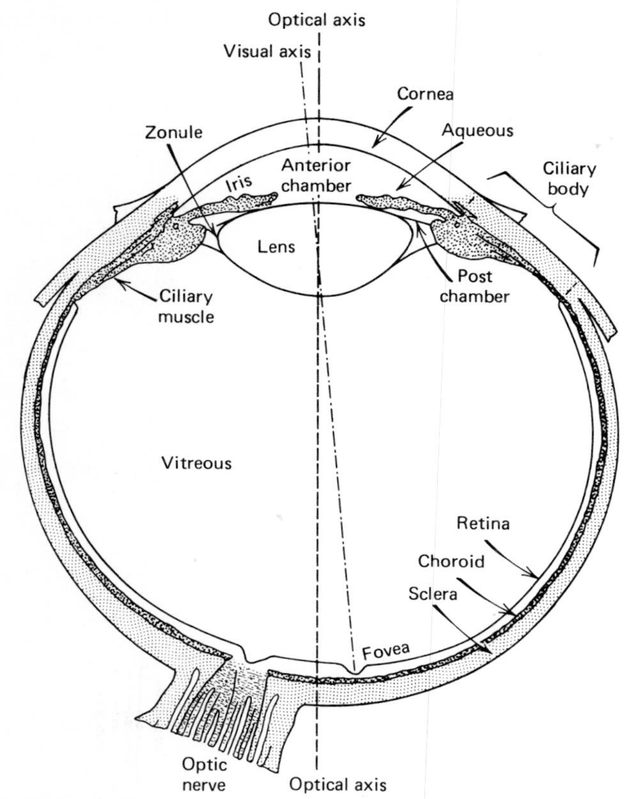

Figure 4.4 Detailed anatomy of the eye. (Adapted from Wolff, *Anatomy of the Eye and Orbit,* 6th ed., H. K. Lewis, 1968)

of our field of view. When we examine something closely we are using the fovea. Because of the close spacing of the cones, in this region, foveal vision is very sharp. The density of cones becomes progressively less at increasing distances from the fovea.

The distribution of rods is somewhat different. On the central 1° of the fovea there are no rods at all. Rods begin to appear toward the outer part of the fovea, and their concentration is greatest about 30° from the fovea. Then, like the cones, the density of rods drops off toward the periphery of the retina.

It is normally assumed that the rods and cones form two generally distinct visual systems. At high light levels (characteristic of normal viewing conditions) the more sensitive rods are saturated and do not contribute to vision. Under these conditions the cones operate effectively and produce normal color vision (photopic vision). At low light levels the cones are not sensitive enough to be stimulated. Thus the rods take over and produce night vision (scotopic vision). Scotopic vision is colorless or achromatic because the rods contain only a single

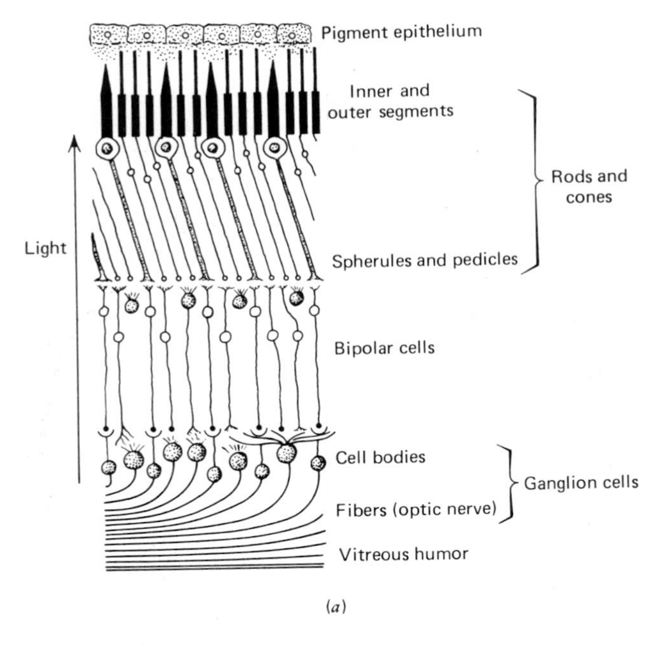

(*a*)

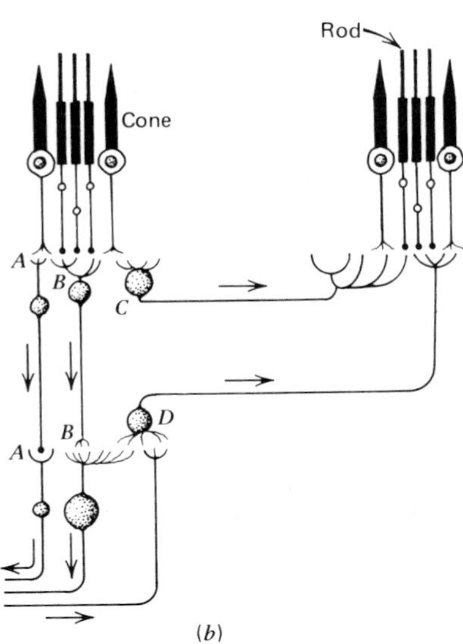

(*b*)

Figure 4.5 Nerve pathways on the retina. (*a*) General organization of nerves. (*b*) Various detailed pathways. (Adapted from Le Grand, *Light, Color and Vision,* Dover, 1967) form of visual material. Also, since the central 1° of the fovea contains no rods, scotopic vision suffers from a "blind spot" at the very center of the visual field. On a dark night after you have allowed your eyes to adapt to the low light level (that is, allowed the rods to unsaturate), try looking at a very faint star. If the star is sufficiently faint, you will not be able to see it when you look straight at it. By looking slightly to one side of the star, however, it will become visible because the image of the star will now be focused just outside the fovea where the concentration of rods is greater. This technique of using "averted vision" is one that people who normally work under low light conditions try to develop.

In addition to looking at the distribution of rods and cones on the retina, it is worth examining the network of nerves that leads from the rods and cones to the optic nerve. Some of the details of this network are shown in Figure 4.5. The retina consists of three layers, one containing the rods and cones, one containing the bipolar cells, and one containing the ganglion cells. Perhaps surprisingly, the layer containing the rods and cones lies at the *back* of the eye. In front of this layer are the bipolar cells that receive signals from the rods and cones. Except in the fovea, bipolar cells are often connected to several rods and cones, and likewise each rod or cone may be connected to more than one bipolar cell. Thus adjacent parts of the retina are connected together. In the fovea, however, it appears that each cone is uniquely connected to one bipolar cell. In front of the bipolar cells is another retinal layer containing the ganglion cells. As with bipolar cells, outside the fovea each ganglion cell receives signals from several bipolar cells. In the fovea, however, each ganglion cell receives information from just one bipolar cell. Thus the cones on the fovea send their signals through the visual network in a much more direct fashion than the cones or rods on the rest of the retina.

The ganglion cells pass on the information they receive by means of individual nerve fibers that form a network running across the retina and joining together to form the optic nerve, much like a large telephone cable containing a bundle of many individual wires. The optic nerve passes out the back of the retina, and the optic nerves from the left and right eye meet at the optic chiasma (see Figure 4.2). At the optic chiasma each optic nerve divides in half, one half corresponding to the right side of the field of view, the other half to the left side of the field of view. The halves of each optic nerve containing information from the right side of the field of view join together and proceed to the left lateral geniculate body. Here the original optic nerve fibers end, but the message is passed on to other nerves that proceed to the left half of the visual cortex in the brain. The entire system is repeated in a symmetric fashion for the left side of the field of view.

THE ZONE THEORY

From what we have said about the structure of the visual system it is clear that at the level of the retinal receptors, cones, the component theory of Young and

Helmholtz seems to be vindicated. However, once we begin to look at how the nerve cells farther along the visual pathway actually operate, we find direct evidence that information originating in the cones is processed in an *opponent* fashion. The possibility of this kind of processing was not appreciated until fairly recently because of a mistaken impression about how nerves operate. It used to be assumed that nerves were essentially passive conduits waiting to conduct an active stimulus from one point to another. We now know that some nerves are in fact always in an active state. Even though unstimulated, these nerves may send a continuous signal comprised of a series of pulses at a definite frequency. Such nerves may either be stimulated (pulse frequency increases) or inhibited (pulse frequency decreases). Thus two *opposing* types of information can be transmitted by such a nerve. Modern zone theories take the view that although the individual receptors behave in a component fashion, nerve cells farther along the visual pathway process the information in a way that resembles Hering's opponent theory. Figure 4.6 is a schematic comparison of the operation of the visual system according to component, opponent, and zone theories. In particular, Figure 4.6*c* shows how the different kinds of cones are connected to the Ganglion cells that act as opponent cells. Let us discuss this illustration in greater detail.

For the average color observer, each pure wavelength of the visible spectrum evokes a perception that consists of two basic hues plus a certain amount of white. Thus, wavelengths between 400 and 475 nm appear as combinations of blue, red, and white. At 475 nm, the color appears as just blue plus white. Between 475 and 500 nm the color is blue, green, and white. At 500 nm the color is green plus white. Between 500 and 580 nm the color is green, yellow, and white. At 580 nm the color is yellow and white. And finally, between 580 and 700 nm the color is yellow, red, and white. No pure wavelength produces a sensation of red alone (or red plus white). Thus at each pure wavelength there are always two chromatic hues plus some white (except at the wavelengths 475 nm, 500 nm, and 580 nm, which are unique blue, green, and yellow, respectively). Now let us consider what happens when we observe pure light of, say, wavelength 425 nm. According to Figure 4.3, the short wavelength receptors are the ones mostly responsible for absorbing light at this wavelength. Our perception of the color of this light tells us that it appears as a combination of blue, red, and some white. Thus the shortwave cones must be connected to the blue side of the blue-yellow opponent cells, the red side of the red-green opponent cells, and white side of the white-black opponent cells. In Figure 4.6*c* the connections have been drawn in this way. Similarly, light of wavelength 610 nm evokes a perception of red, yellow, and some white. But Figure 4.3 shows that at this wavelength it is the longwave cones that are most stimulated. Thus these cones must be connected to the red side of the red-green opponent cells, the yellow side of the blue-yellow opponent cells, and the white side of the white-black opponent cells. Finally, in the wavelength range where the middlewave cones are most sensitive, the perceived color tends to be combinations of

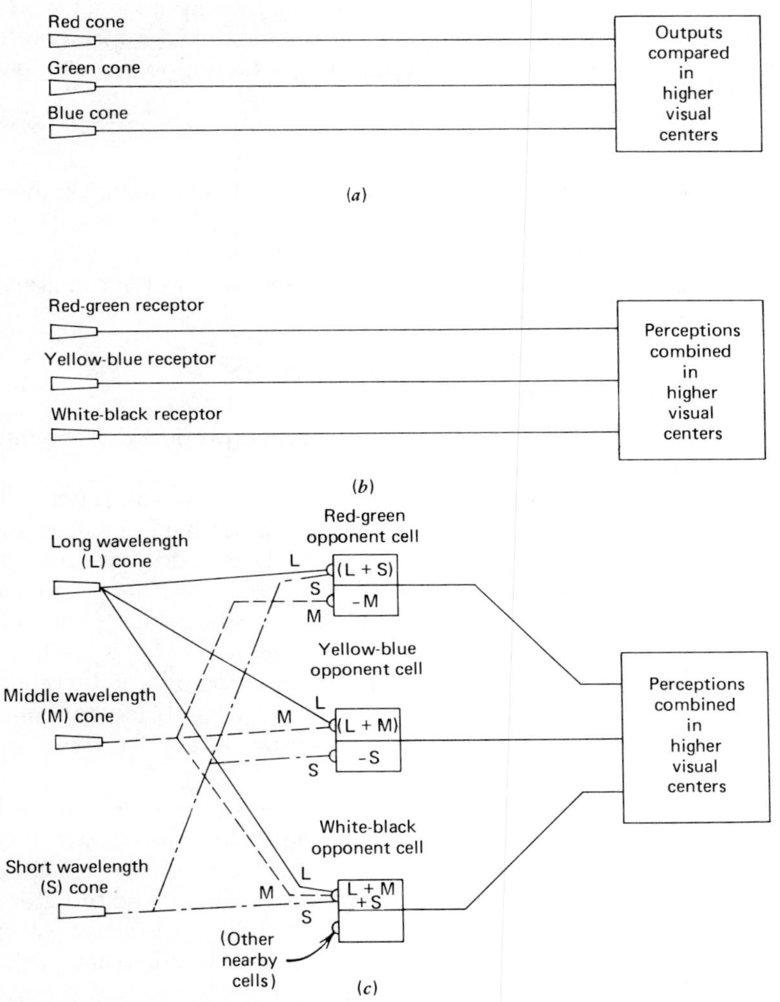

Figure 4.6 Schematic diagrams of (*a*) component theory, (*b*) opponent theory, and (*c*) zone theory of color vision.

green, yellow, and white. Thus we assume that the middlewave cones are connected to the green side of the red-green opponent cells, the yellow side of the blue-yellow opponent cells, and the white side of the white-black opponent cells. Figure 4.6*c* illustrates all these presumed connections.

Now let us examine the opponent cells shown in Figure 4.6*c*. First consider the red-green opponent cell. Imagine that the top half of the cell represents the red half, and the lower half represents the green half. Now into the top half (red) comes the output from the L and S cones. Let us assume that this output tends to stimulate the cell. This stimulation we shall represent algebraically as L

$+ S$. Into the lower (green) half of the cell comes the output from the M cones. We assume that this output tends to inhibit the cell, and we represent this inhibition by $-M$. Thus the net activity of the red-green opponent cell is given by

$$\frac{\text{Red }(+)}{\text{Green }(-)} = (L + S) - M \quad (1)$$

If the right side of eq. (1) is positive (i.e., if $L + S$ is greater than M), then a red (stimulation) signal is produced and sent to higher visual centers. If $(L + S) - M$ is negative, then a green (inhibition) signal is produced.

The blue-yellow opponent cells work in the same way as the red-green cells. Thus, in place of eq. (1) we would have

$$\frac{\text{Yellow }(+)}{\text{Blue }(-)} = (L + M) - S \quad (2)$$

If the right side of eq. (2) is positive a yellow signal is produced. If the right side is negative a blue signal is produced.

Finally, we come to the white-black opponent cell. It would seem that since the L, M, and S cones are all connected to the upper (white) half of this cell, only varying degrees of white could be produced. How does the black half of this cell come into play? The answer lies in lateral inhibition. There is now a great deal of evidence that activity in cells on one part of the retina tends to induce the *opposite* kind of activity in similar cells on adjacent parts of the retina. Thus, if white-black opponent cells are highly stimulated in one part of the retina, they will send an inhibiting signal to the black half of the white-black opponent cells that are nearby. Black is therefore totally a contrast effect. No direct external stimulation produces black.

The white-black opponent cells are not the only ones affected by lateral inhibition. Red-green and yellow-blue opponent cells are affected in the same way, although the lateral connections are not specifically shown in Figure 4.6*c*. Thus, if one area of the retina is sending a green signal (i.e., the red-green cells are inhibited), then the red-green cells in adjacent areas of the retina will receive a signal that tends to stimulate them. This signal will be combined with the $L + S$ stimulation in the red half of the adjacent red-green cells and thus make the output of these cells more red (or less green) than would otherwise have been the case.

The zone theory represented by Figure 4.6*c* is superior in explanatory power to either the component or opponent theories individually. We have, as it were, the best of both worlds. Thus negative afterimages can still be explained by retinal fatigue, while simultaneous contrast can be explained by lateral inhibition. In addition, the peculiarities of red-green color blindness can now be understood in terms of the zone theory, whereas neither the component or opponent theories provided a satisfactory explanation. To see this, imagine two possibilities. In Figure 4.6*c* either the L cones have been replaced by M cones (protanopia), or the M cones have been replaced by L cones (deuteranopia). In the first case, eq. (1) becomes

$$\frac{\text{Red }(+)}{\text{Green }(-)} = M + S - M = S$$

and eq. (2) becomes

$$\frac{\text{Yellow }(+)}{\text{Blue }(-)} = M + M - S = 2M - S$$

Thus the red-green cell has lost its ability to achieve inhibition entirely, and has effectively been nullified, whereas the yellow-blue cell can still be stimulated or inhibited and thus function more or less normally. Loss of the L substance will also leave the person abnormally insensitive to the long wavelength portion of the spectrum.

Now consider the second possibility, M cones replaced by L cones. Then eq. (1) becomes

$$\frac{\text{Red }(+)}{\text{Green }(-)} = L + S - L = S$$

and eq. (2) becomes

$$\frac{\text{Yellow }(+)}{\text{Blue }(-)} = L + L - S = 2L - S$$

Thus again it is the red-green cell that can no longer function. However, loss of the M substance leaves the person less sensitive to the middle wavelengths.

THE RETINEX THEORY

Although the zone theory is extremely elegant and powerful, and is in basic accord with the anatomical and physiological evidence, there remains much to be discovered about exactly how visual information is processed. An interesting theory, called the retinex theory of color vision, has been put forward by Edwin Land of the Polaroid Corporation.

Land considers color constancy to be fundamental to color vision. The retinex theory accounts for color constancy in an elegant fashion. The theory begins by assuming the existence of three types of receptors, the long wavelength (L), middle wavelength (M), and short wavelength (S) cones. But Land proposes that the cones of each type are organized into independent systems, which he calls retinex systems. Thus there is an L retinex system, an M retinex system, and a S retinex system. Each system forms an *independent* achromatic picture of the visual field. This is most easily understood with the aid of Color Plate 10. One very important step takes place in the formation of these independent pictures. Take, for example, the picture formed by the L retinex system. Some area in the colored mosaic of Color Plate 10 will appear, to the L retinex system, as brighter than the rest. Let us arbitrarily assign to this brighter area a brightness of 1.0, regardless of its absolute luminance. All other areas in the mosaic are assigned brightnesses that are proportionately lower. For example, an area appearing half as bright as this brightest area is assigned a value of 0.5, and so on. This step is important for the following reasons. If the illumination changes so that the *absolute* luminance of the entire mosaic changes, the *relative* brightness will remain essentially constant so that the numerical values assigned to each area of the mosaic by the *L* retinex system will stay the same. The relative values are shown on Figure *A* of Color Plate 10. Now exactly the same kind of thing is presumed to occur for the M retinex system and the S retinex system. The relative values assigned to various areas of the colored mosaic by these systems are shown on Figure *B* and *C*, respectively, of Color Plate 10. Notice that every area in the colored mosaic ends up being assigned a unique triplet of relative brightness values by the three retinex systems. Furthermore, these values will stay nearly constant even if the illumination is changed substantially. Thus color constancy is explained as an automatic function of the visual system. Ideas such as memory color are no longer required as the primary explanation. Notice that the retinex theory is entirely consistent with opponent processing. The retinex systems could come either before or after the opponent cells in the visual network. The retinex theory also explains simultaneous contrast. Suppose we have a white area surrounded by an area that contains a little extra blue. Figure 4.7*a* shows the *absolute* amount of L, M, and S light coming from each area. Figure 4.7*b* shows how the three retinex systems will assign the relative brightnesses. Notice that the "blue" area ends up with an assignment of (ℓ = 1, m = 1, s = 1) while the "white" area is assigned values of (ℓ = .97, m = .95, s = 0.8). Thus the "blue" area will appear white and the "white" area will appear yellow.

Brightness constancy and black-white contrast are also naturally explained by the retinex theory. In brightness constancy, it is clear that increasing the illumination in the field of view will have very little effect on the apparent brightness of objects, since each retinex system assigns the brightest area in the field of view a brightness of 1.0 regardless of absolute illumination. In the case of black-white contrast, consider a white area adjacent to a gray area. Suppose that the light coming from white area is increased while the light coming from the gray area remains constant. Each retinex system will continue to assign the white area a brightness of approximately 1.0 while the gray area will be assigned an ever-decreasing value. Thus the gray area will be perceived as increasingly black.

In our description of Land's theory we have said that each retinex system assigns the brightest area in the field of view a relative brightness of one. This is not quite true. Actually, in Land's theory the relative brightness assigned by a given retinex system to the brightest area in the field of view will not be exactly 1.0, but will depend weakly on the absolute luminance of light. This is illustrated in Figure 4.8, which shows that very great changes in absolute luminance result in very small changes in the relative brightness assigned to the brightest area by a retinex system. Nevertheless, these small differences can be important in a situation where the eye is presented with a single uniform stimulus. In this case

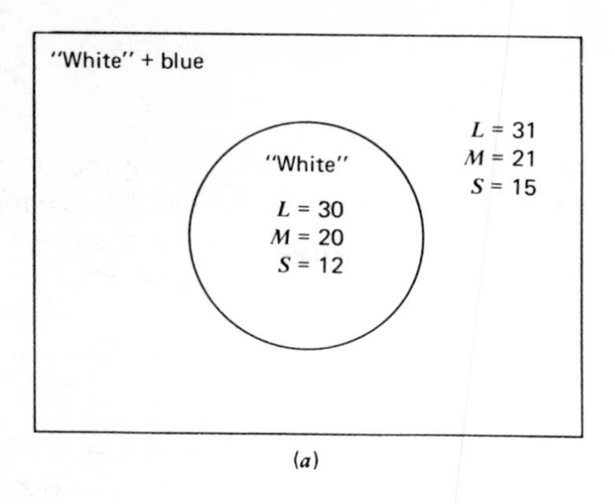

(*a*)

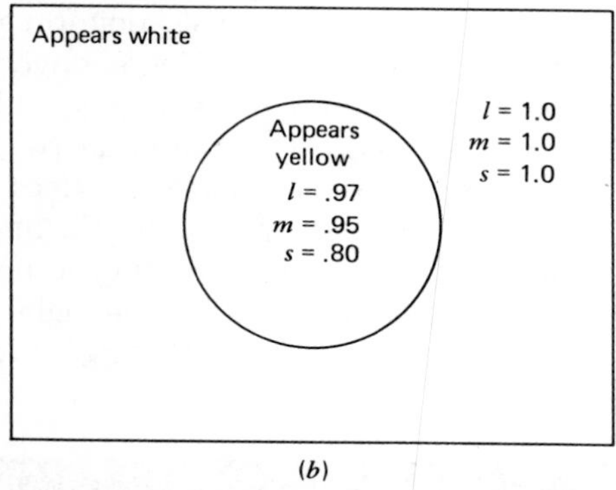

(*b*)

Figure 4.7 (*a*) Light from a "white" projector surrounded by an area to which a weak blue light has been added. Values of L, M, and S give absolute amounts of stimulation of the three retinex systems. (*b*) Relative brightnesses assigned by the three retinex systems to the areas in upper figure. Note that the "bluer" area is assigned values of 1.0, 1.0, 1.0, and thus appears white, while the "white" area is assigned values of 0.97, 0.95, 0.80, and thus appears yellow.

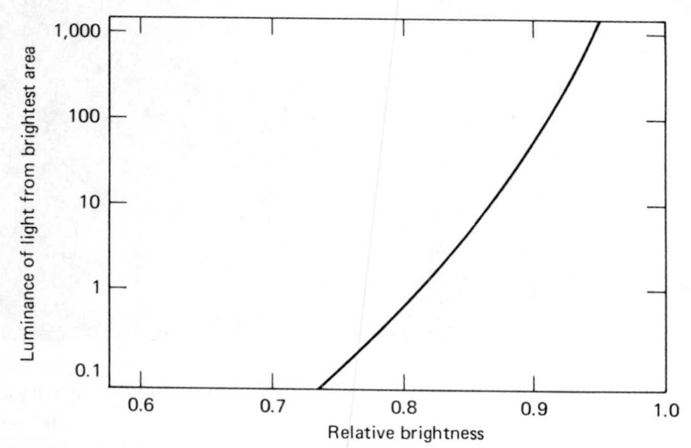

Figure 4.8 Relative brightness assigned by a typical retinex system to the brightest area in the field of view.

the only differences in assigned brightnesses among the three retinex systems will be those resulting from differences in absolute luminance as indicated in Figure 4.8. This will be sufficient, however, to allow the color of the single stimulus to be perceived.

THE IMPORTANCE OF EDGES

Until now we have assumed that the visual system, however it is organized, uses information from every point on the retina to form our view of the visual world. There is evidence that this might not be the case. The field of view is basically made up of a large number of essentially uniform areas that abut each other along sharp boundaries or edges. It seems increasingly likely that the visual system primarily uses information from these edges, rather than from the regions within the uniform areas, to produce our visual perceptions.

One phenomenon that suggests the importance of edges is the effect of nonuniform illumination discussed previously in the context of a black, white, and gray mosaic. It was remarked that the areas within the mosaic retained their "true" brightnesses even though a dark area on the highly illuminated side of the mosaic might actually reflect more light than a bright area on the dimly illu-

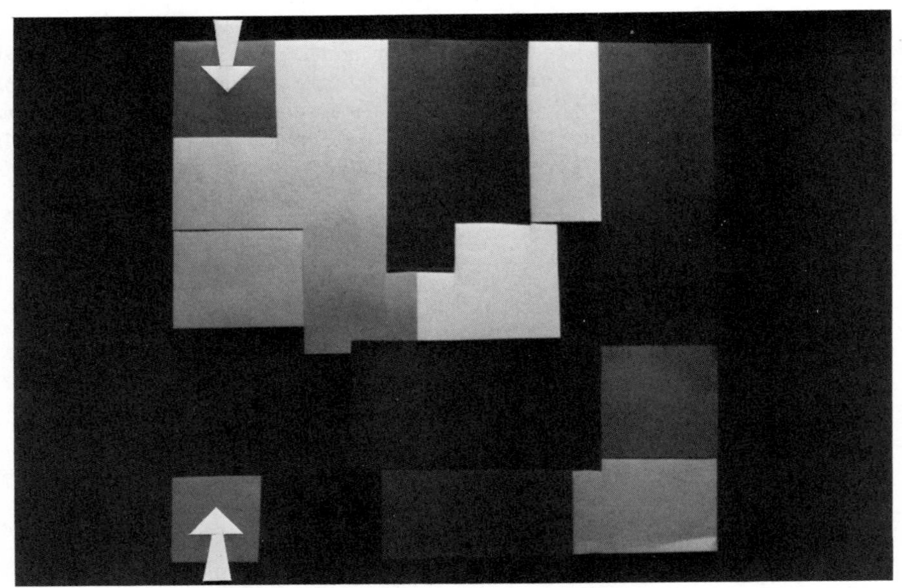

Figure 4.9 Black and white mosaic illuminated nonuniformly. The top is more strongly illuminated than the bottom. Consider the two areas indicated by arrows. The upper appears clearly darker than the lower area even though both reflect the same amount of light.

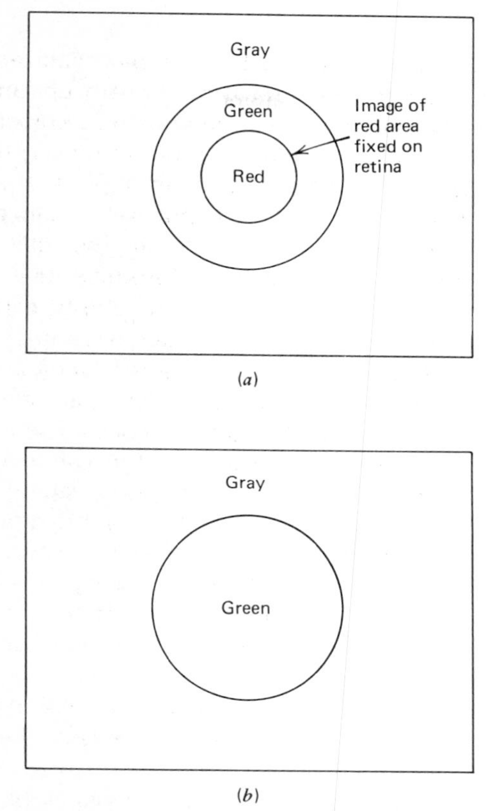

Figure 4.10 (*a*) Actual image on the retina, with red area fixed at a constant retinal location. (*b*) The image seen by the observer.

minated side (Figure 4.9). This suggests that the areas of the mosaic are compared along their edges where true relative brightnesses can be determined, and that the absolute illumination within an area is unimportant.

Another classic illustration of the importance of edges occurs when the field of view is artificially fixed at a constant location on the retina. Normally such fixation does not occur because our eyes involuntarily change their fixation point continuously. The effect of such artificial fixation is to cause the field of view to fade quickly into a uniform gray. Now, envision a colored circular area surrounded by a second circular area of a different color surrounded in turn by a field of gray (Figure 4.10). If the image of the *inner* circular area is fixed on the retina, the image of the inner area soon fades and is replaced by the uniform *color* of the second colored area. Thus it seems that without edge information, the visual system simply cannot see the inner area, and replaces it with the surrounding color. This demonstration strongly suggests that the information from *inside* the central colored area simply plays no role in perception.

KUFFLER UNITS

How, it might be asked, could the visual system be organized to be sensitive to edges and yet ignore the interiors of uniform or nearly uniform areas? One suggestion proposes the existence of what are known as Kuffler units. These units are small areas of the retina in which the central region acts to excite (or inhibit) a nerve while the surrounding region acts in an opposite manner on the same nerve (Figure 4.11). The unit as a whole is precisely balanced so that uniform illumination produces no net effect on the nerve. Thus uniform illumination of any intensity leaves the nerve leading from the Kuffler unit in an unstimulated condition. Only when a boundary across which contrast exists falls on the unit will there be a response. This type of organization discards information about absolute luminance levels, and only passes on information about luminance *differences* to higher levels in the visual system. Such organization clearly can form the basis for brightness contrast and brightness constancy.

What about color constancy? The same receptors can simultaneously be part of several types of Kuffler units by means of various lateral connections. Thus one type of Kuffler unit could be responsible for brightness constancy and brightness contrast, while other types of Kuffler units might exist to process color information. One possibility would be to organize the output of the *R-G* and *Y-B* opponent cells into Kuffler units in the manner shown in Figure 4.12. Consider the *R-G* Kuffler unit. Under uniform illumination of *any* spectral composition, this unit will be balanced so that no output results. However, if an edge across which a difference in red-green content exists falls on the unit, an output will result. That is, the unit will be sensitive to *differences* in red-green content rather than the absolute amount of red or green in the original illumination. Likewise, the *Y-B* Kuffler unit will sense differences in yellow-blue content. Clearly these *R-G* and *Y-B* Kuffler units are capable of passing on precisely the kind of information required to explain color constancy and simultaneous color contrast, and could also be used as a basis for Land's retinex theory.

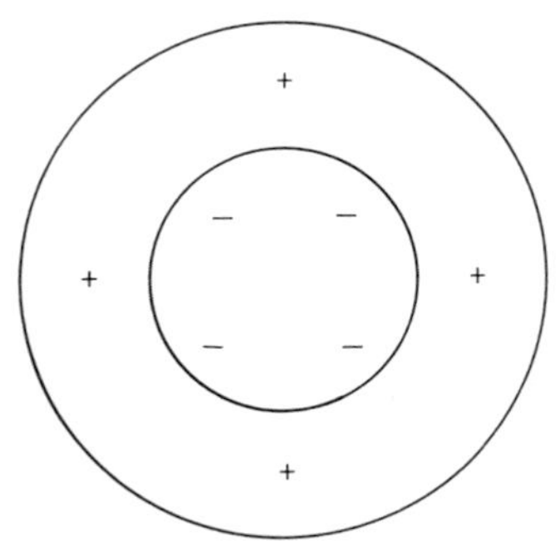

Figure 4.11 Brightness Kuffler Unit; surround and center are in opposition.

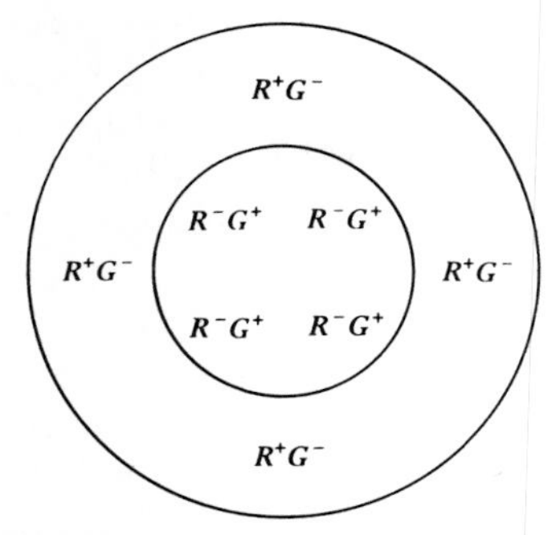

Figure 4.12 Red-Green Kuffler Unit; surround and center consist of red-green opponent cells acting in opposition.

4.6 IN CONCLUSION

Our discussion of color vision has led us from the early rather simple ideas of Newton into the complexities of modern color theory. Today, of course, we have the advantage of detailed anatomical, physiological, neurological, and biochemical evidence as well as very sophisticated techniques of manipulating visual stimuli. Nevertheless, our explanations must remain tentative and open to doubt. The more we learn about the visual system, the more we are struck by its complexity and sophistication. Like most other branches of science, our knowledge of color vision deepens but does not reach an end.

PROBLEMS AND EXERCISES

1. Refer to Figure 4.2. Explain why severing the right optic nerve does not cause blindness in half the visual field. Explain why severing the right optic tract causes blindness in the left half of the field of view.
2. With the aid of Figure 4.3, estimate the relative stimulation of the red, green, and blue receptors produced by monochromatic light of the following wavelengths:
   (a) 600 nm
   (b) 550 nm
   (c) 475 nm
   (d) 450 nm
3. Explain the two varieties of red-green color blindness in terms of
   (a) the Young-Helmholtz theory
   (b) the Hering theory
   (c) the Zone theory
4. On the diagram below are shown the absolute stimulation of the *L*, *M*, and *S* cones by light coming from three different colored surfaces.

(a) According to the simple component theory of color vision, what colors would you expect to see in each case?

(b) Using Land's theory, compute the relative brightness of each area as seen by the three retinex systems (that is, compute *1, m, s* for each area). Now what colors would you expect to see?

$L = 50$	$L = 25$	$L = 50$
$M = 10$	$M = 25$	$M = 15$
$S = 5$	$S = 8$	$S = 13$

Figure 4.13

5. If you stare at a magenta circle and then look away at a white background, what color afterimage will you see? What about a yellow circle? Red? White? Try some of these.

6. Get a package of different colored papers. Try staring at a paper of one color for a few seconds and looking at a paper of a different color. How is the color of the second paper affected by the color of the first? This is a successive contrast effect.

7. Roll up a piece of black paper into a tube about one inch in diameter. Look through this tube at objects of various colors in such a way that only a *uniform patch of color* is visible through the tube. How does the color of an object as seen through the tube compare with the color of the object seen in its normal setting?

5 THE APPEARANCE OF OBJECTS

Why do things look the way they do? A piece of green paper looks different from a glossy piece of green plastic. Metallic surfaces look different from nonmetallic surfaces. Some objects appear transparent, others translucent, and still others opaque. How are we able to recognize these differences? Clearly, the answer must lie in the nature of the light that reaches our eyes.

In general, the light reaching our eyes from an object has two major characteristics. First, the light has a certain spectral composition. That is, it consists of a mix of wavelengths of various intensities. This is the attribute of light that determines the *color* of the object. We have already discussed this at some length in the four previous chapters, and will return again to it presently.

The second major characteristic of the light from an object is the light's *geometrical organization*. That is, most objects do not appear to send to our eyes a uniform amount of light from each portion of their surface. Thus, a glossy object is perceived with definite highlights that are much brighter than the remainder of the object's surface. It is the presence of these highlights that alert us to the glossy nature of the surface. Likewise, a transparent object transmits fairly sharp images of other objects and thus gives itself away as transparent. We may place most objects into one of four categories based on the way that most of the light from the object reaches our eye.

(a) *Nonmetallic Opaque Objects*. Light reaching our eyes from these kinds of objects has been diffusely reflected from the surfaces. Diffuse reflection is illustrated in Figure 5.1*a* and Color Plate 11. Light striking a diffusely reflecting surface scatters in all directions, so that the object appears generally the same if the location of the observer's eye or the light source is changed somewhat.

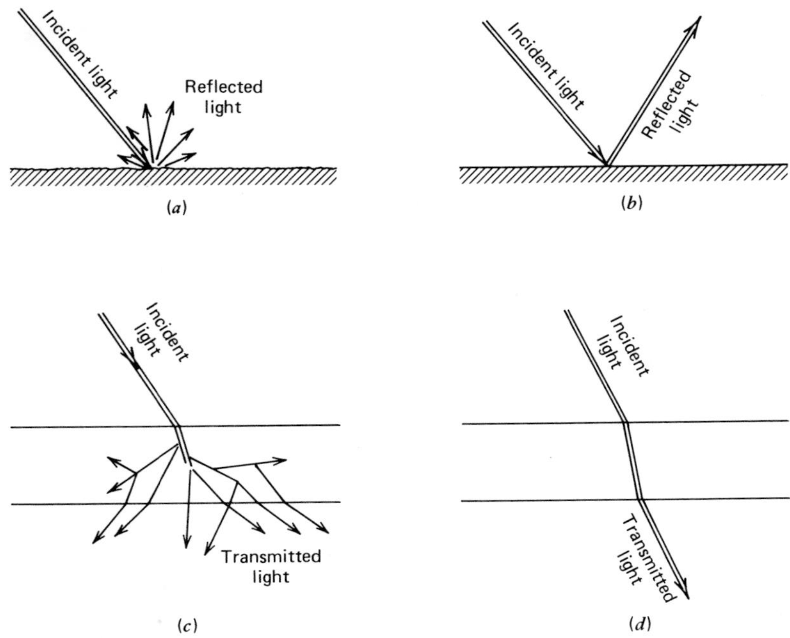

Figure 5.1 (*a*) Diffuse reflection, (*b*) specular reflection, (*c*) diffuse transmission, (*d*) specular transmission.

(b) *Metallic Objects.* Light striking a smooth metallic surface is mostly *specularly reflected.* This is illustrated in Figure 5.1*b* and Color Plate 11. Specular reflection obeys the law of reflection discussed briefly in Chapter 1 and in greater detail in Chapter 6. The location of both the eye and the light source plays a critical role in how much light is seen from a metallic object.

(c) *Translucent Objects.* Light coming from these objects has been primarily *diffusely transmitted* through the objects. This is illustrated in Figure 5.1*c* and Color Plate 11. In this situation, light passes through the object, but is scattered in all directions. Thus the location of both the observer's eye and the light source can be changed somewhat without dramatically changing the appearance of the object.

(d) *Transparent Objects.* These kinds of objects *specularly transmit* most of the light that reaches them (Figure 5.1*d* and Color Plate 11). Although the shape of the transparent object may cause light to bend somewhat, very little light is scattered. Thus most of the light that enters the object along a given line will exit together in a common direction. For this reason, the images of other objects can be seen through a transparent object.

In the rest of this chapter we are going to be primarily interested in the nonmetallic opaque class of objects. Most particularly, we would like to know how to manipulate opaque surfaces to produce a wide range of desired effects. Since the most common way of controlling the appearance of such surfaces is through the use of inks, dyes, paints, and other surface applicants, it is these that we shall be most concerned with. We have already encountered a few of these substances briefly in Chapter 2, but now we can apply a great deal more background to the discussion of their properties.

At the end of the chapter we shall return briefly to a discussion of the four classes of objects with an eye toward how an artist might portray them on canvas. This will allow us to see how a knowledge of paints and pigments can be used to create three-dimensional visual effects on a two-dimensional canvas.

5.1 REFLECTION FROM OPAQUE SURFACES

METALS

Light reflecting from the surface of a smooth metal does so in a specular fashion. Thus a highly polished metal surface makes a good mirror. If the surface is somewhat rough, the tiny imperfections in the metal surface produce a certain amount of diffuse reflection, and the effective specular reflection is thus reduced. However, the spectral composition of the reflected light will be the same regardless of the nature of the surface (provided it is clean). When light strikes the surface of a metal it is *selectively reflected.* That is, some wavelengths are reflected more efficiently than others. This selective reflection is different for each type of metal; thus, gold appears different from silver, which appears different from copper, and so on. The important point is that the reflection takes place right at the surface of the metal. All of the light reflected from the metal will undergo the same selective process. This is a distinguishing characteristic of metals that they do not share with most nonmetallic opaque objects

NONMETALLIC SURFACES

For nonmetallic surfaces, the process of reflection is more complex than for metals. The vast majority of objects fall in this category. Most of these objects have an outer surface that reflects *nonselectively.* That is, the extreme outer surface reflects a certain fraction of the incident light *without* changing its spectral composition. The remainder of the light penetrates below the surface where it is scattered about and selectively absorbed before finally reemerging from the material with its spectral composition changed. The color of the object is primarily determined by the selective absorption that has occurred below the surface, although diffuse nonselective reflection from the outer surface can serve to desaturate the color substantially. Since our main interest centers on the behavior of materials applied to surfaces for the purpose of modifying their appearance, we need to look more closely at this situation.

Figure 5.2 shows a cross section of a surface that has been coated with a paint or other applicant. Three elements of this surface need to be considered.

(a) *The Outer Surface.* Some of the light incident on the outer surface will be reflected nonselectively. Thus, if the incident light is white, the reflected light will be white. If the outer surface is very glossy, this light will be reflected in a specular fashion and, unless the eye is in just the right place (Figure 5.1*b*), the reflection will not be seen. If the outer surface has a matte finish, the nonselective reflection will be highly diffuse. This diffusely reflected light will be seen no matter where the eye is placed, and will have the effect of desaturating the color of the underlying layer of material.

(b) *The Colorant Layer.* Below the outer surface is the colorant layer. This layer may consist of a transparent material that selectively transmits light to and from the support surface below, or the colorant may contain particles of pigment suspended in a binder. If the colorant contains pigment, this pigment may be opaque or transparent, and either colored or white. These various possibilities will be discussed later in detail.

(c) *The Support.* The surface upon which the paint has been applied is called the support. If the colorant layer is sufficiently thin or transparent so that some light reaches the support, this surface will behave like any other opaque surface. Some light will be reflected nonselectively, while some light will penetrate the surface and be selectively scattered back out. The amount of nonselective reflection can be enhanced or reduced by proper selection of the transparent medium (or binder) in the colorant layer.

One special case is worth mentioning here. Suppose the support is a piece of smoothly sanded wood. In this condition, the wood will look quite light in color because the surface is microscopically rough, which produces a great deal of diffuse reflection. Suppose a layer of clear varnish or wax is applied over the wood. Now the outer surface has become quite smooth, making the nonselective reflection specular. Thus the diffuse light from the outer surface is virtually eliminated. In addition, if the varnish has an index of refraction (see Chapter 6) very similar to that of wood, almost no *nonselective* reflection will occur from the

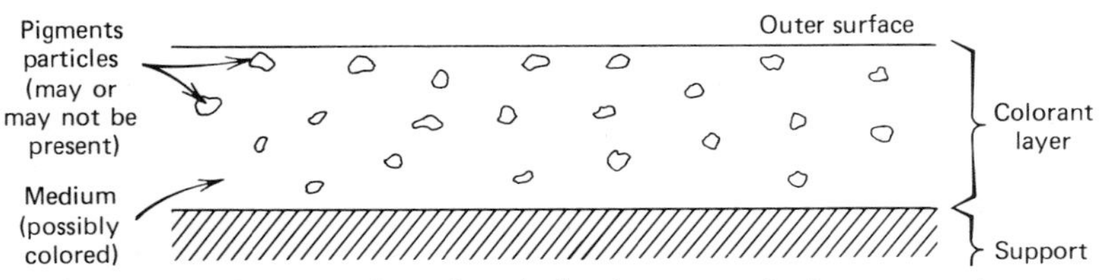

Figure 5.2 Cross section of typical colorant applied over a surface.

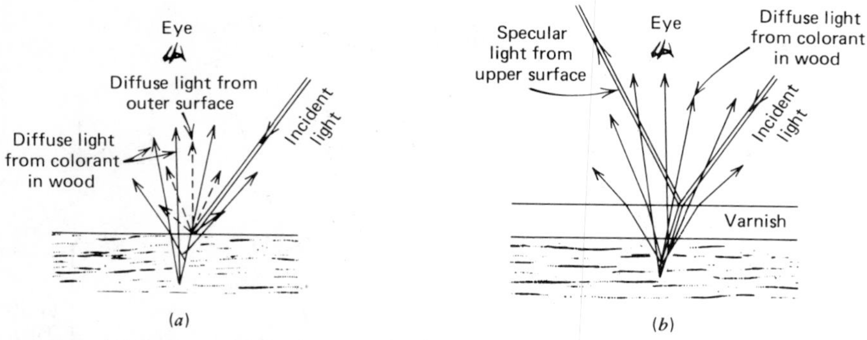

Figure 5.3 (*a*) Reflected light from unfinished wood surface is seen by the "eye" as a mixture of nonselective diffuse light from the outer surface plus selective diffuse light from the colorant in the wood. (*b*) Varnish layer eliminates most of the diffuse nonselective reflection so that the "eye" sees mostly the selective diffuse light from the colorant in the wood.

underlying wood surface. Thus most of the light reflected from the wood will be due to the natural pigments in the wood itself, and the wood surface will look richly colored. This is illustrated in Figure 5.3.

Within the general framework just outlined in (a), (b), and (c) above, the two most interesting and useful examples are provided by transparent colorants on the one hand and pigmented paints on the other.

5.2 TRANSPARENT COLORANTS

Transparent colorants such as many inks, dyes, and watercolors are essentially filters. The basic principles underlying the transmission of light through transparent filters were discussed in Chapter 2. These principles allow us to calculate accurately the transmission and reflection characteristics of various combinations of transparent colorants. However, only by examining typical colorants can we truly appreciate the consequences of these basic principles.

It will be helpful to refer again to Figure 5.2. We are interested in the case where no pigment particles are present and where the medium is colored. Light incident on the surface will be subjected to three modifications. First, some of the light will be nonselectively reflected from the surface of the colorant layer. Depending on how glossy the surface is, this will cause more or less desaturation of the color. Second, light that is not reflected will pass through the colorant and impinge on the underlying support. During this first pass through the colorant the light will be filtered. The transmission curve for the colorant determines the spectral composition of the light that gets through the colorant layer to the support. At the support, reflection will take place. If the support itself is colored, the net reflection will be selective, so that the relative spectral composition of

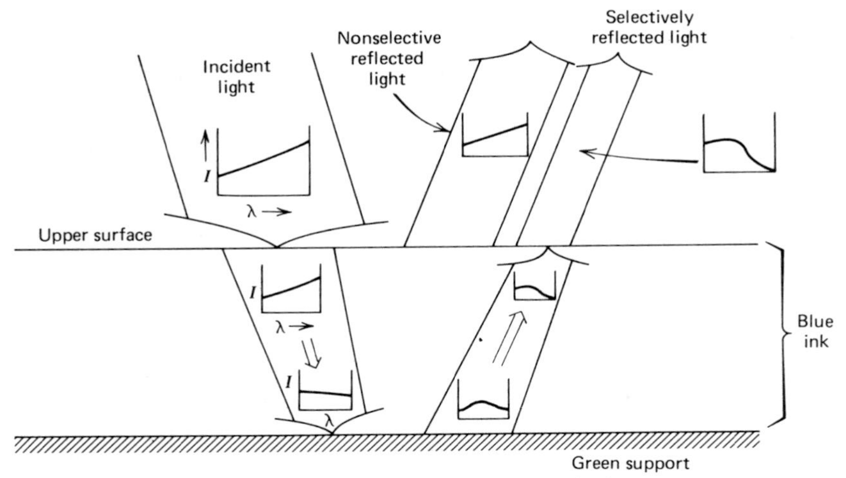

Figure 5.4 Details of light reflection from blue ink applied over a green support.

the light will be changed again. Finally, this reflected light will pass again through the colorant layer and undergo additional filtering. Figure 5.4 represents schematically the entire process just described. Clearly, a complete description of the reflected light requires consideration of the nonselective reflection at the upper surface and the reflectance characteristics of the support in addition to knowledge of the transmission characteristics of the colorant. Usually, the nonselective reflection at the upper surface can be taken into account by adding a small (5–10%) uniform reflectance to the reflectance curve of the colorant and

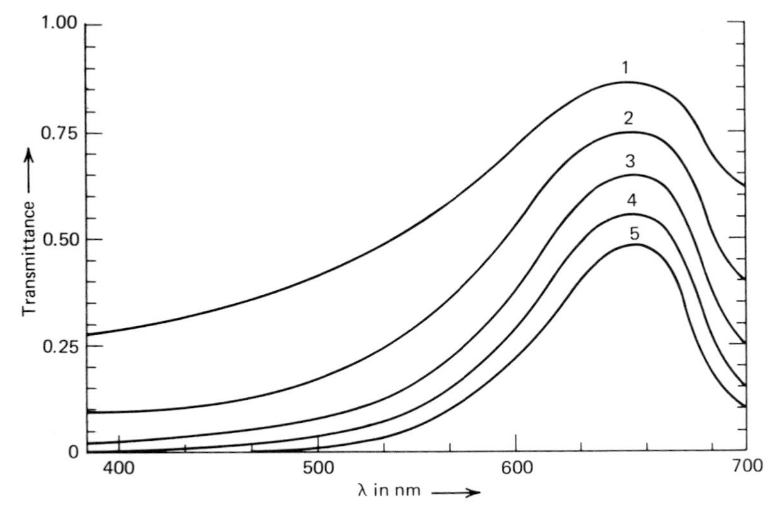

Figure 5.5 Transmission curves of an orange filter showing the effect of increasing thickness or concentration of colorant layer by up to a factor of five.

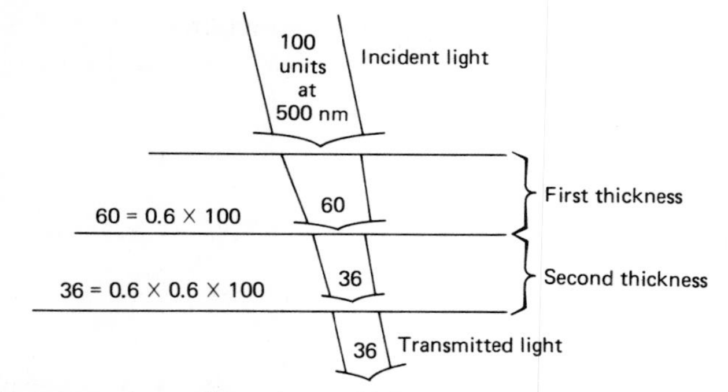

Figure 5.6 The successive action of two thicknesses of filter on 100 units of light (λ = 550 nm).

support. In addition, the support is often nearly white so that most of the selective action comes from the colorant layer. In the rest of this section, therefore, we shall concentrate on the properties of the colorant itself, and not consider the effects of nonselective reflection from the upper surface, or selective reflection from the support. However, in a real situation these must be considered.

SINGLE COLORANT: EFFECT OF THICKNESS AND CONCENTRATION

For a single layer of colorant, the nature of the *reflected* light is really determined by the *transmission* characteristics of the colorant layer. The best way to examine the transmission properties of a material is to examine the transmittance curve. Figure 5.5 is the transmittance curve for a particular colorant layer of *specific thickness and colorant concentration* (curve 1 in Figure 5.5). If either the layer thickness *or* the dye concentration is changed, the transmittance curve will also change. For example, doubling the layer thickness can be viewed as two separate successive transmissions of the light through two identical filters. Consider, for example, light of 500 nm wavelength incident on a double thickness of a colorant layer (Figure 5.6). Suppose a single thickness transmits 0.6 of the light at this wavelength.

If 100 units of light are incident on the first layer, then 0.6 of this or 60 units will make it to the second layer. Of these 60 units, 0.6 or 36 units, will get through the second layer. Thus for the double layer as a whole, the net transmittance is 36 units out of the original 100, or 0.36. The transmittance of the double layer is given by the *product* of the transmittances for each layer (i.e., $0.6 \times 0.6 = 0.36$). We can state this result in a succinct algebraic form.

Let T_0 = transmittance of colorant layer of thickness d_0 (at some particular wavelength).

T = transmittance of colorant layer of some other thickness d.

If we think of a thickness d_0 of a colorant layer as a "single" thickness, then the ratio d/d_0 gives the "number of layers" of colorant. If T_0 is the transmittance for one layer, then, as we have seen, the transmittance for two layers is T_0^2, for three layers T_0^3, and so on. We can generalize this to the number of layers given by the ratio d/d_0.

Then

$$T = T_0^{(d/d_0)} \tag{1}$$

Example 1

A colorant of a particular thickness has a transmittance of 0.7 at a wavelength of 550 nm. What is the transmittance of a layer of this colorant that is three times as thick?

We have

$$T_0 = .7$$

and

$$\frac{d}{d_0} = 3$$

Thus

$$T = (0.7)^3 = (0.7)(0.7)(0.7) = 0.343 \blacksquare$$

Example 2

Suppose in the previous example, the original layer had been made one-half as thick instead of three times as thick?

Then

$$T_0 = 0.7$$

and

$$\frac{d}{d_0} = \frac{1}{2}$$

Thus

$$T = (0.7)^{1/2} = \sqrt{0.7} = 0.837$$

Equation (1) tells us how to deal with different thicknesses of a given colorant. The same approach is used with different concentrations of dye within the colorant layer. That is, doubling the dye concentration has exactly the same effect as doubling the layer thickness. This makes sense, since in both cases the light has to pass through twice as much coloring material. Thus, we may expand upon eq. (1) and now write:

T_0 = transmittance of colorant layer of thickness d_0 and concentration c_0.
T = transmittance of colorant layer of thickness d and concentration c

The product $(d_0 c_0)$ represents the effective amount of colorant in a layer of thickness d_0 and colorant concentration c_0. Likewise the product (dc) represents the effective amount of colorant in a layer of thickness d and colorant concentration c. Thus the ratio $dc/d_0 c_0$ gives the effective "number of layers" of the combination (dc) compared to the combination $(d_0 c_0)$. Thus eq. (1) becomes

$$T = T_0^{(dc/d_0 c_0)}$$

or

$$\boxed{T = T_0^{(d/d_0)(c/c_0)}} \quad \blacksquare \qquad (2)$$

Example 3

A colorant of a particular thickness and concentration has a transmittance of 0.6 at a wavelength of 550 nm. The thickness is quadrupled, but the concentration is cut in half. What is the resulting transmittance?

We have

$$T_0 = 0.6$$

$$\frac{d}{d_0} = 4$$

$$\frac{c}{c_0} = \frac{1}{2}$$

Thus

$$T = (0.6)^{(4)(1/2)} = (0.6)^2 = 0.36$$

Use of eq. (2) allows us to calculate the effect of changes in both layer thickness and dye concentration. In Figure 5.5, the curves labeled 2 through 5 illustrate the effect of increasing either concentration or layer thickness by a factor of 2 through 5, respectively. Actually, these curves are even more general than this. Thus, if curve 1 in Figure 5.5 represents the transmittance of a colorant of some thickness d_0 and concentration c_0, then curve 2 represents the transmittance of the colorant for *any* combination of d and c for which

$$\left(\frac{d}{d_0}\right)\left(\frac{c}{c_0}\right) = 2$$

Curves 3, 4, and 5 in Figure 5.5 likewise represent situations for which

$$\left(\frac{d}{d_0}\right)\left(\frac{c}{c_0}\right) = 3, 4, \text{ or } 5, \text{ respectively} \quad \blacksquare$$

Although the curves in Figure 5.5 accurately represent the changes in transmittance with increasing layer thickness or dye concentration, it is hard to visualize the resulting color changes. Two things occur as we move from curve 1 toward curve 5 in Figure 5.5. First, the CIE chromaticity coordinates of the color change (recall Chapter 3); in addition, the color becomes darker. Figure 5.7 is CIE diagram that illustrates this process. Point *C* is the chromaticity of CIE Standard Source *C* that we presume is used to illuminate our colorant. Assuming the colorant has been placed on a perfectly white surface, if the colorant had *zero* thickness, then all the light from Source *C* would be reflected. The chromaticity point would be at point *C*, and the resulting value of Y(%) (the CIE measure of reflectance) would be 100%. For the combination of layer thickness and concentration represented by curve 1 in Figure 5.5, the chromaticity of the reflected

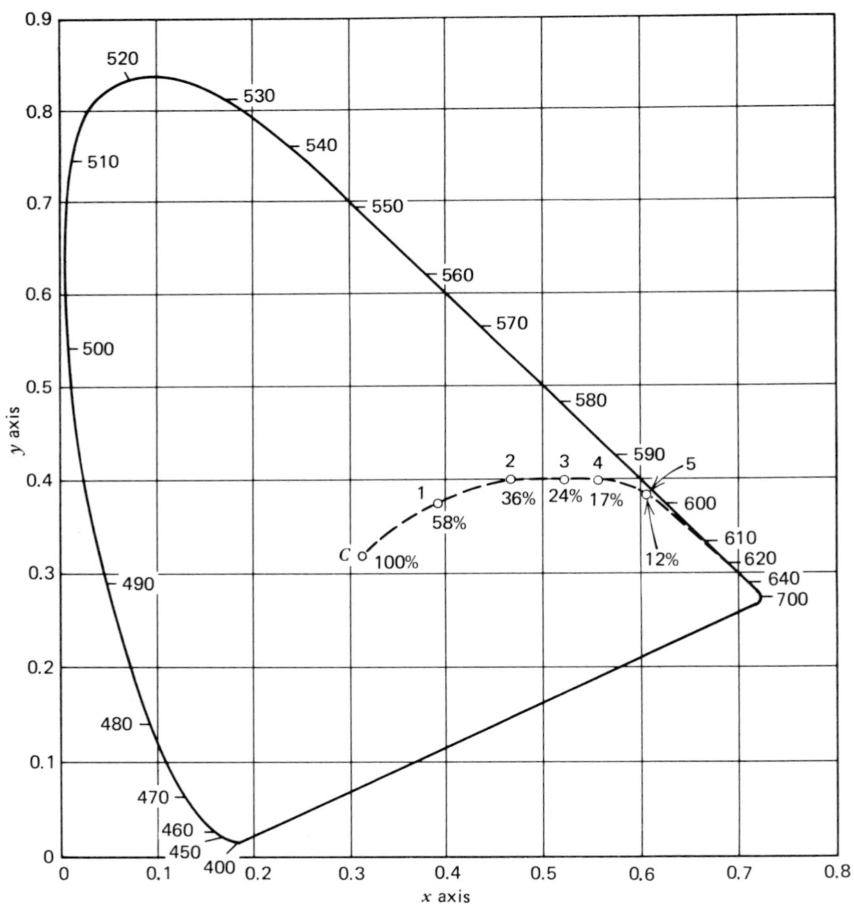

Figure 5.7 Chromaticities of curves shown in Figure 5.5 assuming illumination by Standard Source *C*.

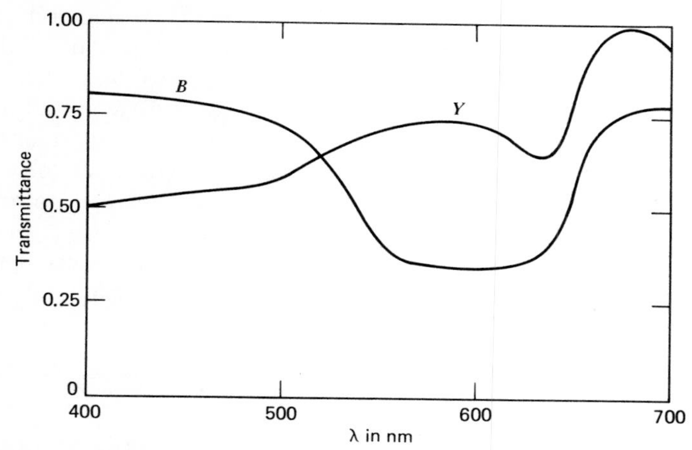

Figure 5.8 Transmittance curves of a yellow (Y) and blue (B) colorant at some fixed thickness, d_o, and concentration, C_o.

light has shifted on the CIE diagram to the point labeled 1. Next to this point is the corresponding value of Y(%) as a percentage. Points 2 through 5 on the CIE diagram represent the chromaticities corresponding to curves 2 through 5 in Figure 5.5. Also on the CIE diagram is a dashed line representing the additional changes in chromaticity that occur as the colorant is made even thicker or more concentrated. At first glance it seems that an extremely saturated pure color can be achieved by making the colorant layer sufficiently thick or concentrated. This is misleading, however, since it does not take into account the greatly reduced brightness indicated by the very small values of Y(%) that result. The actual appearance of the color tends toward brown and then black if the colorant becomes too thick or concentrated. Thus there is a practical limit to the purity of color that can be reached.

TRANSPARENT COLORANT MIXTURES

If two transparent colorants are mixed, the effect is the same as placing two filters one after the other. Light must now pass through both kinds of colorants, and will undergo selective modification by each. Use of two colorants in various proportions greatly increases the chromaticities that can be achieved. Consider, for example, the two transmittance curves labeled Y and B in Figure 5.8. The Y curve represents a yellow colorant of some thickness and concentration, while the B curve represents a blue colorant of the same thickness and concentration. In the remainder of this discussion we shall assume that the colorant thickness is always the same. This will enable us to examine the effect of varying the relative and absolute concentrations of both colorants in a mixture. For the two colorants whose transmittance curves are shown in Figure 5.8, a whole series of chromaticities can be achieved. To begin with, either colorant can be prepared and used alone in a variety of concentrations. The curves labeled Y and B in Figure 5.9 show the chromaticities of increasing concentrations of the yellow and blue colorants individually, assuming illumination by CIE Source C. The other lines show chromaticities resulting from mixtures of the two colorants where the *ratio* of concentrations is held fixed (shown by the fraction labeling each curve) while the total concentration is increased. The entire range of chromaticities contained within the area covered by these curves can be achieved in principle with some mixture of these two colorants. However, in practice this range is limited by the very low values of reflectance that result when concentrations get too high. Nevertheless, a large variety of hues can be produced.

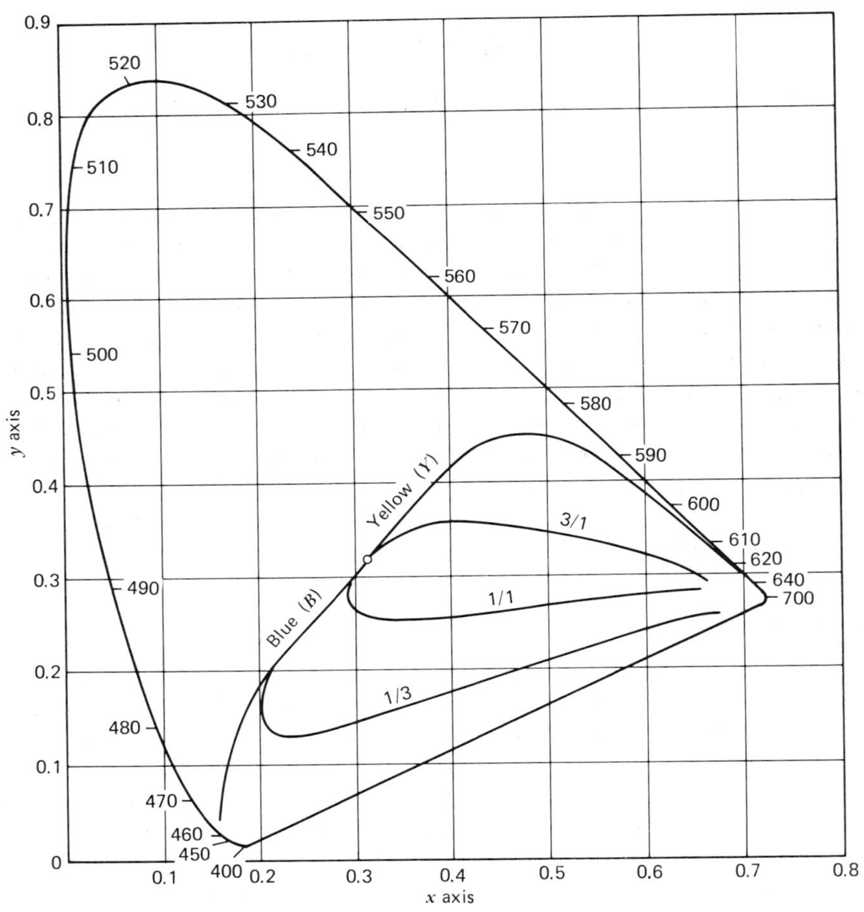

Figure 5.9 Chromaticity traces of various concentrations of the blue and yellow colorants of Figure 5.8. Light lines show chromaticity traces of various mixtures of these two colorants.

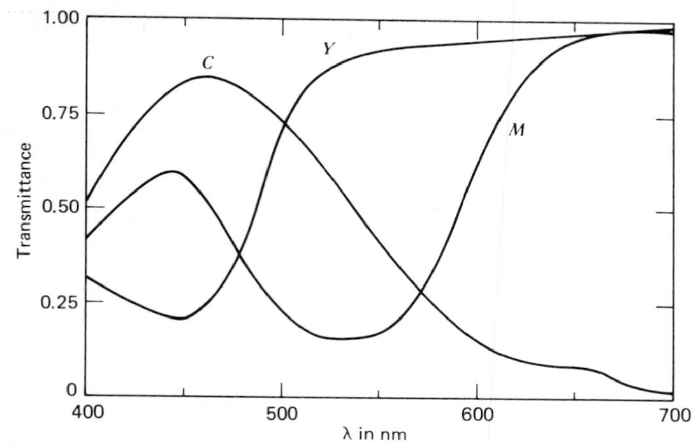

Figure 5.10 Transmittance curves for a particular set of dyes, Cyan (*C*), Yellow (*Y*), and magenta (*M*), at a fixed thickness and concentration.

Mixtures of three colorants can be used to produce virtually any hue required. Figure 5.10 shows the transmittance curves for three standard colorants often used in color reproduction: magenta, yellow, and cyan. Figure 5.11 is a CIE diagram illustrating how these three colorants can be used to produce a wide range of chromaticities. At first glance it might seem as if only two of the three colorants are required at any given point on the CIE diagram. This ignores the possibility of different values of reflectance having the same chromaticity coordinates. There is a unique *relative* combination of the three colorants that produces an achromatic gray. Various concentrations of this combination can be added to any mixture illustrated on Figure 5.11 to *reduce* the reflectance of the mixture without changing the chromaticity. Thus, when all three color variables x, y, and Y(%) are taken into account, it will be found also that all three colorants are required. This more realistic three-dimensional situation is shown in Figure 5.12, which is a CIE diagram with a vertical axis added to illustrate the possible values of Y(%) for each chromaticity. Notice that for high values of Y(%), chromaticities cluster near the white point. This makes sense, because high reflectances require very little selective absorption and thus imply highly desaturated colors. Extremely saturated colors can only occur if a great deal of light is selectively absorbed. Thus high purity implies rather low values of reflectance as expressed by Y(%).

5.3 OPAQUE COLORANTS

As we have seen, in the case of transparent colorants it is possible to make fairly exact calculations of the chromaticities of various colorant mixtures. Unfortunately, the situation is not so pleasant when it comes to opaque materials. Such materials as house paints, and artists' tempera, pastel, and most oils fall into this

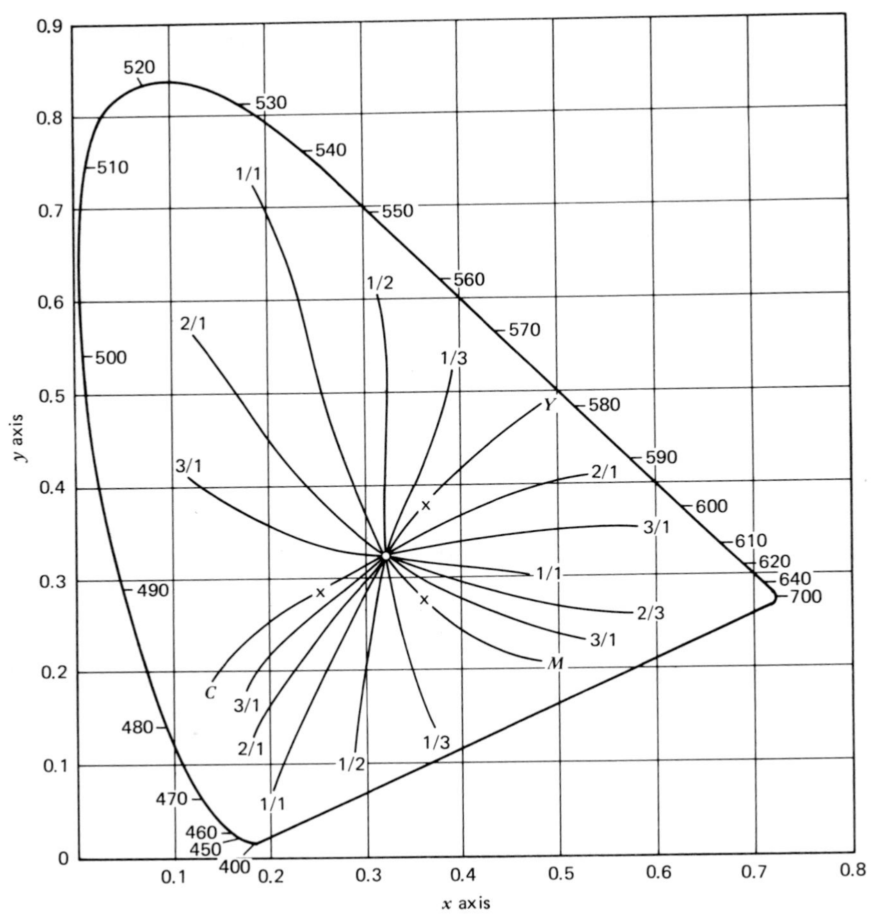

Figure 5.11 Chromaticity traces for the dyes of Figure 5.10. Dark curves labeled *C, Y,* and *M* are for increasing concentrations of dye (points marked *x* are for concentrations shown in Figure 5.10). Lighter solid lines show increasing dye concentrations for fixed proportions of various combinations of two dyes.

category. For these compounds, exact calculations are virtually impossible. The best we can do is describe qualitatively the basic principles that govern their behavior. A firm grasp of these principles will allow the practicing artist to understand why various mixtures produce the results that they do, and will give the artist guidance in preparing his or her materials.

THE PIGMENT PARTICLES

The particles of pigment in an opaque paint are themselves opaque. The surface of an opaque particle acts like any other opaque surface and reflects some light nonselectively. Light that penetrates the particle's surface will be selectively reflected back into the medium in which the particle is suspended. Generally speaking, the more reflections the light makes from different particles in the colorant layer, the more saturated the color of the light becomes. Thus the color will usually become more saturated as the particle concentration is increased. If the concentration becomes too great, however, the particles at the paint's surface virtually touch each other and light is reflected back outside the colorant layer after only one or two reflections from a particle of pigment. An extreme example of this is provided by pastels, which are all pigment and no medium. Because light will usually be reflected only from one or two particles of the pastel pigment, pastels are typically quite desaturated in color. Figure 5.13 illustrates the effect of increasing particle concentration on the reflection of light.

THE SUPPORT

For some opaque paints in sufficiently thin layers, some of the light incident on the surface will penetrate through the colorant layer to the support below. Here it will be reflected to some extent depending on the nature of the support. If the support is white, the light reaching the support will be almost totally reflected back into the colorant layer where it will suffer additional reflections with pigment particles before escaping from the paint. These additional reflections tend to increase the saturation of the paint's color. If the support is black,

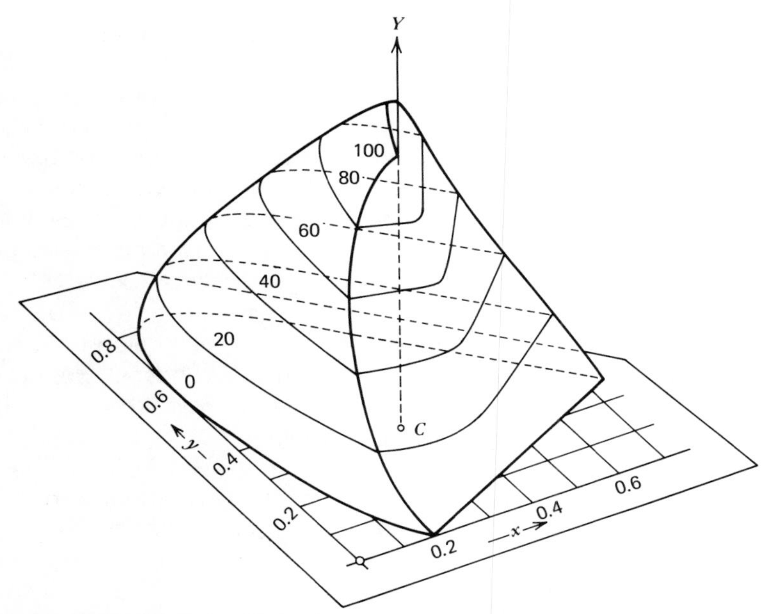

Figure 5.12 Three-dimensional CIE chromaticity diagram showing that the high purity in a reflectance spectrum is necessarily accompanied by a low value of Y.

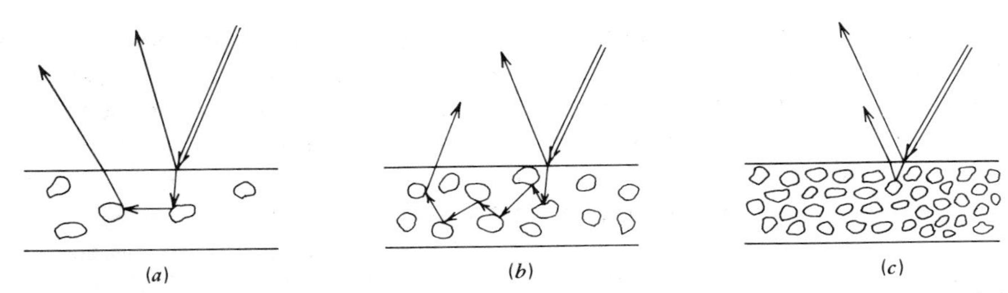

Figure 5.13 (*a*) Low pigment concentration results in relatively few reflections and thus limited selectivity. (*b*) Increased pigment concentration leads to more reflections in the colorant layer and thus the light becomes more selectively reflected. (*c*) Very high pigment concentrations may result in reduced selectivity because light will now reflect only from the upper most particles of pigment.

no such increase in saturation will occur because little or no light will be reflected back into the colorant layer. A colored support will tend to distort the color of the pigment layer by "showing through." That is, the light reflected from the support will contain a mix of wavelengths different from the original incident light and will thus change the mix of wavelengths reflected from the painted surface. If a paint has very good "hiding power," the effect of the support is negligible.

THE OUTER SURFACE

In Section 5.1 the effect of the outer surface was briefly discussed. It was pointed out that matte surfaces and glossy surfaces behave quite differently under most circumstances. The nonselective reflection from a matte surface is diffuse and will thus be seen as if mixed with the selectively reflected light from below the surface regardless of the location of light sources or the viewing angle. The effect of this added nonselective diffuse light, which may amount to 40 to 50% of the incident light in extreme cases is to desaturate the color of the surface. On the other hand, if a glossy surface is illuminated by a single light source, except in the direction of specular reflection, the surface color will be much more saturated. This same glossy surface can appear quite desaturated, however, if it is placed in a brightly lit room with lightly colored walls. In this case light will strike the glossy surface from virtually every angle so that a great deal of specular glare is seen regardless of viewing angle. This should be kept in mind when exhibiting oil paintings for example.

Another point to be kept in mind is the effect of the angle of the incident light as it strikes the surface of the paint. For most oil paints, about 4 to 5% of the light that strikes the surface vertically will be nonselectively reflected while the rest will penetrate the colorant layer. As the angle of incidence becomes more oblique, an increasing percentage of light will be nonselectively reflected, until at a grazing angle virtually all the incident light will suffer this type of reflection.

This has two effects. First, the increase in nonselective reflection directly reduces the color saturation. Second, the more light reflected from the upper surface, the *less* that is available to penetrate the colorant layer and produce selective reflection. The worst possible situation occurs when a painting is illuminated at an oblique angle and viewed so that the reflected glare is visible. In this case the actual colors of the painting might be virtually unobservable. Figure 5.14 illustrates the effects of various angles of illumination and observation. An artist should keep in mind that the colors he *sees* on the canvas are dictated not only by the selection of pigments but also by the angle of the illumination used while the painting is in progress. In exhibiting, the same illumination angle must be used if the same colors are to be seen.

COLORANT MIXTURES

The effect of colorant mixtures involving opaque pigments is quite complex. Such mixtures do not follow the simple subtractive laws that govern the behavior of transparent colorants. In general, only practical experience with the specific colorants of interest can lead to satisfactory results. Mixtures of opaque colorants will follow *approximately* the same rules as transparent colorants, but the detailed results of various mixtures cannot be accurately predicted. For example, a mixture of the two pigments whose reflectance curves are shown in Figure 5.15 will produce a green, but exactly which green cannot be predicted. Figure 5.16 shows the variations in chromaticity produced by mixing a number of opaque colorants.

An interesting situation occurs when opaque white pigment is added to a transparent colorant. If the colorant is very dark to begin with, the introduction of the highly reflecting white pigment will increase the amount of light reflected back from the colorant layer and may also increase the extent of the selective

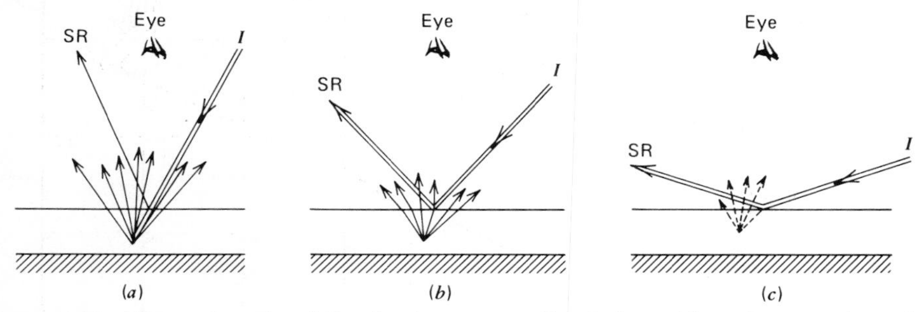

Figure 5.14 Effect of angle of illumination on an oil painting with a glossy surface. (*a*) Best results are obtained when illumination is slightly oblique. This minimizes specular reflection (SR) and maximizes diffuse reflection from within the colorant layer. (*b*) and (*c*) show effects of increasing the obliquity of the illumination. This results in more specular reflection and less diffuse reflection from the colorant layer.

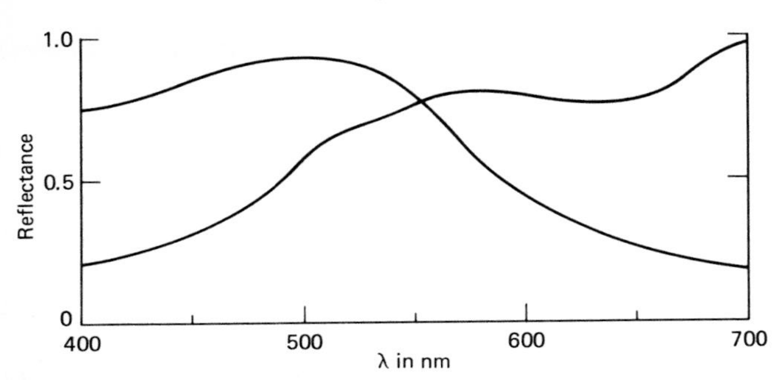

Figure 5.15 Reflectance curves of two opaque pigments. A mixture consisting of equal parts of these two colorants produces a green. The exact chromaticity of the mixture could not be predicted from these curves however.

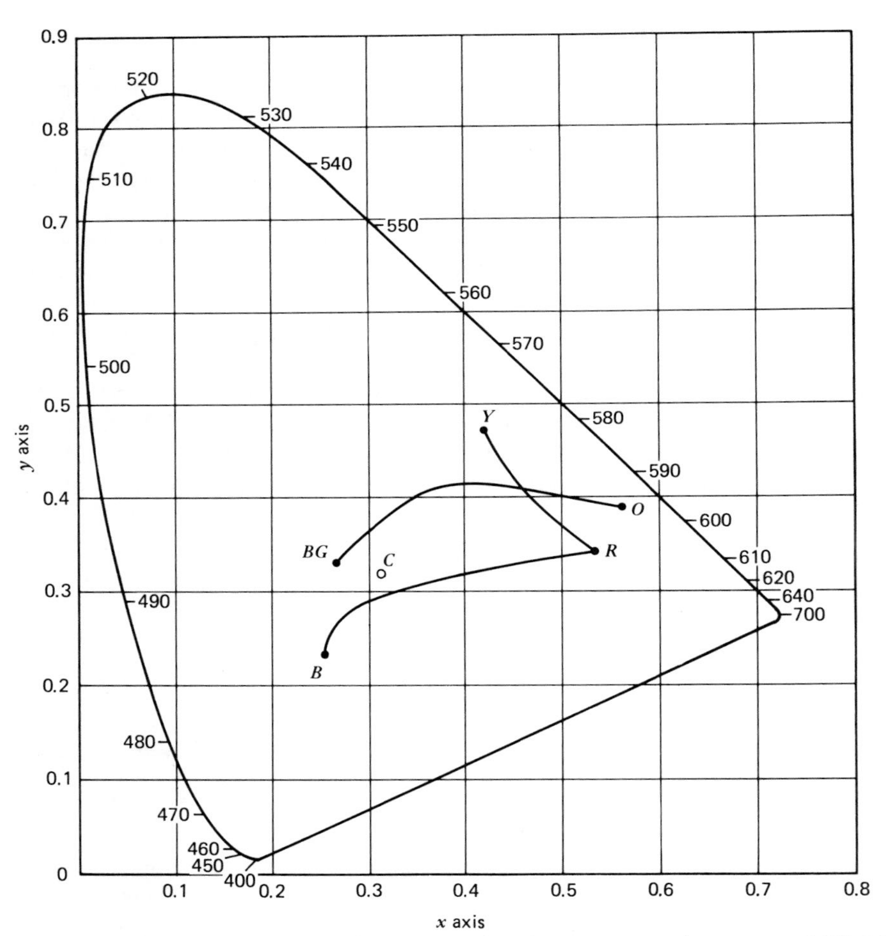

Figure 5.16 Chromaticity traces of mixtures of various opaque pigments, red (*R*), blue (*B*), yellow (*Y*), orange (*O*), and blue-green (*BG*).

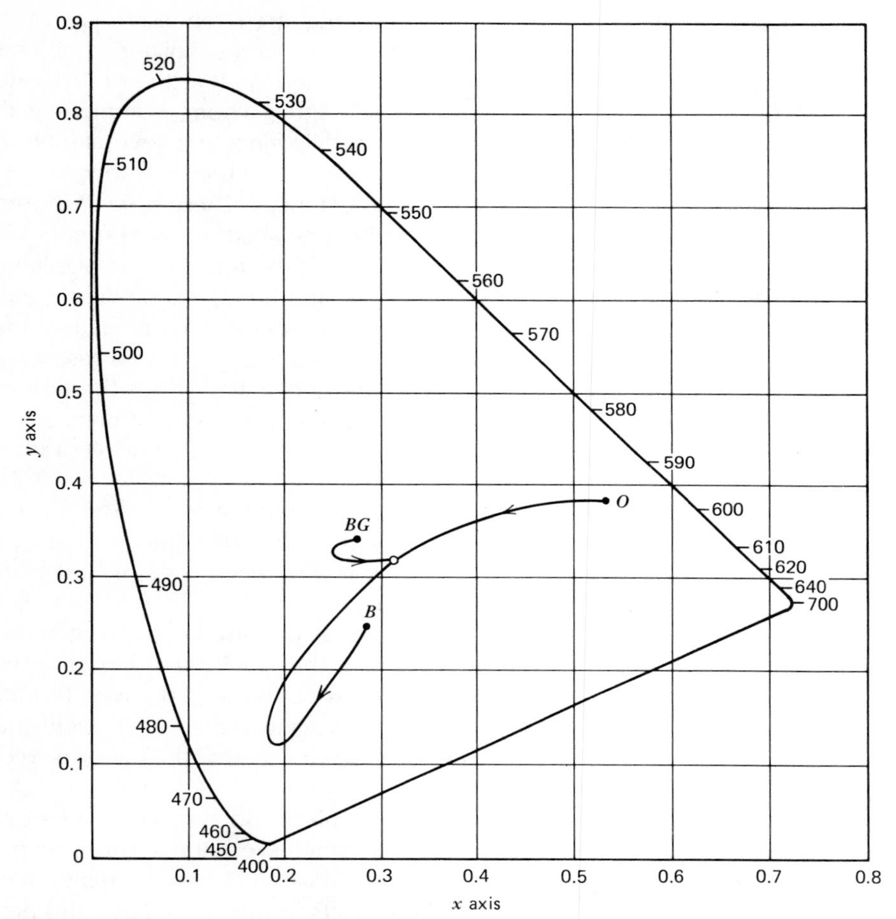

Figure 5.17 Effects of adding opaque white pigment to several colorants, blue (*B*), blue-green (*BG*), and orange (*O*). Note that, up to a point, the addition of white to the darker colorants, *B* and *BG*, increases their purity.

action of the colorant. This increased selective action occurs because of the multiple reflections of the light from the white particles and the resulting increased distance traveled by the light in the transparent colorant. Thus, adding opaque white pigment to a dark transparent colorant may increase, up to a point, the saturation of the color. Figure 5.17 shows the effect of adding white opaque pigment to several transparent colorants.

5.4 THE USE OF COLOR IN PAINTING

As any artist knows well, color has many uses in art. Most of these are beyond the subject matter of this text, and we will not attempt to discuss them. However, we can make a few points about paintings that are intended to portray a realistic scene. In this case, the artist usually wishes us to recognize real objects in a real setting of some sort. For example, let us suppose that the objective is to portray the quiet elegance of a gentleman's sitting room, complete with crackling fire, well-stocked book shelves, subdued lighting, and several men of means engaged in leisurely discussion. The actual scene, in the flesh so to speak, will contain objects having a wide variety of reflectances. These objects in turn will be illuminated nonuniformly. Some will be in deep shadow while others will be directly illuminated by a nearby light source. Thus the range of surface luminances (the absolute amount of light reflected by an object) will be very great. On the other hand, since paintings are displayed under uniform illumination, the range of luminances associated with different areas of our painting is restricted to the range of surfaces reflectances achievable by oil paints. Thus the task of the artist is to create on the canvas the *illusion* of the vast range of luminances which exist in the original scene. This is done by applying artistic principles rather than those of light and color.

Another problem faced by artists is the fact that in the real world, the surface of a uniformly colored object will frequently reflect a different spectrum of light from different areas of the surface. This is due to the reflectance of light from other nearby colored objects and the frequent occurrence of nonuniform illumination. We see the object's surface, however, as uniform in color regardless of these nonuniform reflectances. In his painting the artist must duplicate this effect. Thus the surface of a blue bowl must *look* uniformly blue even though in the painting a range of hues will be necessary to achieve this effect. Highlights and shading must be represented on canvas in such a way that the uniform color of the bowl is still apparent.

Often in viewing a painting we are struck by the predominance of a certain color. This effect also occurs in the real world when we enter a room with a specific color scheme or a room with colored illumination. A predominant coloristic effect does not mean that most of the objects in the scene have the predominant hue. What is more common is to find the average color of the scene shifted away from white in a certain direction. The solid oval on the CIE diagram in Figure 5.18 is centered on the chromaticity of Standard Source *C*, and represents the range of chromaticities we might expect in a room with no predominant color that is illuminated by Source *C*. Now, if the same room were illuminated by some colored source, *C'*, all the colors would shift in that direction. The chromaticities in the scene would now occupy the area inside the dashed oval in Figure 5.18. Our perception would be that of a normal range of hues illuminated by a colored illuminant. A similar effect occurs when Source *C* illuminates a scene in which the objects themselves have the chromaticities represented by the dashed oval in Figure 5.18. If direct evidence of the nature of the light source is not included in the paintings (or in the actual room being viewed), the two cases cannot be distinguished by the eye, and our perception

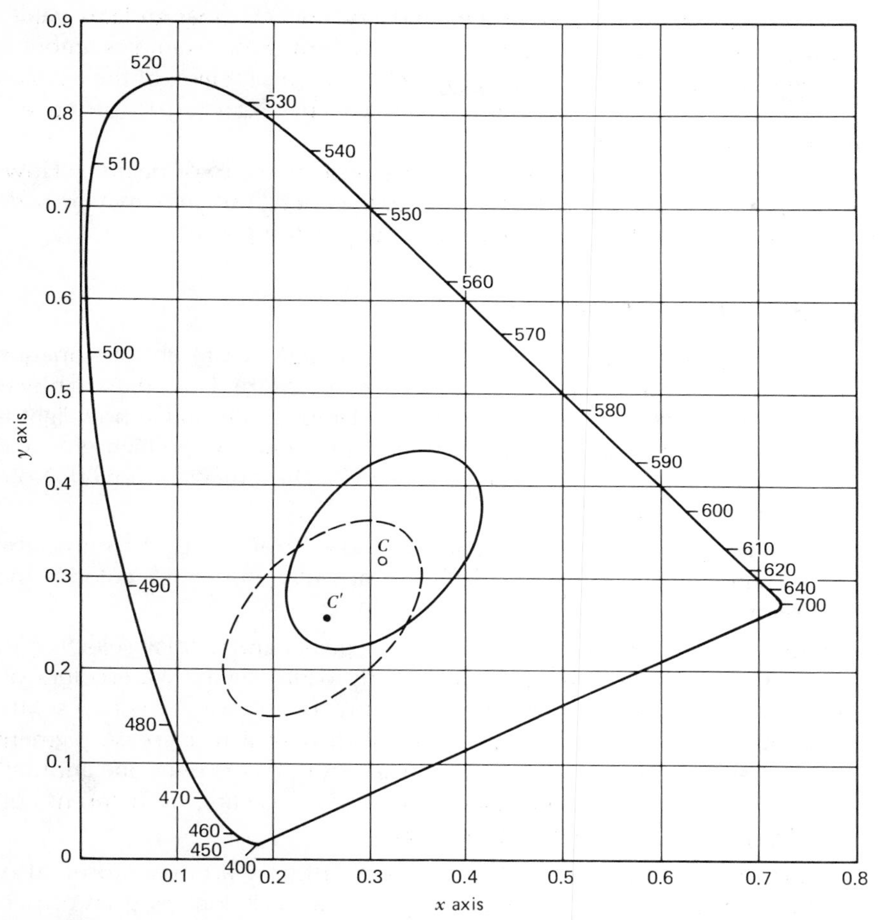

Figure 5.18 Solid curve shows range of chromaticities usually obtained using Source *C* and a balanced range of pigments. Shifting the choice of pigments to the range of chromaticities in the dashed oval produces the perception of a colored illuminant (*C′*).

alternates between the two possibilities. Normally, however, the light source is visible, and the two cases are easily distinguished.

In addition to the possibility of a predominant color in a scene there may occur a restriction in the perceived range of values and chromas (review Chapter 3: Munsell system). This kind of restriction can control the mood of the painting. For example, a full range of hue, value, and chroma is characteristic of direct sunlight on a multicolored scene. Restricting the range of values to the low end of the scale would produce a completely different effect, for example, one of quality and elegance in the painting of a room. The same room painted in low chromas and high values would look light and airy. Low chromas and low values would make the room seem shabby and dull. Each of these examples amounts to restricting the colors in the painting to a relatively small region of the Munsell or Ostwald color trees. Different restrictions can be expected to produce a variety of special moods and effects.

All of the effects discussed in this section are well known to oil painters. However, looking at these effects in the light of our knowledge of color may provide a different perspective that will be useful to the practicing artist.

PROBLEMS AND EXERCISES

1. Obtain a glossy opaque object (such as a small vase) and an opaque object with a matte finish (such as a piece of cloth). Hold these objects near a single light source (such as a table lamp) and examine how light is reflected from them. In particular, notice the nature of the highlights and how they change when either your head or the object is moved. Note the color of the highlights.
2. Obtain a metallic object, and compare how it reflects light to how the objects in question 1 reflected light. In particular, notice the color of the highlights.
3. Obtain an inexpensive set of watercolor paints and a small selection of colored papers. First, try painting a fairly dilute set of watercolors on white paper. Now increase the concentration of watercolor. Also try allowing the paint to dry and then painting over it to increase pigment thickness. Try painting different combinations of colors over one another. Can you account for the resulting colors? Try painting on different colored papers. Can you predict the results?
4. Obtain a small package of food coloring from a grocery store. Also, obtain a small transparent container such as a small drinking glass or perhaps a test tube. Fill the container about $\frac{3}{4}$ full of water. Now add a single drop of one of the food dyes. Note the color of the fluid. Now add more coloring, one drop at a time. Carefully note how the color (chroma, value, *and* hue) changes as the dye concentration increases. Try making a variety of mixtures of different colors and concentrations. See what kind of variety of colors you can produce.

6 GEOMETRIC OPTICS

Our discussions of vision up to this point have been primarily concerned with the question of what color is and how we interpret the color stimulations that reach our eyes. At this juncture let us divert our attention to a different aspect of seeing: how light is collected and brought to focus on the retina. To facilitate this understanding we shall begin by investigating the laws of geometric optics and then use these ideas to understand the methods by which lenses and mirrors can collect light and cause it to produce images on the retina of the eye or the film of a camera. The basic principles we shall investigate also are the building blocks of various types of devices, such as telescopes, projectors, optical fibers, and microscopes. These instruments extend the eye's versatility, making it possible to see many additional things it would otherwise not be able to see.

6.1 REFLECTION

Most of the objects around us are not emitters of light but instead reflect part of the light that is incident on them. Reflected light is responsible for most of our visual perceptions. Although most objects reflect primarily by diffuse reflection, certain highly polished objects are primarily specular reflectors. Such objects can be used to form optical images of various kinds. The formation of such images can be understood on the basis of a law that was well known in the Greek era some two thousand years ago.

This basic law of specular reflection can be stated as: *The angle at which a ray is incident to a surface must equal the angle at which it is reflected from the surface.* This law is illustrated in Figure 6.1. The angle of incidence is defined as the angle between an incoming or incident ray of light and a line that is perpendicular to the reflecting surface at the point of incidence. The perpendicular line is called the *normal* line. The angle of reflection must lie in the same plane as the normal and the incident ray.

The formation of images by specular reflection requires the use of materials, such as metals, which inherently reflect by specular reflection. Even with such

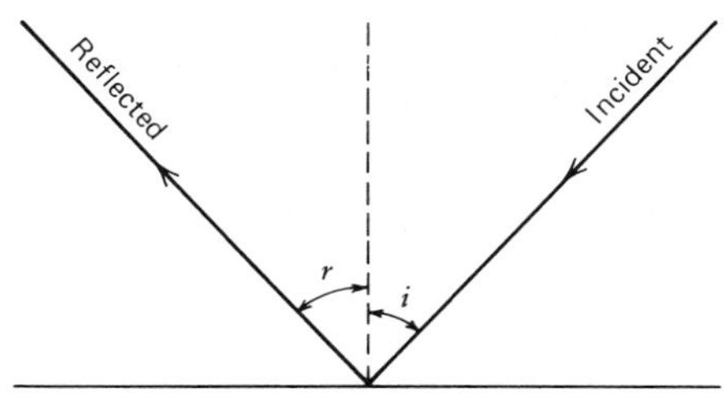

Figure 6.1 The angle of incidence equals the angle of reflection.

materials, however, if the surface of interest is rough the multitude of miniature bumps it contains results in the light being scattered in all directions or being diffusely reflected. Figure 6.2 shows light rays hitting such a surface and obeying the law of reflection on a microscope scale, but not on a large scale. Regardless of how smooth a surface may be polished it will still diffuse some portion of the light incident on it. The dust that has settled on the surface or small microscopic irregularities are sufficient to diffusely scatter some of the incident light reaching the surface. However, if these imperfections are kept to a minimum, it is possible to construct high quality reflecting devices.

If a surface is very smooth in relation to the size of the electromagnetic waves that are striking it, the surface can be considered to be polished for that type of wave. When the distance between successive bumps or facets of the surface is one-half wavelength or less of the incoming radiation the area will act as a uniform reflector or mirror surface. Depending on the shape of the mirror surface, different types and sizes of images (visual reproduction of an object) are produced of the objects whose light is reflected.

6.2 PLANE MIRRORS

Daily experience with plane or flat mirrors has led us to know that the images produced by these reflecting surfaces are most similar to the object that provides the light. When light leaves a point of an object it spreads out in all directions. If some of the light strikes a nearby mirror, an image will be formed according to the law of reflection. To understand the nature of the image and its method of formation, consider what happens when you look into a mirror.

The tip of your nose reflects light that is striking it and this light spreads in all

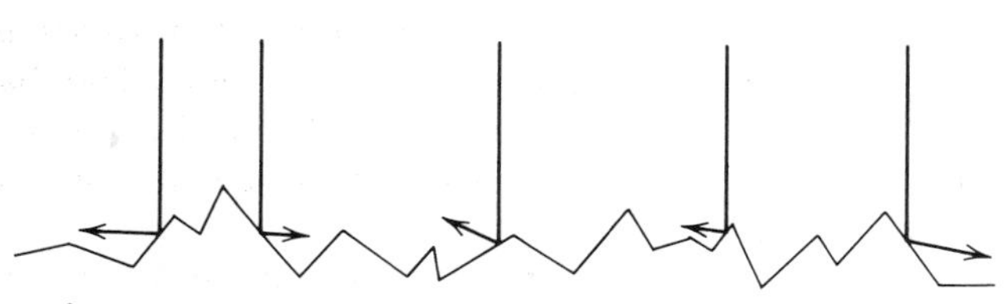

Figure 6.2 A rough surface results in diffuse reflection.

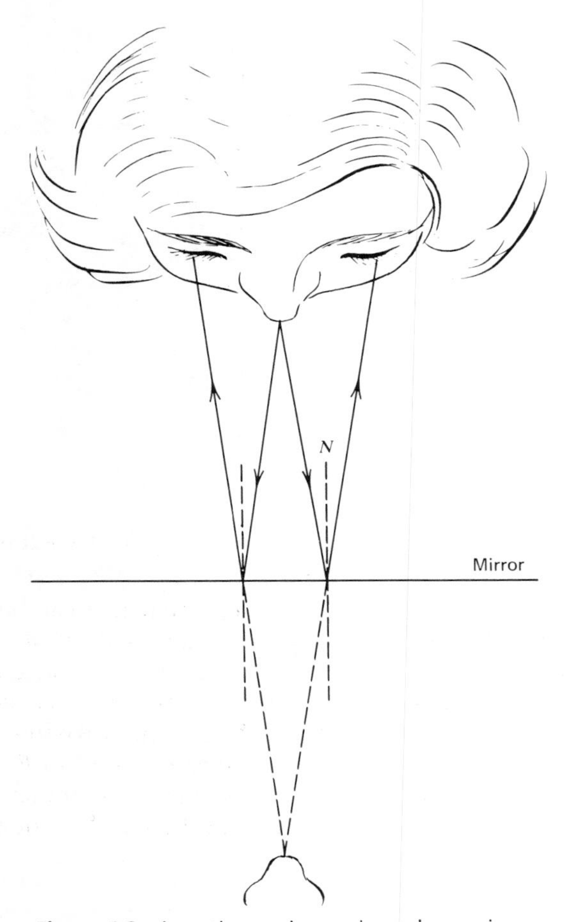

Figure 6.3 Locating an image in a plane mirror.

directions. Some of the light that leaves your nose will travel in the proper direction to strike the mirror, which will reflect it. Certain of the rays (Figure 6.3) will bounce off the mirror and reflect back into your eyes. Your eyes trace these rays of light to the point where they apparently originate. This point is where the image of the tip of your nose appears to be. As can be seen from the illustration, the position of the image (point where the light rays come together or appear to meet) is on the opposite side of the mirror from where your nose is. An examination of the diagram will also show that the image's distance from the mirror is equal to the distance of your nose from the mirror. Images in which the rays of light do not really intersect, but instead are projected by your brain to do so, are referred to as *virtual images.* That is, no light actually need exist at the location of a virtual image. The image is located where the light *seems* to come from.

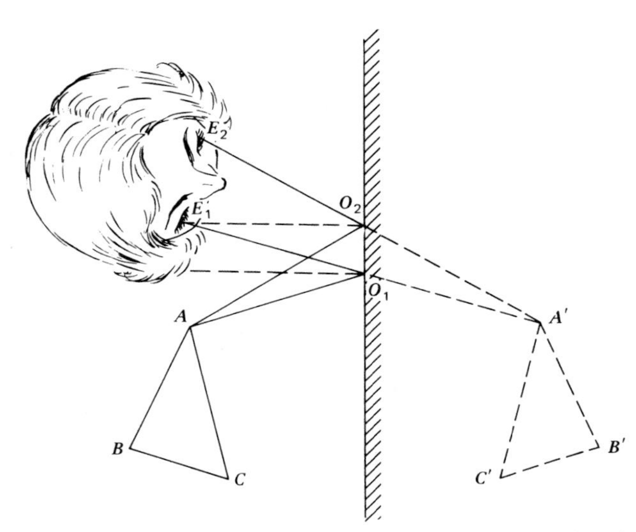

Figure 6.4 Locating the mirror image of a triangle. Note the left-right reversal.

Another quality of the image formed by a plane mirror can be seen by examining Figure 6.4, which illustrates an experimental method of constructing the image formed by the mirror. A triangle, *ABC,* drawn on a piece of paper serves as the object whose image is to be located. A pin is then placed at angle *A* of the triangle. Light rays AO_1 and AO_2 leave the pin and strike the mirror reflecting to the right and left eyes of the observer along lines O_1E_1 and O_2E_2. The observer's brain projects these two lines as having come from behind the mirror and intersecting at *A′*, which is the location of the image of point *A*. The same

Figure 6.5 Mirror image showing left to right reversal.

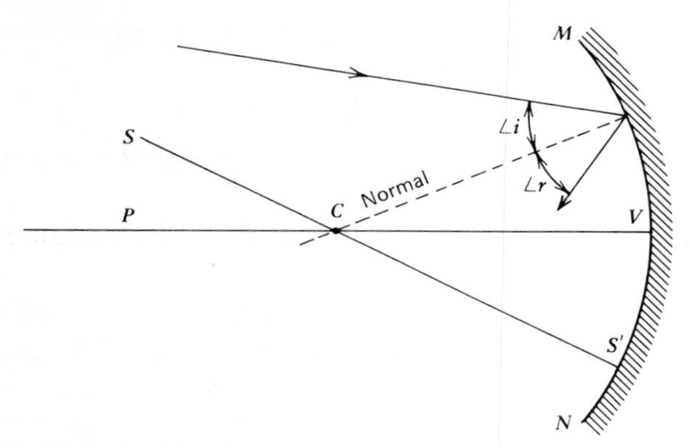

Figure 6.6 The basic geometrical features of a concave spherical mirror.

procedure is then repeated for points *B* and *C* and their images are found to be located at *B′* and *C′*. In each instance we note that the images appear to be as far behind the mirror as the object is in front. Construction of the triangle from three image points shows that the size of the object and the image are identical. Examination of triangle *A′B′C′* makes it appear that the image in the mirror has been reversed from right to left. This is more easily seen in Figure 6.5 where it appears that the "image" is holding up his left hand whereas the real person is holding up his right hand. Actually, each part of the image is located directly across from the corresponding point of the object. Thus the image of the upraised right hand is in fact on the right side of the mirror. But now it looks like a left hand because a real person standing behind the mirror would have to *turn around* in order to face out, and this would reverse the location of right and left. It is in this sense that a plane mirror is said to create an image that has reversed right and left.

6.3 CURVED MIRRORS

The reflection of light from curved mirrors follows the same fundamental principle of reflection as the light from the plane mirror we have been discussing. However, the shape of the surface allows for differences in the types of images that are formed. With concave or convex mirrors it is possible to obtain images that are enlarged or reduced in size, are erect or inverted in their direction, and that can be virtual or real. The method of determining the characteristics of the image one will get from a particular object placed at various distances from a mirror will be shown in this section.

As an aid to understanding the formation of these images we need to define a series of terms for concave mirrors. If, in Figure 6.6, we assume that *MN* represents a concave spherical mirror in two dimensions, then we need the following definitions.

1. The *center of curvature, C,* is the center of the sphere of which the mirror *MN* is a part.
2. The *vertex, V,* is the center of the mirror itself.
3. The *principal axis, PV,* is the line drawn through the center of curvature and the vertex.
4. A *secondary axis* is any other line drawn through the center of curvature. Example *SS′*.
5. A *normal* to the surface of the mirror is any line drawn through the center of curvature to the point of incidence of an incoming light ray. All such lines would be radii of the sphere and thus would be perpendicular to the surface at the point of incidence. In a convex mirror, a normal is a radius produced or extended beyond the mirror.

Three parallel rays of light are shown incident on a concave mirror in Figure 6.7. Ray *PV,* which is along the principal axis, will strike the mirror and be reflected back in the same direction from which it came since its angle of incidence is zero degrees. Ray *AB* and *CD* will not hit the mirror perpendicularly but will hit at some angle to their normals. As can be seen in the diagrams these rays will reflect and pass through a common point on the principal axis known as the *principal focus.* The distance of that point from the mirror is known as the focal length of the mirror. This point is where all rays parallel to the principal axis will converge when reflected from a concave mirror. Conversely, if a light source is placed at the principal focus of such a mirror the rays striking the mirror will be reflected in a parallel beam. This idea is the basic principle used in automobile headlights, which produce beams of light that are nearly parallel.

If a spherical mirror is made from too large a fraction of a sphere, the parallel rays of light deviate from the principal focus and instead focus at points closer to the vertex (Figure 6.8). This imperfection is known as *spherical aberration* and results in fuzzy images. In large mirrors it can be avoided by reducing the

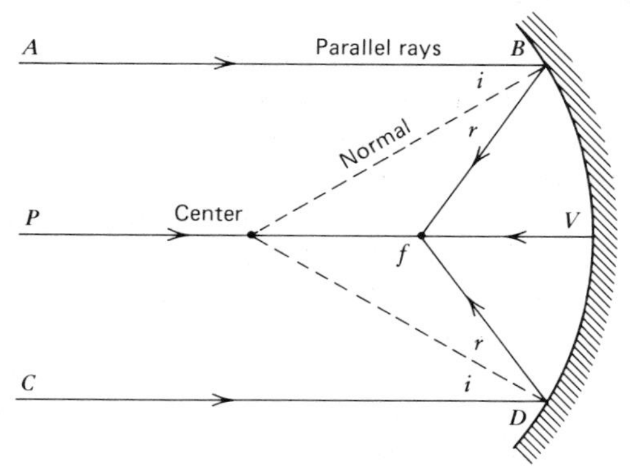

Figure 6.7 Locating the focal point of a concave mirror.

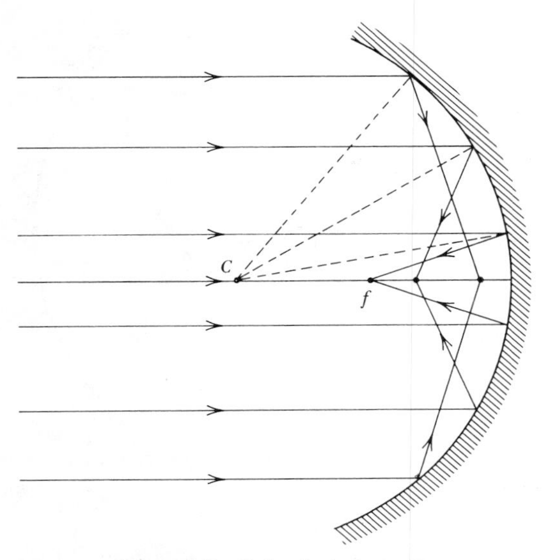

Figure 6.8 Spherical aberration.

curvature of the mirror (in effect cutting the mirror from a sphere of larger radius) but this increases the focal length of the mirror. The problem can be avoided by making the mirror parabolic in shape rather than spherical. Parabolic mirrors do not suffer from spherical aberration regardless of size, and are frequently used in large reflecting astronomical telescopes (which will be discussed in Chapter 7).

Now let us assume that our concave spherical mirror is small enough so that there is essentially no spherical aberration. How does such a mirror form an image? We must distinguish two different situations; an object outside the focal point of the mirror, and an object between the focal point and the mirror itself. The first possibility is illustrated in Figure 6.9, the second in Figure 6.10. Let us consider the situation in Figure 6.9 first, that of the object outside the focal point. The "object" to be considered is the arrow *SP*. In the diagram, three light rays are shown leaving the head of the arrow, point *S*, reflecting from the mirror, and then coming back together at point *S′*. Actually there will be many rays of light that leave point *S* and strike the mirror. All of these will converge at point *S′*, which is the *image* of point *S*. Light will leave other points on the arrow and will converge again to form a series of images. In this way an image of the entire arrow will be formed. As can be seen from Figure 6.9, the image is *inverted* and *real*. We call the image *real* because light actually converges at the location of the image, unlike the case of the plane mirror.

In order to locate the image and to determine its size, we need only locate point *S′*. Figure 6.9 makes it clear that once this point is located, the size and position of the image are determined. To find point *S′* we can follow *any* two

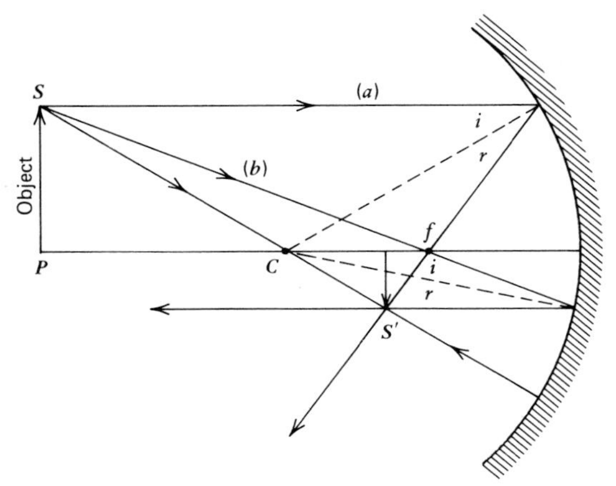

Figure 6.9 Locating an image formed by a concave mirror of an object placed outside the focal point of the mirror.

rays that leave point *S* and strike the mirror until they intersect. However, to draw the angle of reflection equal to the angle of incidence for most rays requires careful use of a protractor and even with care it would be easy to make an error. Fortunately, there are two rays that behave in a very simple, easily drawn fashion. These two rays, which are illustrated in Figure 6.9, are (*a*) the ray that leaves *S parallel* to the principal axis and therefore, reflects from the mirror through the focal point, and (*b*) the ray that leaves *S* and passes through the focal point and, therefore, reflects from the mirror parallel to the principal axis. The third ray shown in Figure 6.9 can also be used to help locate the point *S′*. This ray leaves *S* and passes through the center of curvature of the mirror, point *C*. Of course, this ray simply reflects back on itself since it strikes the mirror normally.

As long as the object is outside the focal point, the image will always be real and inverted for a concave spherical mirror. A little experimentation with a real mirror or with drawing graphical constructions such as Figure 6.9 reveals the following results. As the object is removed farther from the mirror, the image moves toward the focal point and gets smaller. Conversely, as the object is brought closer to the focal point, the image moves away from the focal point and gets larger.

Now let us turn our attention to the case where the object is located inside the focal point, as in Figure 6.10. The construction proceeds just as before. That is, we draw (*a*) a ray that leaves *S* parallel to the principal axis and reflects through the focal point, and (*b*) a ray that leaves *S* in a direction *as if* it originally came from the focal point and therefore reflects parallel to the principal axis. The third ray in Figure 6.10 leaves *S* in a direction as if it came from the center of curvature, *C*. This ray thus reflects directly back on itself. All three rays shown in the diagram never actually converge again after they leave point *S*. After

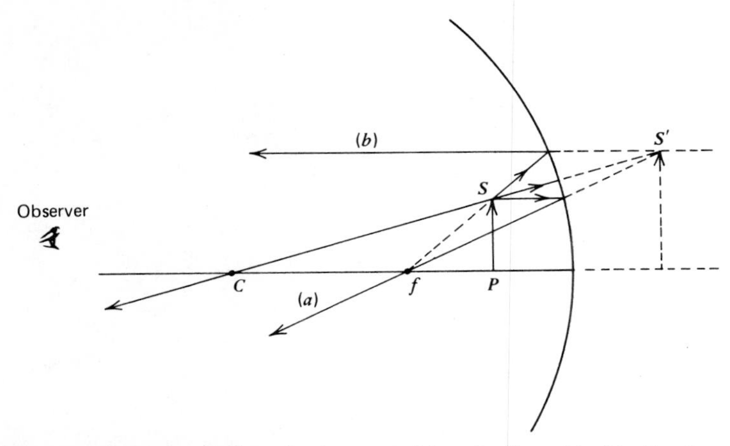

Figure 6.10 Virtual image produced when an object is placed inside the focal point of a concave spherical mirror.

striking the mirror, the rays continue to diverge from each other. However, an observer would see the rays apparently originating from point S', which is called the image of point S. As before, the location of point S' determines the location and size of the entire image of the object. In this case, the image is erect, enlarged, and virtual. As long as the object is inside the focal point, the image will always have these characteristics. One common use of a concave spherical mirror in this fashion is the so-called beauty or shaving mirror. At normal viewing distance the face is *inside* the focal point so that an erect, magnified image results. If, however, you were to back up a few feet the image of your face would become inverted.

Parallel rays of light reflected off a convex mirror are shown in Figure 6.11. As can be seen, these rays diverge and appear to come from a point on the opposite side of the mirror. This point of apparent divergence is the focal point for the convex mirror. The center of curvature of the mirror is also located on the side of the mirror opposite the incident parallel rays. Such a mirror can also collect light from diverse directions and reflect it into a parallel beam. Often convex mirrors are found being used in this manner in shopping areas where it is desired to have a view of a wide area channeled to one store employee. When the rays of light from divergent areas of the store strike the mirror, they reflect from it to a nearly parallel beam giving the observer the ability to watch a large area from one position (Figure 6.12). These mirrors are also frequently found attached to the bottom of a plane rearview mirror on the side of a truck. Used in this fashion the rays of the light from several lanes of traffic can be seen by the driver giving him wider view of the traffic behind him.

The image of an object that will be formed when the light reflects from the convex mirror is shown in Figure 6.13. As the illustration indicates the rays will not actually meet but will form a virtual, upright, and reduced image on the side

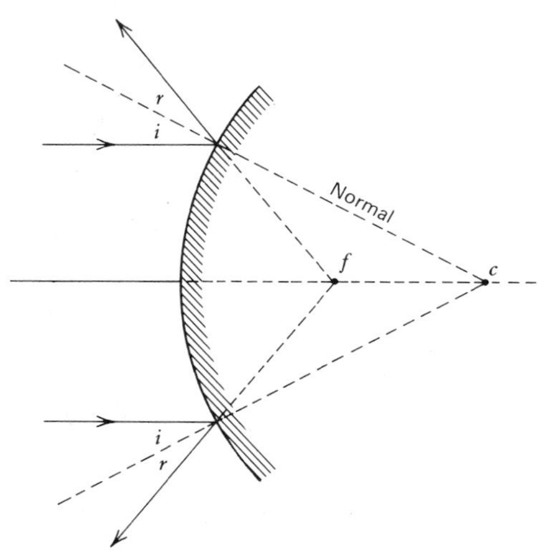

Figure 6.11 Parallel rays of light reflected from a convex mirror.

of the mirror opposite the object. Regardless of the placement of the object relative to the mirror, the resulting image will always have the same characteristics as the one in the diagram. Moving the object closer to the mirror will result in its image becoming larger but it will never reach the size of the object.

Uses of the mirrors described in this section are found in many different types of optical instruments that use lenses and in other optical devices. In the next

Figure 6.12 Observing a wide area using a convex mirror.

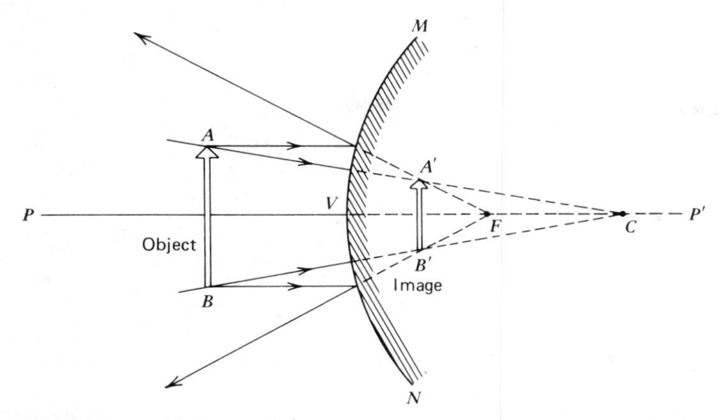

Figure 6.13 Locating the image of an object as observed in a convex mirror.

section we shall discuss refraction, which is the basis of understanding lenses. Armed with this additional knowledge we shall be able to better comprehend the operation of additional optical devices and the process of vision.

6.4 REFRACTION

The term refraction refers to the bending of a wave as it moves from one medium in which it is traveling into a second medium that propagates the wave at a different speed. This bending occurs for waves on water, sound waves, and light waves, and has been observed and written about for thousands of years. The early Greeks studied this phenomenon and understood it from a qualitative standpoint; that is, they knew the way or direction a wave would bend but they were unable to determine the amount or degree of bending that would take place.

Perhaps the most common example of refraction which people readily notice, is the bending of light from an object that is partially immersed in water or some other liquid. The change in speed of the light as it leaves the liquid and enters the air causes the light to refract and thus the object appears to be bent or distorted at the boundary position (Figure 6.14).

A complete understanding of the law describing refraction was not achieved until approximately 1600 A.D. Named in honor of Willebrod Snell, who experimentally discovered the relationship between the incident ray and the refracted ray, this law correctly predicts the extent to which light will be refracted as it moves from medium to medium. The most common form of Snell's law is:

$$\frac{\sin \measuredangle i}{\sin \measuredangle r} = \frac{\eta_2}{\eta_1}$$

In this equation the value η_1 (eta) stands for the index of refraction of the first medium the light is traveling in, and η_2 represents the index for the medium into

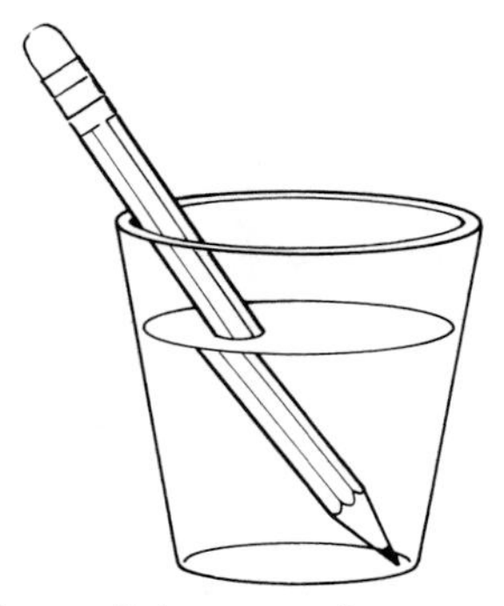

Figure 6.14 Refraction as light passes from one medium to another.

which the light will be moving. The *index of refraction* for a medium is the measure of how fast light travels in that material relative to the speed of light in a vacuum.

$$\eta = \frac{\text{speed of light in a vacuum}}{\text{speed of light in a medium}}$$

Since the speed of light in a vacuum is unexceeded by the speed of light traveling in any material, the index of refraction for any medium will be greater than one.

Shown below are values for some common light transmitting materials.

Some Indices of Refraction at 590 nm

Medium	Index of Refraction
Water	1.33
Air	1.0003
Fused quartz	1.46
Glass, crown	1.52
Glass, flint	1.66
Diamond	2.40

The table indicates that the values given are for a wavelength of 590 nm. This value is specified because the index for a given material does depend on the wavelength. The importance of this fact will be discussed later.

The $\sin \measuredangle i$ stands for the mathematical relationship between two sides of a right triangle that would have one angle equal to the incident angle of the light ray (the angle between the incoming ray and the normal line — Figure 6.15). It is well known that any triangle has three interior angles that always add up to exactly 180°. A right triangle is any triangle that has one 90° angle — the other two angles must add up to 90°. Figure 6.16 illustrates a right triangle. The triangle

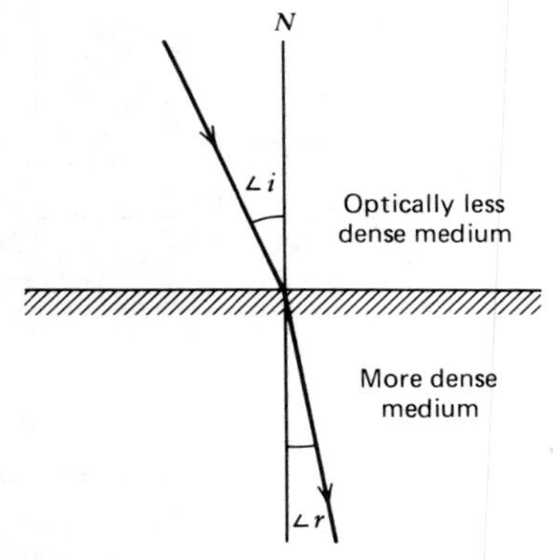

Figure 6.15 The incident and refracted angles.

has one side (opposite the right angle) called the hypotenuse. The other two sides are called the legs of the right triangle. Now consider the angle labeled θ in Figure 6.16, and notice the leg of the triangle opposite the angle θ. If the length of this leg is divided by the length of the hypotenuse, the resulting ratio is known as the *sine of the angle* θ, or sin θ. Each angle has a unique sine, and values of these are available in tables (or on some calculators). Thus, if the value of the angle is known, its sine can be found in a table or vice versa.

For the triangle shown in Figure 6.16 suppose the length of the opposite side was 10 and the hypotenuse 20. This would give a value of $\sin \theta = \frac{10}{20} = 0.5$. A table of sine values would show that the angle corresponding to this value of the sine is 30°.

The sine of the angle of refraction is handled in the same manner. Thus, knowing the indices of refraction for the different materials and the incident angle, it is possible to determine mathematically the angle at which the light will be refracted.

Before attempting to calculate some values for specific materials and angles, let us consider the essence of what the law predicts will happen as light moves from a medium of one index of refraction to a medium of a different index of refraction. There are in effect only two possible situations: (1) where light travels from one medium into another that has a greater index of refraction (greater

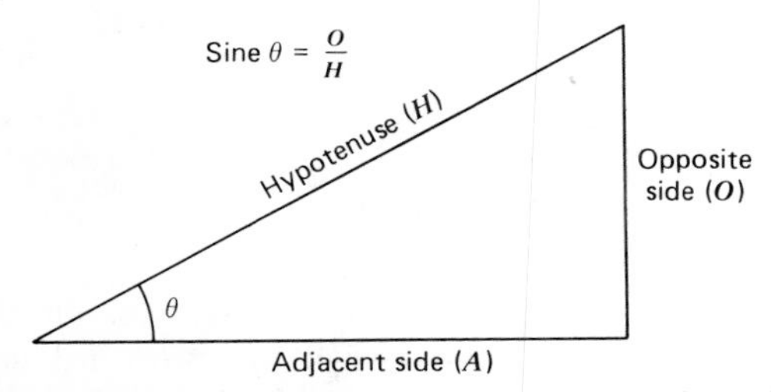

Figure 6.16 Sine of an angle in terms of the sides of a right triangle.

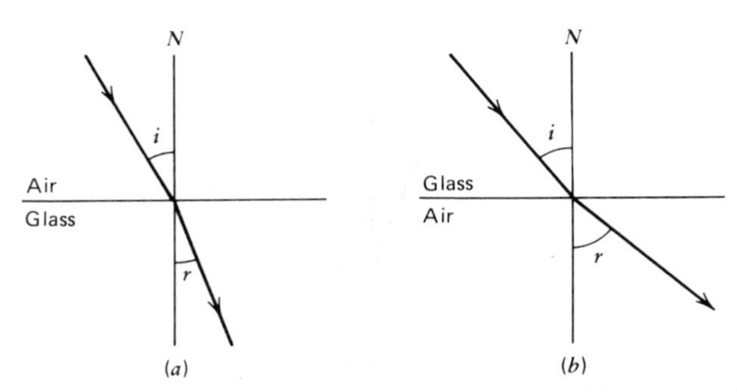

Figure 6.17 Refraction of light as it moves into different media.

optical density) and (2) where light moves from a material of high index of refraction to one of a lower index of refraction. For these two cases Snell's law will show us:

Case 1. When light moves from one medium to another of higher index of refraction it bends toward the normal (Figure 6.17*a*).

Case 2. When light moves from one medium to another of a lower index of refraction it bends away from the normal. (Figure 6.17*b*)

Although these rules enable us to determine the general directions that light will bend under different conditions, we cannot determine how much the light will bend without using Snell's law. Nevertheless, these rules are most helpful in

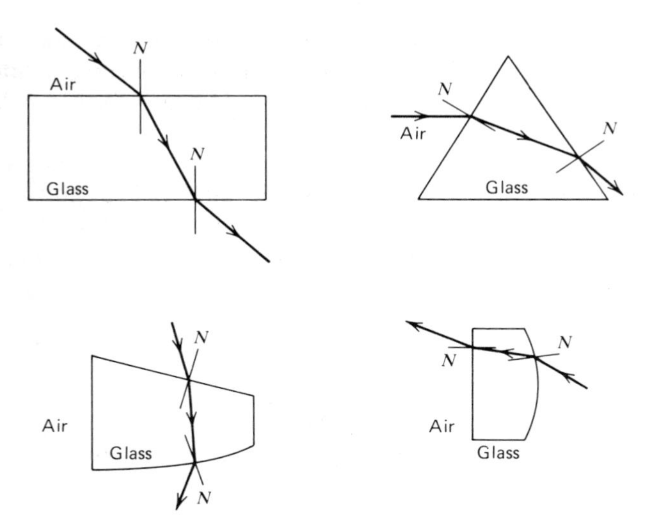

Figure 6.18 The refraction of light as it passes through different shaped objects of an index of refraction greater than air. Note the light bends toward the normal on entering the objects and away from the normal when exiting the objects.

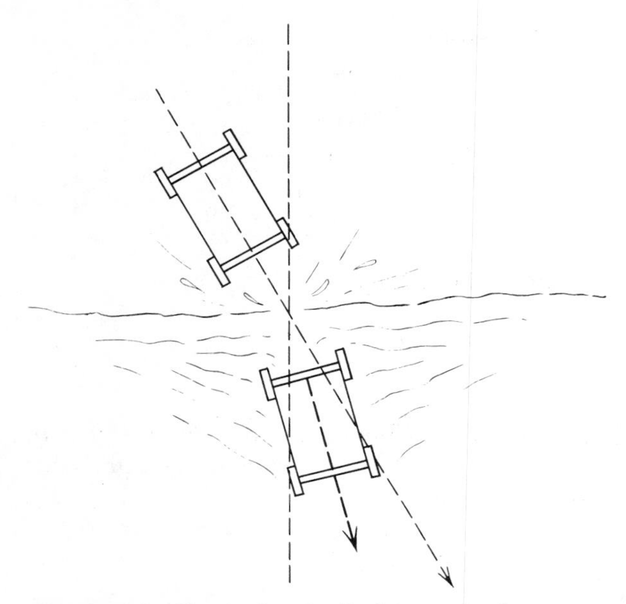

Figure 6.19 The "refraction" of the path of a wagon.

understanding how light refracts through differently shaped surfaces. Figure 6.18 shows the path of light through some different surfaces (the normals to these surfaces are drawn in for clarity).

The manner in which the light rays bend is often explained by considering the following analogy. Suppose you had a wagon rolling across a paved area that was bounded by an area of mud. If the wagon left the paved area and moved into the mud at an angle as shown in Figure 6.19, its direction would be altered as the front wheel left the paved area and moved into the more retarding medium (the mud). In a similar manner a light ray moving from air to glass encounters a greater retardation of its speed and thus its direction is changed.

Let us now apply the mathematical statement of Snell's law and calculate the actual amount of bending that would occur at a given surface. Suppose that a light ray in air was incident on a piece of transparent plastic (refractive index = 1.6) at an angle of 28° as shown in Figure 6.20. At what angle will the light be refracted on entering the plastic? Using Snell's law and a value of 1.0 for the index of refraction for air we have:

$$\frac{\sin \sphericalangle i}{\sin \sphericalangle r} = \frac{\eta_2}{\eta_1}$$

$$^{*}\frac{\sin 28^\circ}{\sin \sphericalangle r} = \frac{1.6}{1}$$

*(*NOTE:* Use the sine tables in Appendix C to find these values.)

Now sin 28° = .47. Therefore,

$$\frac{.47}{\sin \measuredangle r} = 1.6$$

$$.47 = 1.6 \sin \measuredangle r$$

$$\sin \measuredangle r = \frac{.47}{1.6} = .29$$

Thus $\measuredangle r$ = the angle whose *sine = .29

or

$$\measuredangle r = 17°$$

As a second example let us assume that we know the incident and refracted angles and desire to find the index of refraction of the refracting material. Again we shall assume that $\eta_1 = 1$ for air.

$$\frac{\sin \measuredangle i}{\sin \measuredangle r} = \frac{\eta_2}{\eta_1}$$

Suppose we observe that $\measuredangle i = 30°$ and $\measuredangle r = 22°$. Since sin 30° = .5 and sin 22° = .375 we have:

$$\frac{.5}{.375} = \frac{\eta_2}{1}$$

or

$$1.33 = \eta_2$$

$$= \text{index of refraction}$$

Returning to the table giving values of η, we see that this material that is refracting the light could be water. Assuming that it is water, what is the speed of light in that medium? Using the definition given for index of refraction we have

$$\eta = \frac{\text{speed of light in a vacuum}}{\text{speed of light in a medium}}$$

$$1.33 = \frac{3 \times 10^8 \text{ m/sec}}{v_m}$$

$$1.33\, v_m = 3 \times 10^8 \text{ m/sec}$$

or

$$v_m = 2.25 \times 10^8 \text{ m/sec} = \text{speed of light in water}$$

*(*NOTE:* Use the sine tables in Appendix C to find these values.)

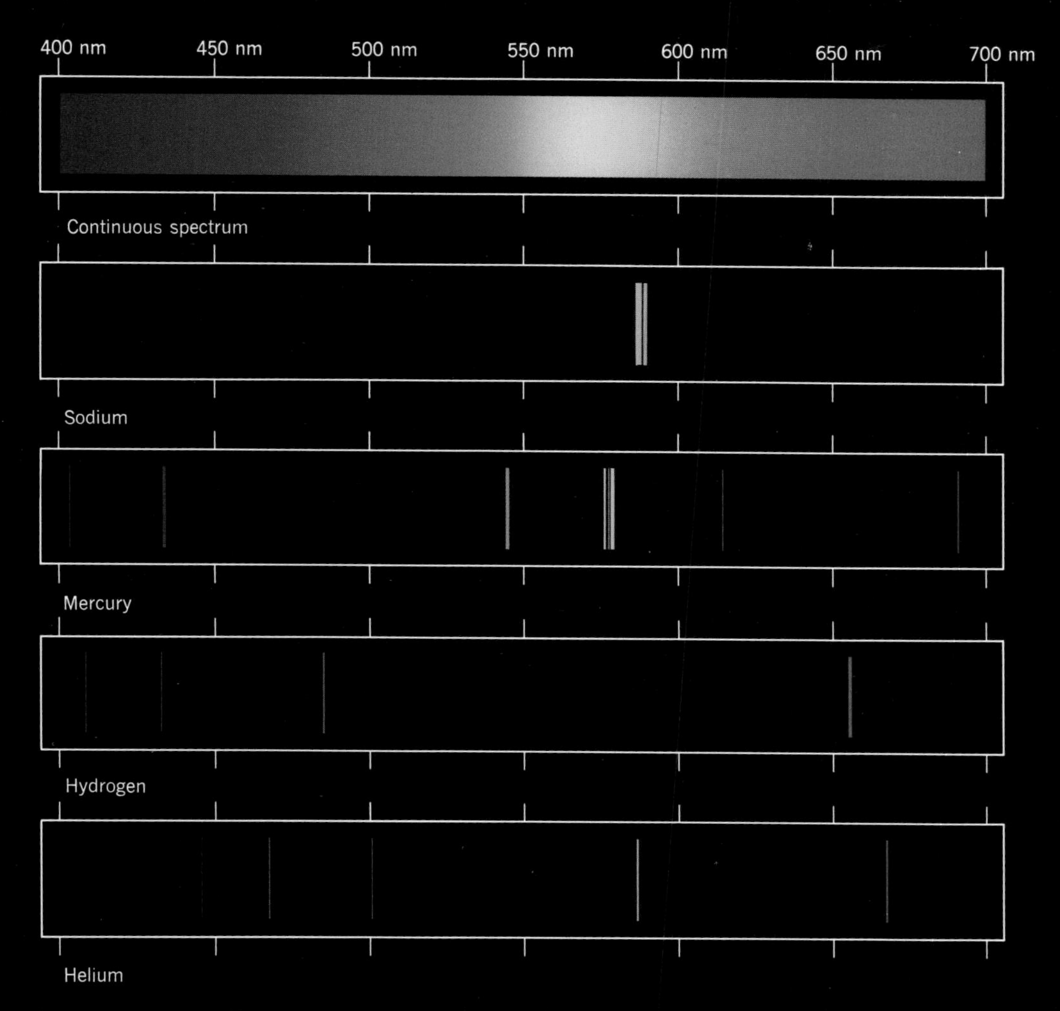

Plate 1 The visible spectrum from 400 nm (violet) to 700 nm (red). This is an example of a continuous spectrum, since there are no gaps or dark lines within it.

Plate 2 Examples of bright-line spectra: sodium, mercury, hydrogen, and helium. These sources produce light at only a limited number of distinct wavelengths.

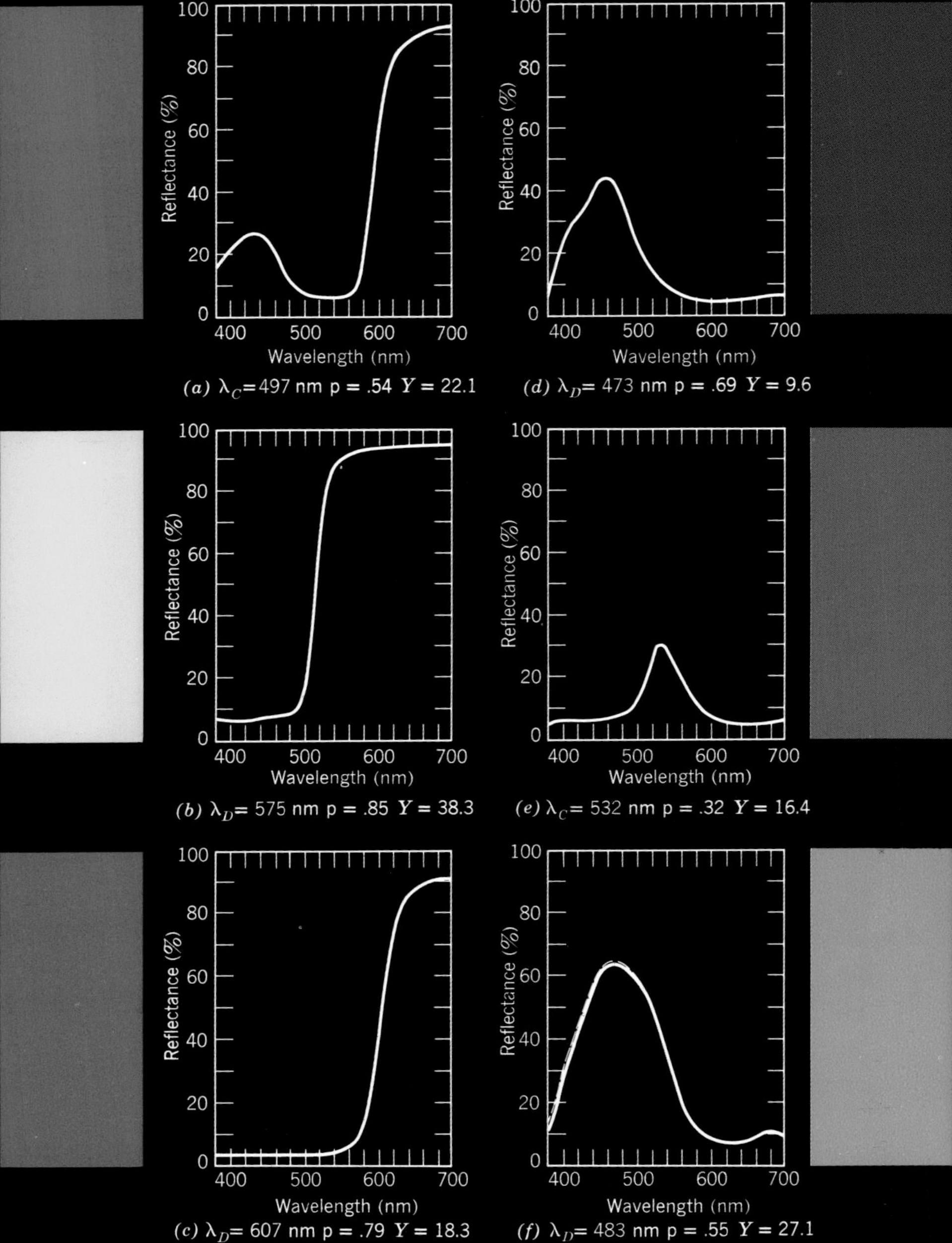
100
80
60
40
20
0
Reflectance (%)
400
500
600
700
Wavelength (nm)
(a) λ_C = 497 nm p = .54 Y = 22.1
(d) λ_D = 473 nm p = .69 Y = 9.6
(b) λ_D = 575 nm p = .85 Y = 38.3
(e) λ_C = 532 nm p = .32 Y = 16.4
(c) λ_D = 607 nm p = .79 Y = 18.3
(f) λ_D = 483 nm p = .55 Y = 27.1

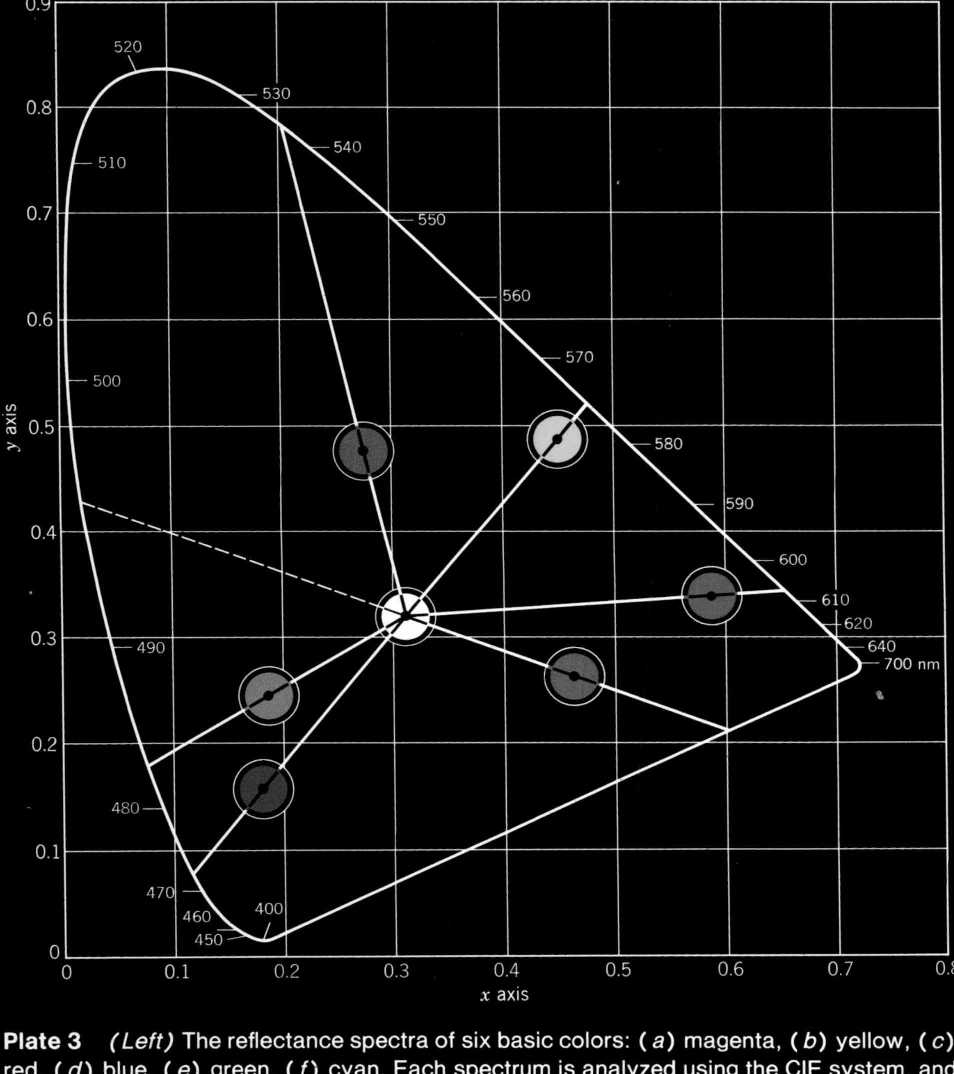

Plate 3 *(Left)* The reflectance spectra of six basic colors: (*a*) magenta, (*b*) yellow, (*c*) red, (*d*) blue, (*e*) green, (*f*) cyan. Each spectrum is analyzed using the CIE system, and the resulting values of λ_D, p, and Y are given. *(Above)* The CIE diagram shows the chromaticity of each color. The spectra (*a–f*) could also represent transmittance spectra rather than reflectance spectra.

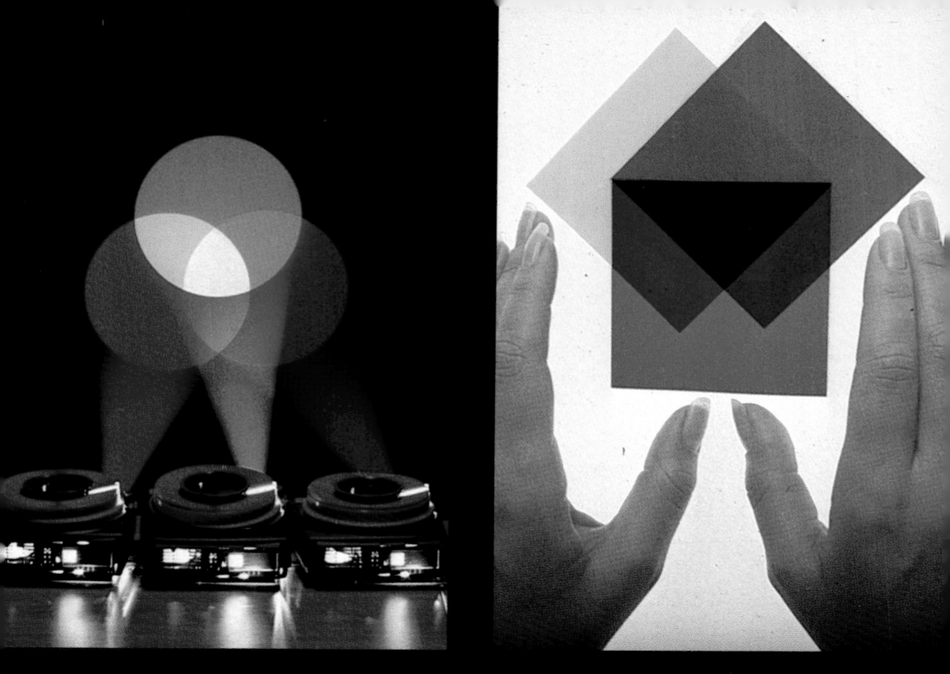

Plate 4 *(Left)* The additive primaries red, green, and blue. Note that when all three are present in equal amounts, white results. *(Right)* The subtractive primaries magenta, yellow, and cyan. When all three are placed over each other, black results.

Plate 5 *(Left)* The painting *Sunday Afternoon on the Island of La Grande Jatte* by Georges Seurat. The use of thousands of tiny dots of different colors to produce a variety of color sensations is known as pointillism. *(Right)* Detail of Seurat's painting. (Collection

Plate 6 Magnified view of the magenta, cyan, and yellow dots used in color printing (Courtesy of Hammermill Papers Group, Division of Hammermill Paper Company)

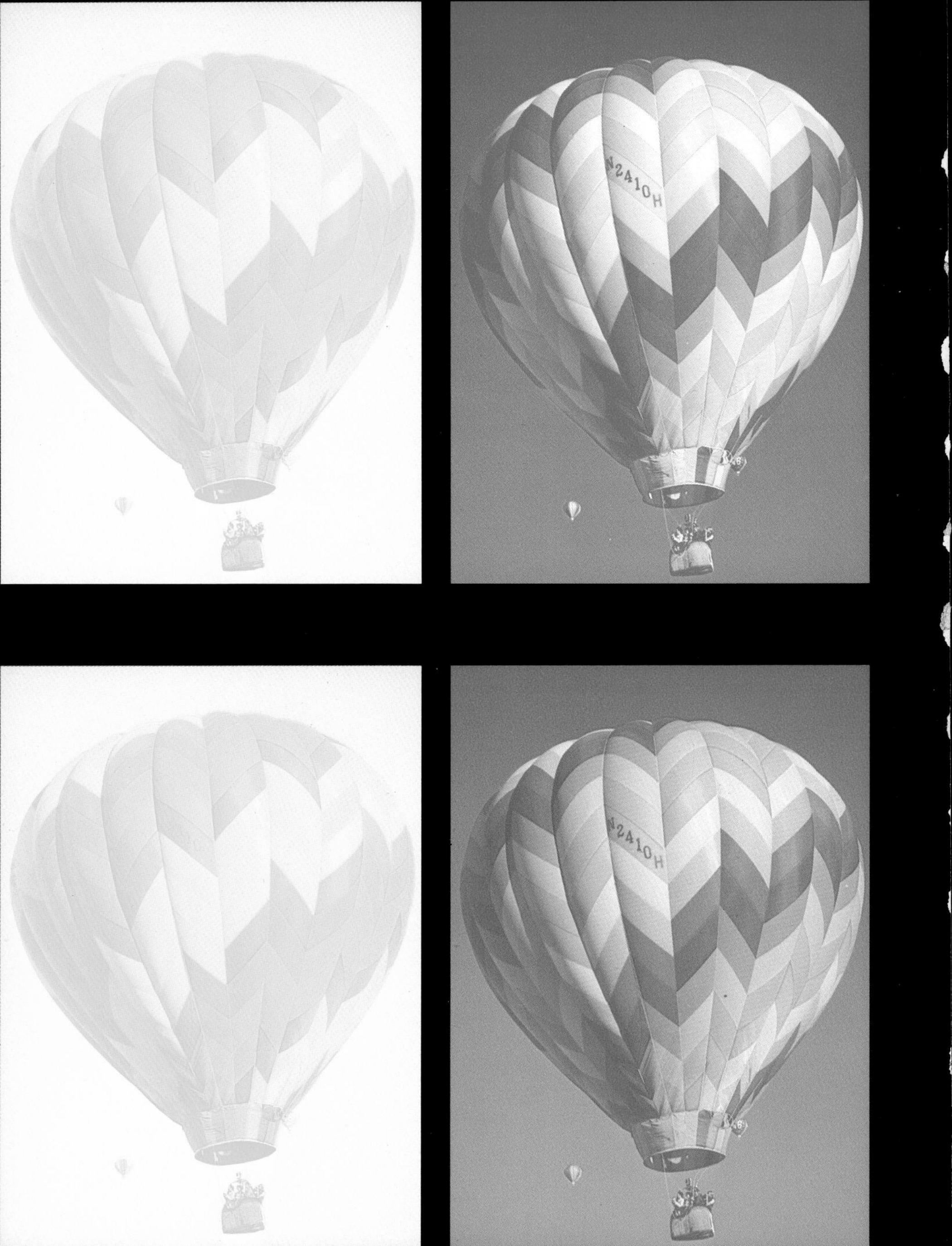
N2410H
N2410H

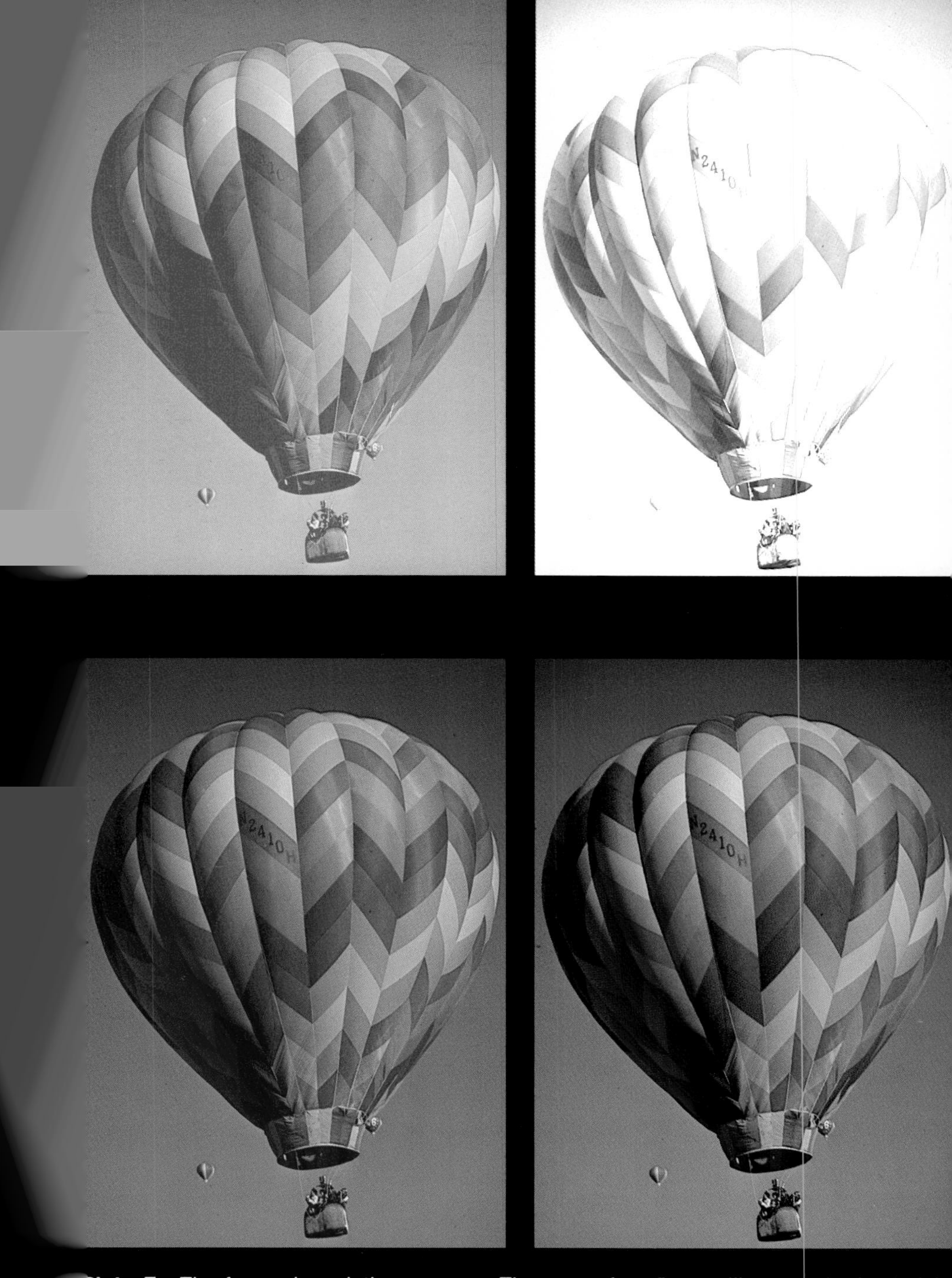

Plate 7 The four-color printing process. The upper four figures illustrate the basic inks used in the process. The lower figures show how the final reproduction is built up. (Photo

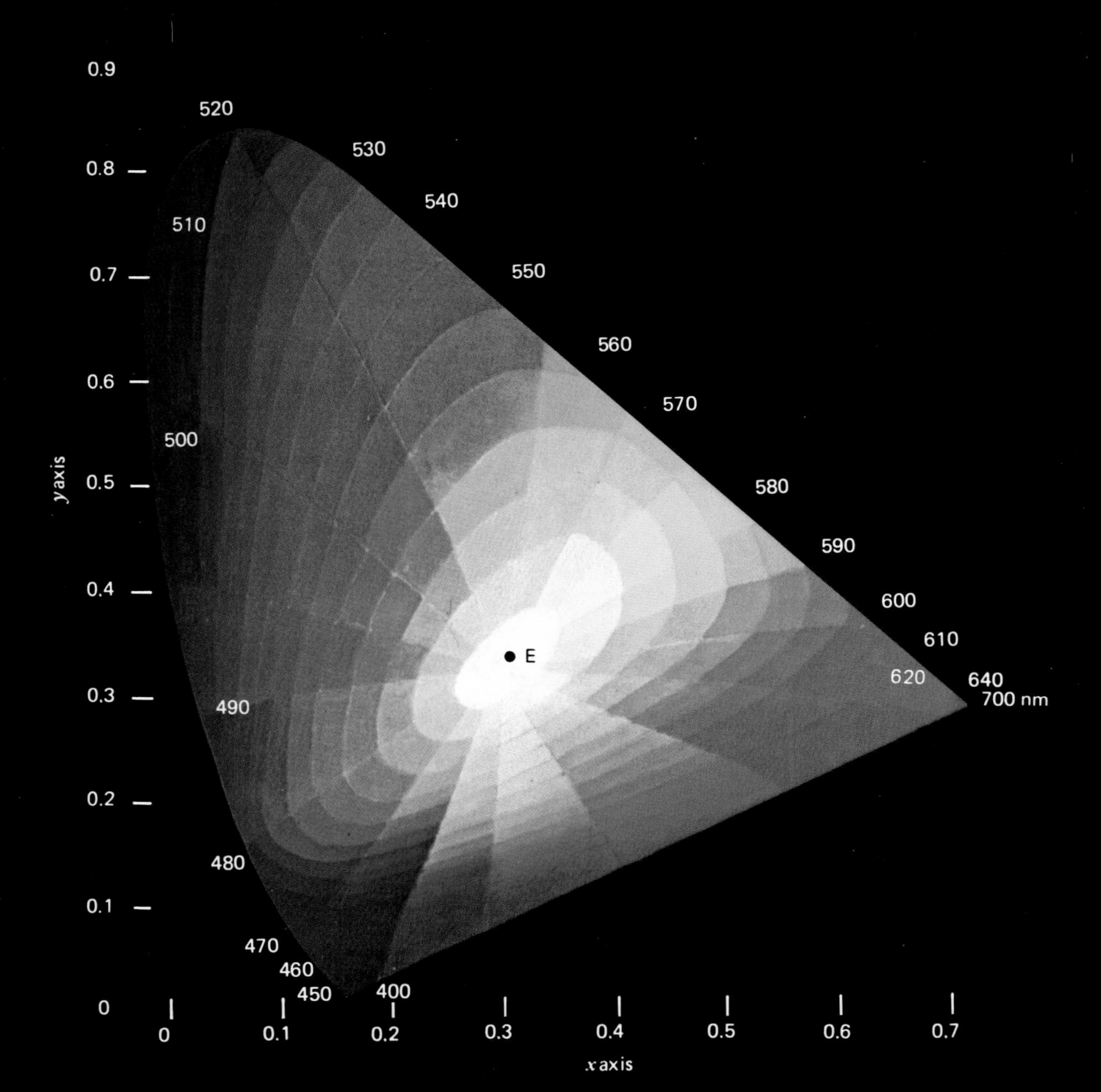

Plate 8 The CIE chromaticity diagram showing the approximate colors associated with each chromaticity. (Science of Color, Optical Society of America, 1963)

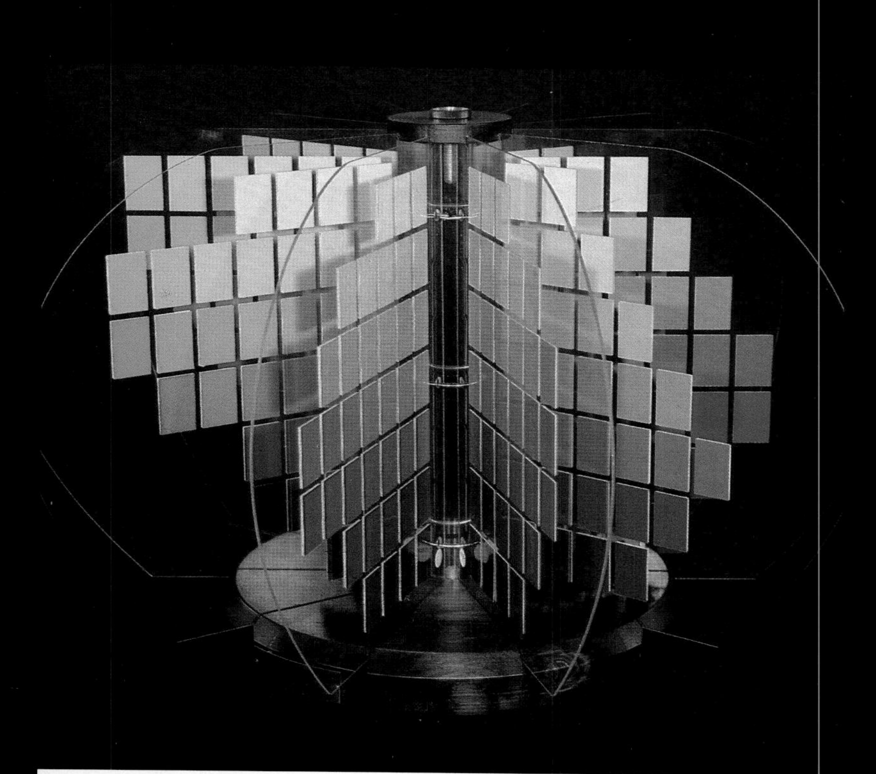

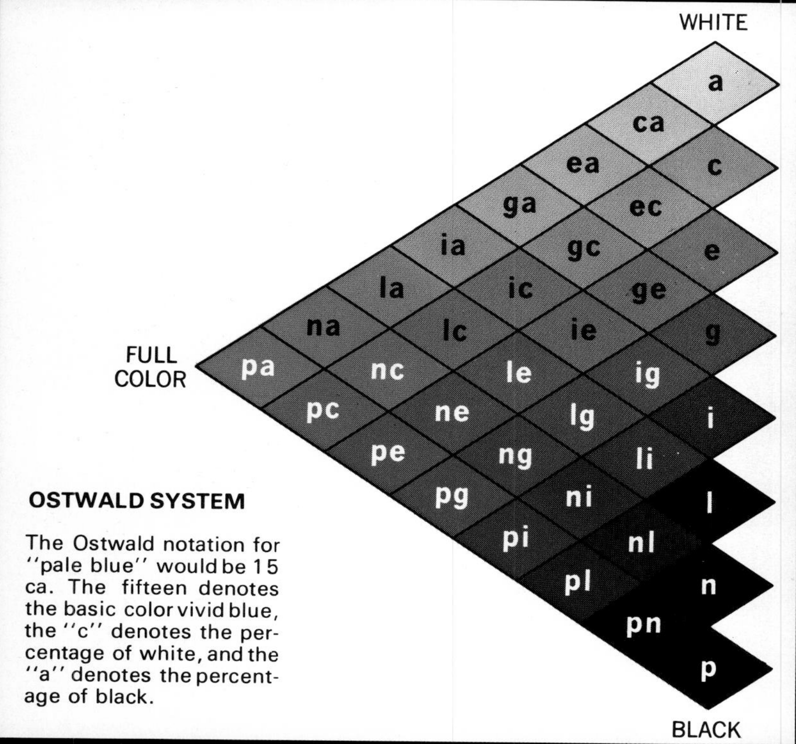

Plate 9 *(Top)* The Munsell color tree. (Courtesy Munsell Color, Macbeth Division, Killmorgen Corporation) *(Bottom)* An example of an Ostwald color chart. (From "Light and Color," published by the General Electric Company, Lighting Business Group)

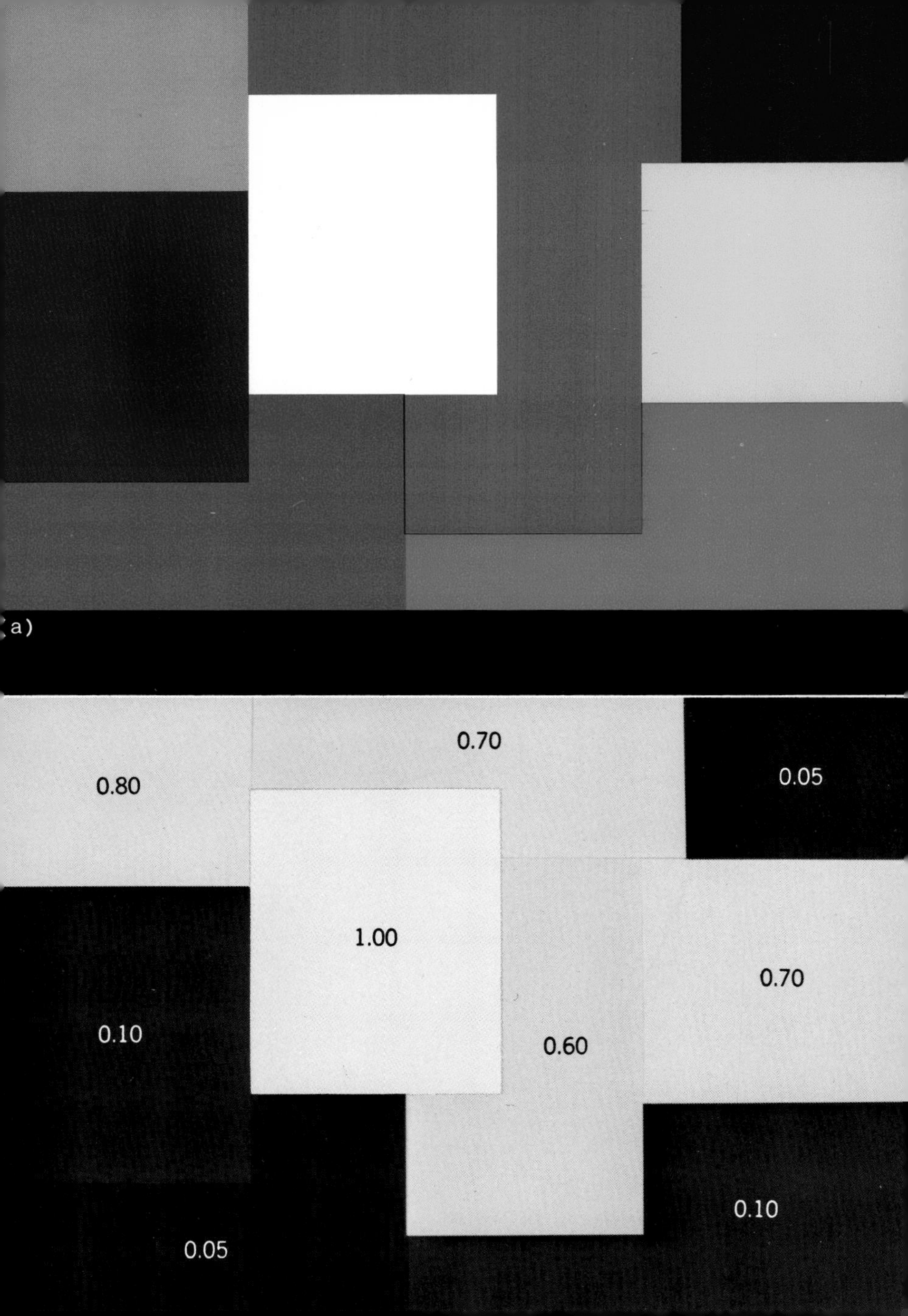
a)
0.80
0.70
0.05
1.00
0.70
0.10
0.60
0.10
0.05

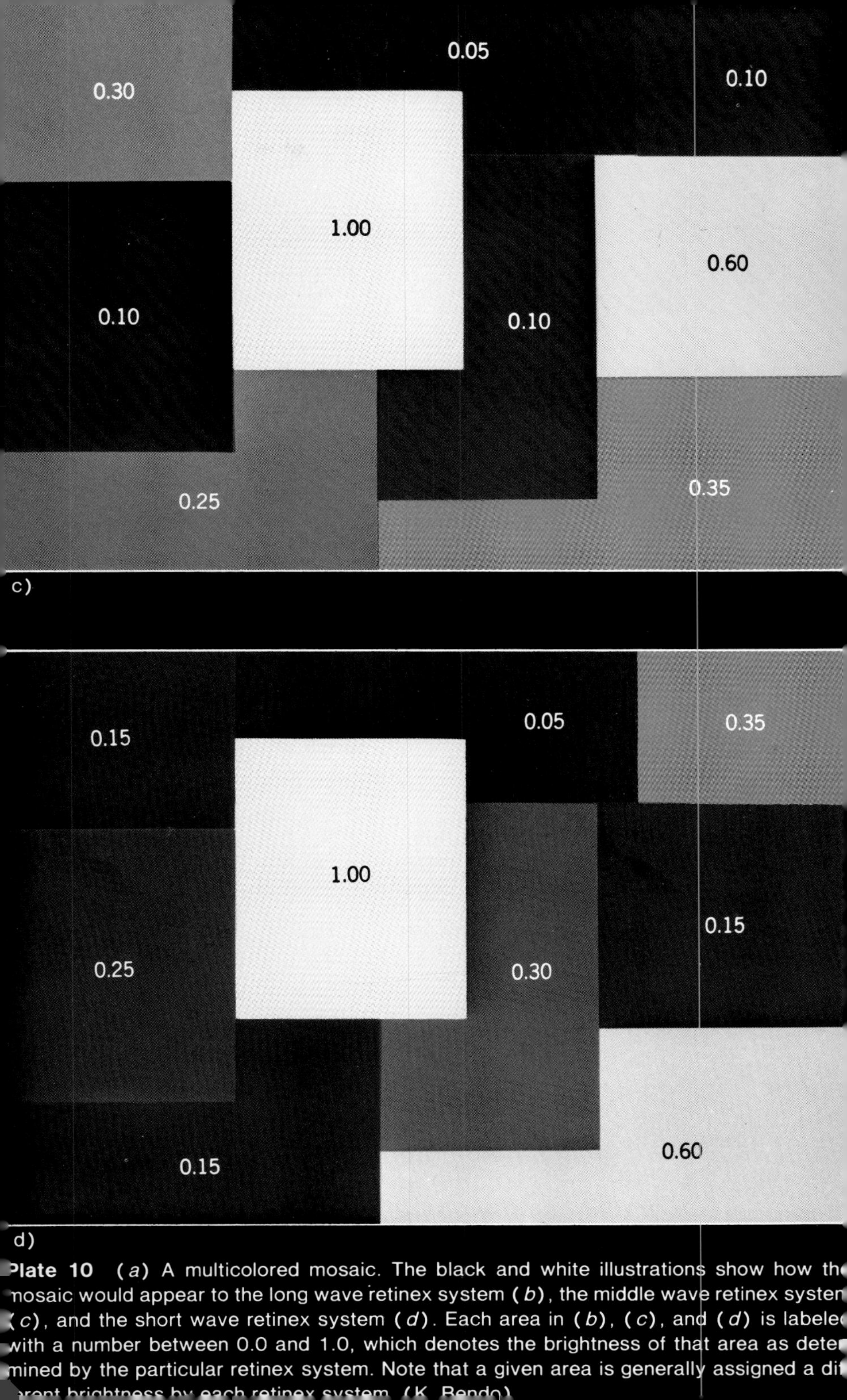

Plate 10 (*a*) A multicolored mosaic. The black and white illustrations show how the mosaic would appear to the long wave retinex system (*b*), the middle wave retinex system (*c*), and the short wave retinex system (*d*). Each area in (*b*), (*c*), and (*d*) is labeled with a number between 0.0 and 1.0, which denotes the brightness of that area as determined by the particular retinex system. Note that a given area is generally assigned a different brightness by each retinex system. (K. Bendo)

Plate 11 Four basic classes of objects: from left to right, nonmetallic opaque, transparent, translucent, and metallic. (K. Bendo)

Plate 12 A double rainbow, Mount McKinley National Park, Alaska. The primary rainbow is on the bottom, the secondary on top. The area between the two rainbows is Alexander's dark band. (Art Wolfe/Photo Researchers)

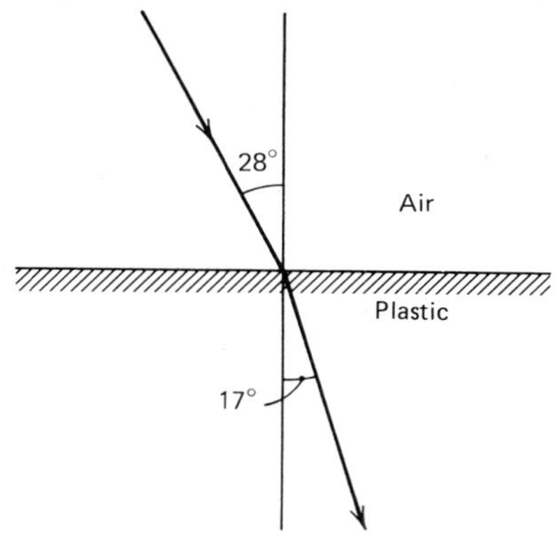

Figure 6.20 Snell's law permits us to calculate the angle of refraction for light entering different media.

As stated earlier, the table of indices of refraction previously given was for light at a wavelength of 590 nm. Experimentally it is found that different wavelengths of light refract by different amounts. The shorter the wavelength (the higher the frequency) the larger the index of refraction. This fact results in the spreading out (dispersion) of multiwavelength light as it passes at an angle from one medium to another of a different index or refraction. Therefore we can understand how a prism disperses the white light that is incident on it into the spectrum of colors. The incident white light contains all the different wavelengths of the visible spectrum. The shorter but higher frequency waves are retarded more when interacting with the medium than the longer low frequency waves. Thus the light spreads out into its many components (Figure 6.21). Shining monochromatic light (such as that which a laser emits) on a prism will cause the light to bend, but there will be no dispersion because all the waves will refract at the same angle.

The ability of a prism to refract the different wavelengths by different amounts is highly desirable for studying the anatomy of light, as is done when one uses a prism spectroscope. However, it presents a real problem to the designer of lenses for optical instruments such as telescopes, microscopes, and

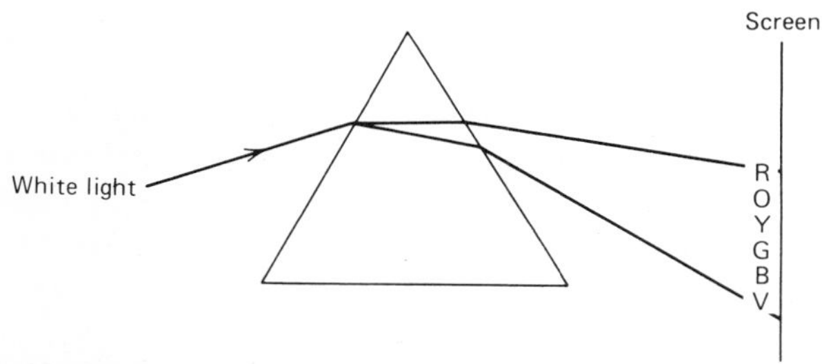

Figure 6.21 The dispersion of white light into a colored spectrum.

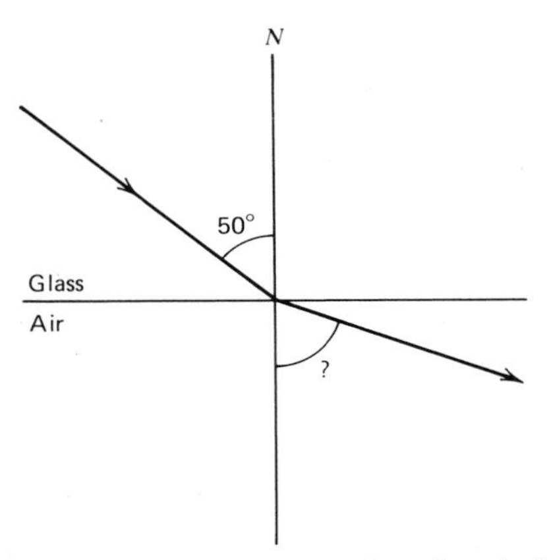

Figure 6.22 Light entering a medium that has a reduced optical density (smaller index of refraction) bends from the normal.

cameras. The dispersion of light in these devices gives rise to imperfections known as chromatic aberration, which will be discussed when we investigate lenses.

Another most interesting principle involving Snell's law and the refraction of light can be demonstrated by the consideration of the following problem. Suppose a light ray in glass ($\eta = 1.5$) is incident on the glass-air boundary as shown in Figure 6.22. At what angle will the light be refracted?

At first observation it appears that this problem is just a routine refraction problem but, as the calculations below show, there is one major difficulty that is encountered when you attempt to find the solution. This difficulty occurs when you endeavor to find the refracted angle for the value of the sine function that has been calculated.

$$\frac{\sin \measuredangle i}{\sin \measuredangle r} = \frac{\eta_2}{\eta_1}$$

$$\frac{\sin 50^\circ}{\sin \measuredangle r} = \frac{1}{1.5}$$

$$\sin \measuredangle r = 1.5 \sin 50^\circ = 1.5 \times .766$$

$$\sin \measuredangle r = 1.15? \text{ (Now what?)}$$

Our table of sine values clearly indicates that *no* angle has a sine function that is greater than one (1). The implication of this is that *no refraction* can take place – all the light will be reflected from the internal glass-air boundary. This phenomenon, which can occur only when light is moving from an optical material to a material of lower index of refraction, is known as *total internal reflection,* and has many uses in today's world.

The smallest angle of incidence at which this can occur depends on the type of material that light is traveling in and what type it is trying to enter. Using

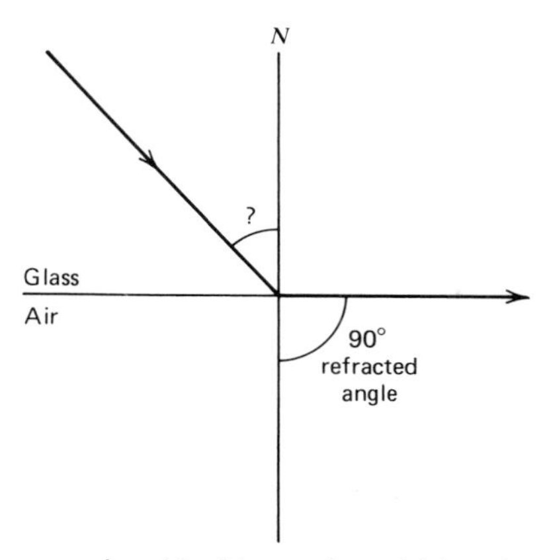

Figure 6.23 The maximum angle of incidence for which refraction can take place. For glass this is approximately 42°.

Snell's law and knowing that the light can be refracted up to a maximum angle of 90° (Figure 6.23), we can calculate the maximum angle of incidence (critical angle) is at which the light can strike our glass surface without undergoing total internal refraction. Shown below is this calculation for glass to air.

$$\frac{\sin \measuredangle i}{\sin \measuredangle r} = \frac{\eta_2}{\eta_1}$$

$$\frac{\sin \measuredangle i}{\sin 90^\circ} = \frac{1}{1.5}$$

$$\frac{\sin \measuredangle i}{1} = \frac{1}{1.5} = .667$$

$$\measuredangle i = 42^\circ$$

Thus any time light strikes this boundry at an incident angle greater than 42°, it is totally reflected from the inner glass surface. This type of light behavior is used in many ways to aid in the development of optical devices. One such use is in binoculars where prisms reflect the light internally and thereby shorten the necessary length of the binoculars (Figure 6.24).

In Figure 6.24 the light enters the prism along a perpendicular or normal and is therefore unrefracted. It then strikes the prism face at a 45° angle and undergoes total internal reflection to the next face where it again is reflected. The process continues until the light exits the prisms to the ocular or eye lens. Although the same thing could be accomplished by using mirrors, the prisms offer two definite advantages: (1) there is no need to silver the surfaces and (2) they are much easier to support than the four mirrors that would be necessary to do the task.

Another very interesting and useful phenomenon dependent on total internal reflection is fiber optics. Fiber optics consists of bundles of long thin fibers in

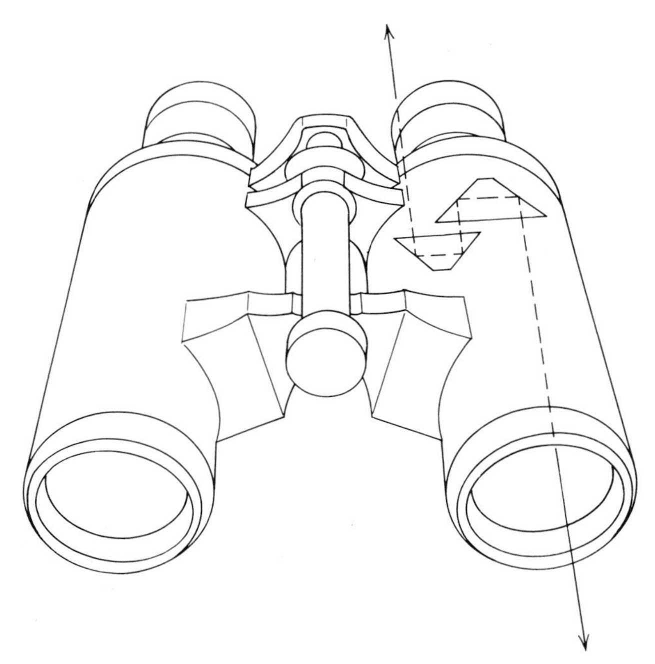

Figure 6.24 The path of light through prism binoculars.

which the light that enters one end is continually reflected internally. Because of the narrowness of the fiber, light entering through its end will strike the sides at an angle that will exceed the critical angle for the material; thus, even though the fiber might be bent the light is still reflected internally until it reaches the other end of the fiber (Figure 6.25*a*).

One of the most important uses of fiber optics is in the medical profession for nonsurgical internal examinations. A bundle of fibers can be inserted into a body cavity and light can be shown through some of the fibers to illuminate the interior area. This light then reflects off the area of interest and enters the end of the remaining fibers to be transmitted out to the examiner where the image can be magnified, photographed, or even shown on a television screen. With this method a doctor can examine the interior of the stomach for ulcers or look for a tumor in the bladder of a patient without the necessity of surgery. Often a small incision can be made in an injured knee allowing a surgeon the opportunity to view and evaluate the damage before undertaking a major surgical procedure.

Most recently some major communication firms have begun to use fiber optics to replace the copper cables that carry electrical signals. The high frequency of light allows for many more messages to be sent through one fiber than can be sent through a copper wire with an electric current. In addition the cost of copper cables is growing much faster than the cost of the optical fibers that are replacing them. Likewise, the light necessary to send the messages is

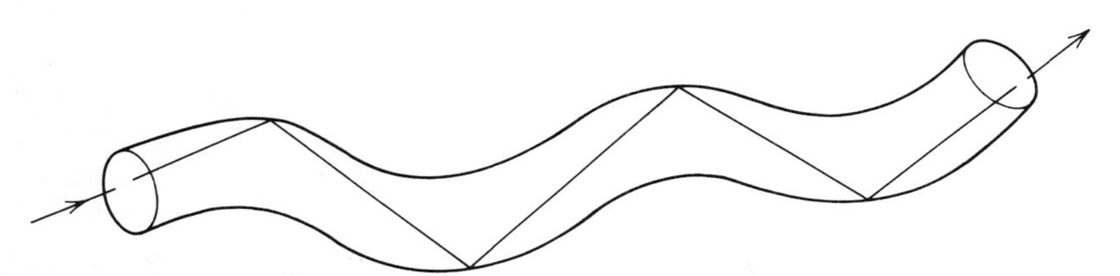

Figure 6.25*a* Light undergoes total internal reflection in fiber optics.

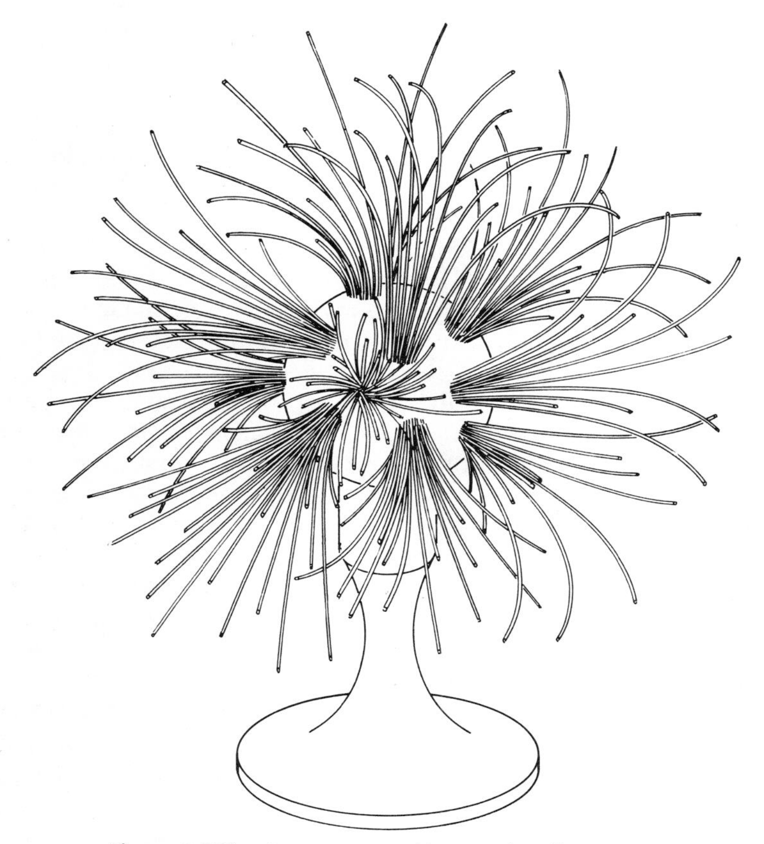

Figure 6.25*b* An ornamental lamp using fiber optics.

less expensive than the electrical energy expended by the conventional technique.

Some ornamental lamps now make use of fiber optics. These lamps consist of a central light source to which many fibers are attached. Light enters the fibers and illuminates them and then exits the other end. Variations of these lamps often include filters that color the central light before entering the fibers (Figure 6.25*b*).

6.5 REFRACTION BY LENSES

As light from a source passes through a lens it is capable of producing images in much the same way images are produced when light is reflected from curved mirrors. These images can be real or virtual, magnified or reduced, and erect or inverted. This image formation can be predicted by applying the principles of refraction to the light as it passes through a convex lens. In Figure 6.26 we have constructed the principal axis of a convex lens and have drawn rays of light parallel to that axis. On entering the lens we observe the light bending toward the normals to the surface and on leaving these rays bend away from the normals as Snell's law dictates they should. As the light exits the lens, we observe that the rays converge and pass through a common point – the focal point. The distance between the center of the lens and the focal point is known as the focal length of the lens. Parallel light rays shining from either the left-hand side of a *thin* lens or the right-hand side of the lens would focus at the same distance from the lens. Thus we have focal points on both sides of the lens that are equidistant from the lens.

Using the fact that the rays of light parallel to the principal axis refract through the focal point, and one additional observation we can locate the image formed by the light from objects by tracing three rays as we did in locating the images formed by mirrors. Figure 6.27 illustrates the method of determining the image of an object located at a distance outside the focal point. In locating the image we shall find the positions where the light from the top and the bottom of the object (the arrow) focus as we did in locating the images with the mirrors. The first ray is drawn from the top of the arrow parallel to the principal axis. As stated above this ray will refract through the lens and pass through the focal point. The second ray is the inverse of the first. It is drawn through the focal point on the object side and thus after going through the lens will be parallel to the principal axis. The third ray passes through the center of the lens with very little if any deviation. This passage without significant deviation can occur because we are only considering thin lenses whose sides at the center of the lens are effectively parallel. In Figure 6.18 we have shown that light passing through a medium with parallel sides continues in a direction parallel to the incident ray. When the medium is thin the deviation from the first ray is minimal; thus since we are considering only thin lenses the ray of light effectively passes

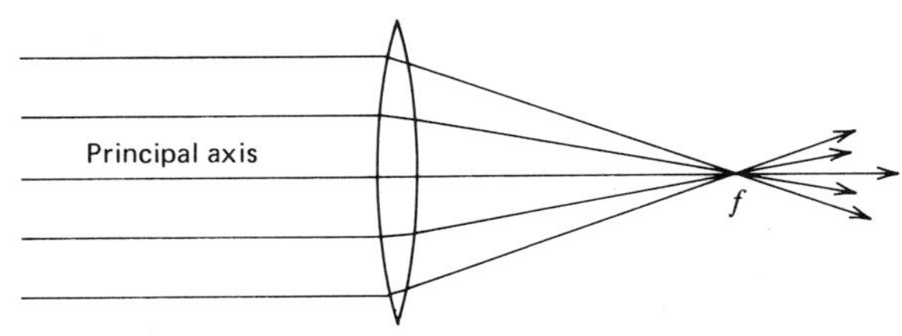

Figure 6.26 The focusing properties of a convex lens.

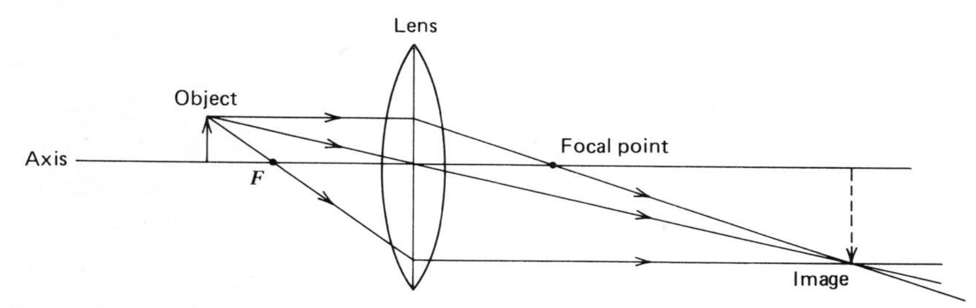

Figure 6.27 A ray diagram showing the three rays useful in locating the image of an object formed by a convex lens.

straight through the center of the lens. The light from the bottom of the object must necessarily focus on the principal axis of the lens. Connecting these two points we observe that the image in this case is inverted, magnified, and real. If the object is moved progressively farther from the lens the magnification is reduced until, when the object is at a distance equal to twice the focal length from the lens, the image and object are equal in size. Further removing the object produces an image that is reduced in size.

Placing an object at a position between the focal point and the lens results in the production of an image that is upright, magnified, and virtual. Figure 6.28 indicates that the light rays leaving the object diverge when they pass through the lens from this position. Here, as we saw with virtual images formed with mirrors, the image is formed when the diverging rays are projected backwards until they appear to meet. This type of image is what one observes when using a magnifying glass. The object to be viewed is placed inside the focal point of the lens and is observed from the opposite side of the lens. If the object is

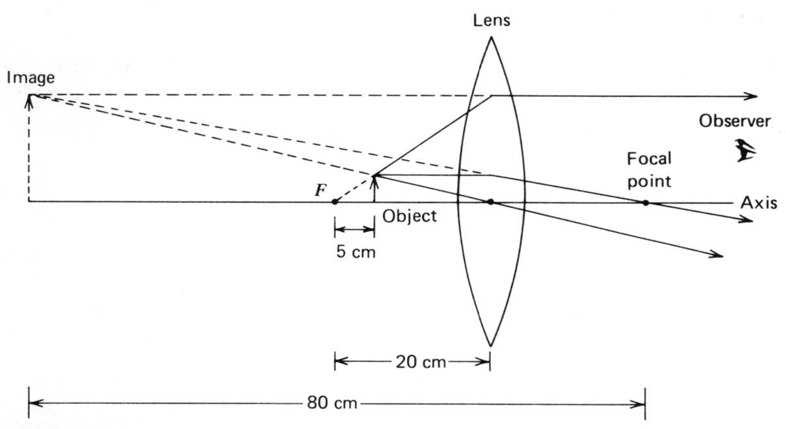

Figure 6.28 When an object is placed inside the focal point of a convex lens, the image is upright, magnified, and virtual.

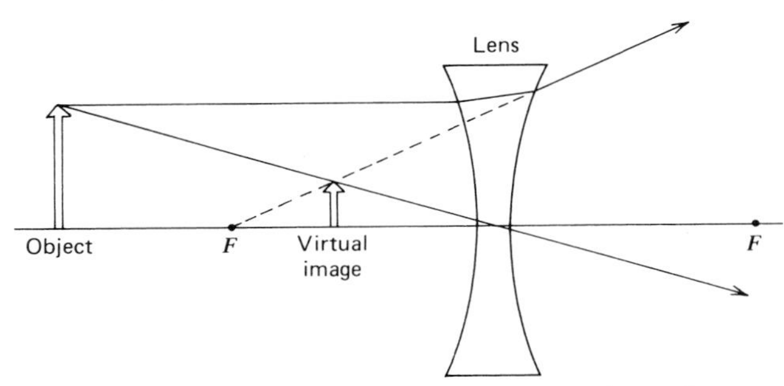

Figure 6.29 The images formed as one views an object through a concave lens are always upright, reduced, and virtual.

placed too distant from the lens (beyond the focal point) the observer will note that the image seen will become inverted – he or she is now viewing it as in Figure 6.27.

Concave lenses refract light and generate images of objects that resemble those produced by convex mirrors. When parallel light rays are incident on a concave lens they refract and diverge as if they were coming from one point (the focal point) on the same side of the lens as the incoming rays (Figure 6.29). Using this knowledge and remembering that the light will pass through the center of thin lens with little deviation, we can locate the image as in Figure 6.29. The image formed in this instance is found on the same side of the lens as the object. Regardless of where the object is placed the image will always be erect, reduced in size, and virtual. This type of lens is frequently used by artists who wish to view a developing work as if from a distance. Using the concave lens will give them an upright view of what they would observe from a distance.

6.6 CHROMATIC ABERRATION

In Section 6.4 we mentioned that since different wavelengths of light refract by different amounts lenses are subject to imperfections known as chromatic aberrations. These imperfections result because for each of the different colors

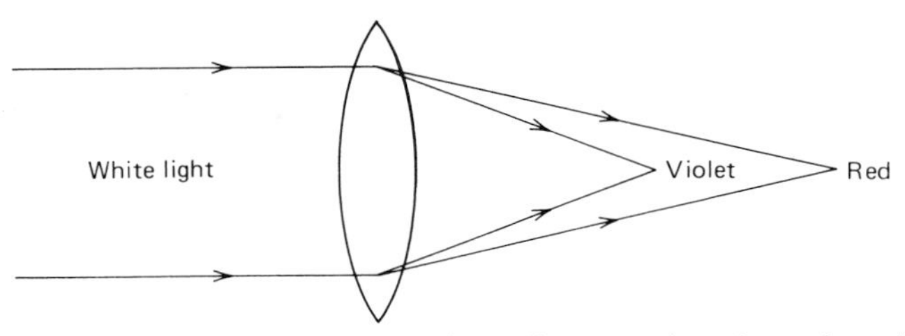

Figure 6.30 Chromatic aberration causes the different colors (wavelengths) of light to focus at different positions when passing through a lens.

of light entering a lens there is a slightly different focal point. Since violet light has the highest index of refraction for the visible light, it bends the most and its image is found closest to the lens, while red light having the smallest index of refraction forms its image farthest from the lens (Figure 6.30). At no position will all the different colors simultaneously be in focus; thus the image at any position will be somewhat "fuzzy" with a slight color fringe around its edges.

To eliminate this problem most lenses in optical devices are color-corrected or achromatic lenses. These achromatic lenses actually consist of a combination of two lenses of different shapes and indices of refraction. They are fitted together so that the combination appears as a single lens. The dispersion produced by one of the lens is counteracted by the other while leaving a net converging or diverging power for the combination.

6.7 LENS EQUATIONS

To understand the operation of different optical instruments such as cameras, telescopes, and microscopes, knowledge of the exact location of images as well as their sizes is necessary and desirable. This information can be acquired by developing an equation that gives these values provided we know the initial size and position of the object. In Figure 6.31 the image of an object has been located by drawing the rays described in Section 6.5. We have labeled the focal length of the lens as f and the object distance (distance of the object from the focal point) as d_o. Likewise the image distance is called d_i, the height of object H_o, and the height of image H_i. By considering triangles formed by the construction rays and the object and image heights, we can find a simple relation between these values.

On the left-hand side of the diagram we note that the two triangles that have been shaded are similar triangles; that is, the ratio of corresponding sides are

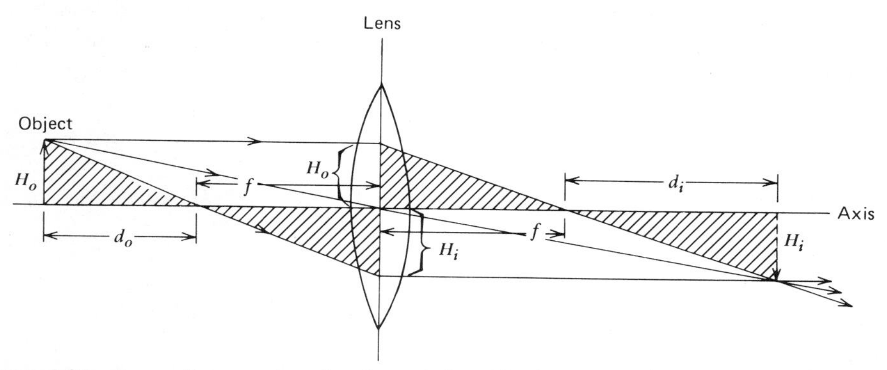

Figure 6.31 A ray diagram showing the relationships between f, H_o, H_i, d_o, and d_i. Analysis of this diagram leads to the equation: $\frac{H_i}{H_o} = \frac{f}{d_o}$ and $d_i d_o = f^2$.

the same. Hence H_i and H_o, which appear as the shorter legs of the similar right triangles, are one set of corresponding sides. The other legs of these triangles, d_o and f (which appear as the longer legs), are another set of corresponding sides. Thus we can write:

$$\frac{H_i}{H_o} = \frac{f}{d_o}$$

The ratio of H_i/H_o gives a comparison of the height of the image that is produced by a thin convex lens to the height of the original object. This tells us how many times larger or smaller the image is than the original object and is referred to as the magnification. For example, if an object was placed at a distance of 15 cm from a lens of focal length 10 cm (making the value of d_o = 5cm) we would find the magnification of the image by

$$\text{Mag} = \frac{H_i}{H_o} = \frac{f}{d_o} = \frac{10 \text{ cm}}{5 \text{ cm}} = 2$$

Thus the image of the object is magnified by a factor of 2 or is twice as high as the original object (it will also be twice as wide). If the object was 2 cm high, then the image height would be:

$$\frac{H_i}{H_o} = \frac{f}{d_o}$$

$$H_i = H_o\left(\frac{f}{d_o}\right)$$

$$H_i = 2 \text{ cm} \times \frac{10 \text{ cm}}{5 \text{ cm}} = 2 \text{ cm} \times 2$$

$$H_i = 4 \text{ cm}$$

Now observing the right-hand side of Figure 6.31 we observe that the two shaded triangles are also similar triangles. Thus, we can again compare corresponding sides and find:

$$\frac{H_i}{H_o} = \frac{d_i}{f}$$

Now we have two expressions for the magnification, H_i/H_o, which are different but equivalent. Equating these two expressions we find:

$$\frac{H_i}{H_o} = \frac{d_i}{f} \quad \text{and} \quad \frac{H_i}{H_o} = \frac{f}{d_o}$$

Thus

$$\frac{d_i}{f} = \frac{f}{d_o}$$

or

$$d_i d_o = f^2$$

This equation, known as the Newtonian form of the thin lens equation, gives the relationship between the location of the object, the position of the image, and the focal length of the lens being used. The use of this equation in conjunction with the magnification equation and a knowledge of where an object is will enable us to locate where an image will be, and to compare the size of the image to the size of the original object.

We have derived the Newtonian lens equation for the specific case of an object placed outside the focal point of a convex lens. This same equation also describes the case of an object placed inside the focal point of a convex lens, and an object placed on one side of a concave lens. In these cases the quantities d_i, d_o, and f must follow the following conventions however.

1. *Object inside focal point of convex lens (Figure 6.32).*
   - **(a)** d_o is to the *right* of the focal point on the *same* side of the lens as the object, and d_o is taken to be *negative.*
   - **(b)** d_i (from the equation $d_o d_i = f^2$) will then compute as negative. This means that d_i should be measured *left* from the focal point on the opposite side of the lens as the object.
   - **(c)** The magnification (from the equation mag. $= f/d_o$) will compute as negative. This means the image should be taken as *upright* and virtual.
2. *Object on one side of concave lens (Figure 6.33).*
   - **(a)** d_o is measured from the focal point on the *opposite* side of the lens as the object, and d_o is taken to be positive.
   - **(b)** d_i is measured to the *right* of the focal point on the same side of the lens as the object, and is taken to be positive.
   - **(c)** f is taken to be intrinsically negative for a concave lens. The magni-

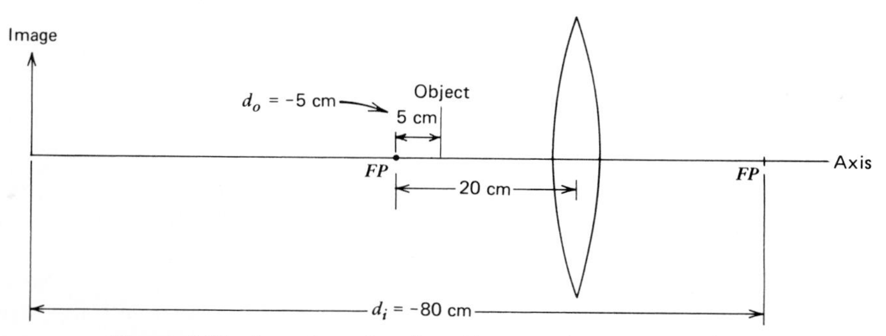

Figure 6.32 Location of a virtual image using a convex lens.

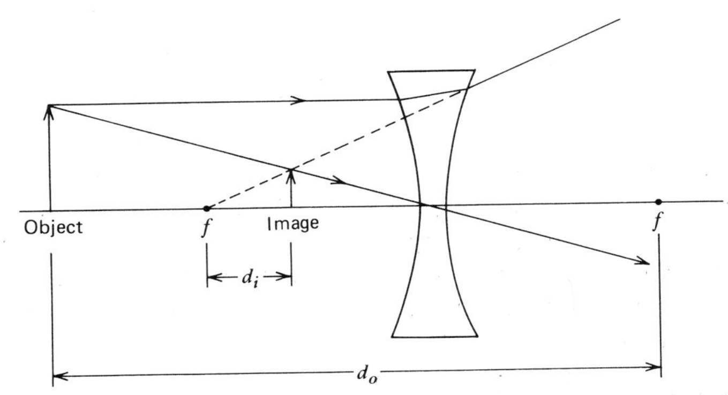

Figure 6.33 Image formed by a concave lens, and the conventions for d_o, d_i, and f.

fication (from the equation mag $= f/d_o$) will then compute negative. This again means that the image should be taken as *upright* and virtual.

With these conventions in mind we can now compute the size and location of the image formed when an object is placed near a lens.

For example, suppose that a candle was burning at a distance of 60 cm to the left of a convex lens of focal length 20 cm. The position of the image of the candle would be found by first calculating the value of d_o and then using the lens equation. You must remember that d_o is always measured from the focal point; therefore in this case $d_o = (60 - 20)$ cm $=$ 40 cm. Hence

$$d_o d_i = f^2$$

$$40 \text{ cm } d_i = (20 \text{ cm})^2 = 400 \text{ cm}^2$$

$$d_i = \frac{400 \text{ cm}^2}{40 \text{ cm}} = 10 \text{ cm}$$

and

$$\text{Magnification} = \frac{H_i}{H_o} = \frac{f}{d_o}$$

$$M = \frac{20 \text{ cm}}{40 \text{ cm}} = 0.5$$

The values found here tell us that the image of the candle would be 10 cm beyond the focal point on the right side of the lens (30 cm to the right of the lens) and would be one-half as high as the candle.

One further example at this point is informative in the use of this thin lens equation. Suppose that using the previous lens of 20 cm focal length the object was placed at a distance of 15 cm to the left of the lens. In this case the object is *inside* the focal point. Thus the object distance is given as negative, in this case -5 cm.

Calculating the position of the image would give

$$d_o d_i = f^2$$

$$(-5 \text{ cm})\ d_i = (20 \text{ cm})^2 = 400 \text{ cm}$$

$$d_i = \frac{400 \text{ cm}^2}{-5 \text{ cm}} = -80 \text{ cm}$$

The minus sign in front of the 80 cm indicates that the image distance should be measured in the *opposite direction* from the focal point than it normally is. Thus, from the focal point to the right of the lens the image would be located 80 cm *left* or a distance of 60 cm to the left of the lens in this instance (Figure 6.32). The negative sign also tells us that the image is virtual. From the placement of the object between the focal point and the lens we realize the lens is being used as a magnifying glass and the results found here agree with those shown in Figure 6.28.

Calculating the magnification of the image gives

$$M = \frac{f}{d_o}$$

$$M = \frac{20 \text{ cm}}{-5 \text{ cm}} = -4$$

The value of -4 is interpreted to mean the image is magnified 4 times but is upright instead of inverted as it would be when the object is located beyond the focal point. This again agrees with the image in Figure 6.28.

The Newtonian form of the lens equation we have developed and illustrated here is simpler algebraically than some other forms of lens equations that might be used. In the next chapter we shall use this equation to develop our understanding of optical instruments, in particular, the eye and the camera.

IN CONCLUSION

In this chapter we have investigated the basic principles of geometric optics that deal with situations in which light can be considered as traveling in straight lines. We have found that the law of reflection enables us to locate images formed when light reflects from plane or curved mirrors. Snell's law of refraction has been used to determine the path of light as it travels from one medium to another of a different index of refraction. Using Snell's law we found that when the angle of incidence of light falling on a boundary between one medium and a second less dense medium is increased, the angle of incidence eventually reaches a critical value where total internal reflection occurs. This phenomenon explains the passage of light in fiber optics.

Using a ray diagram of light passing through a thin lens we have developed equations that can be used to locate the position and size of the images produced by light from objects passing through lenses.

PROBLEMS AND EXERCISES

1. By careful construction of a ray diagram find the image of an object placed 40 cm from a concave mirror of focal length 20 cm. What are the characteristics (size, direction, and nature) of the image?
2. Where would the image of an object very distant from a concave mirror be located? What would the size of such an image be?
3. If you desired to take a photograph of yourself while standing six feet from a plane mirror for what distance would you set the camera focus?
4. Using the law of reflection construct a ray diagram to show that to view your entire height in a plane mirror you need a mirror one-half your height. Does it matter how far you stand from the mirror?
5. The index of refraction for a certain type of plastic is 1.7. Find the speed of light in this plastic.
6. A ray of light traveling in air strikes a glass surface ($\eta = 1.5$) at an angle of 24° from the normal. At what angle will it be refracted in the glass?
7. Examine the index of refraction for a diamond and then give a plausible explanation for the many different colors of light you see from its sparkle. Why don't you see these same colors from a piece of glass cut in the same way?
8. Consider the classic cut of a diamond and suggest reasons why a large amount of light that enters the top of the diamond exits through the top.
9. Explain the following statement: "Looking upward when swimming under water is like looking out of a funnel."
10. An object is placed at a distance of 12 cm from a lens of focal length 10 cm. Where will its image be found and how large will it be?
11. Using a magnifying glass of 25 cm focal length, you look at an object that is 20 cm from the glass. Where and how large will you see the image of this object?

7 APPLIED GEOMETRICAL OPTICS

Man has built machines that have often mimicked what nature has developed in the human body. For example, a system of cables and levers in a crane operates in the same fashion as the bicep muscle and the elbow that work together to move the forearm. Likewise, we might see a pump with a design that works on the same principles as those governing the heart. However, of all man-made instruments none more closely resembles the operation of a part of the human body than a camera does the eye. Numerous comparisons can be drawn between the two because they have developed in response to similar problems. In this chapter we shall invoke the principles learned in the previous chapter to study these similarities and to better understand the processes of vision and photography. Using the same principles we shall also extend our knowledge to the operation of additional optical instruments.

7.1 THE REDUCED EYE

The camera and the eye are constructed and act in many very similar ways. The eye and the camera each consist of an adjustable lens system at one end and a light-sensitive material at the other end with a protective cover around them. In addition, the interior walls of the enclosures are blackened to prevent the scattering of light onto the photosensitive film or retina. Each device also has an opening that is variable in size depending upon the light incident on it and is corrected to reduce chromatic and spherical aberrations. In the following sections we shall examine first the optical properties of the eye and then compare the operation of the camera to the eye.

The diagram of the complete eyeball (Figure 7.1) shows that it is a sphere of an approximate diameter of 24 mm with a transparent bulge (the cornea) through which light enters the front of the eye. The diameter of the cornea is about 12 mm with a radius of curvature of approximate 7.7 mm. The cornea is

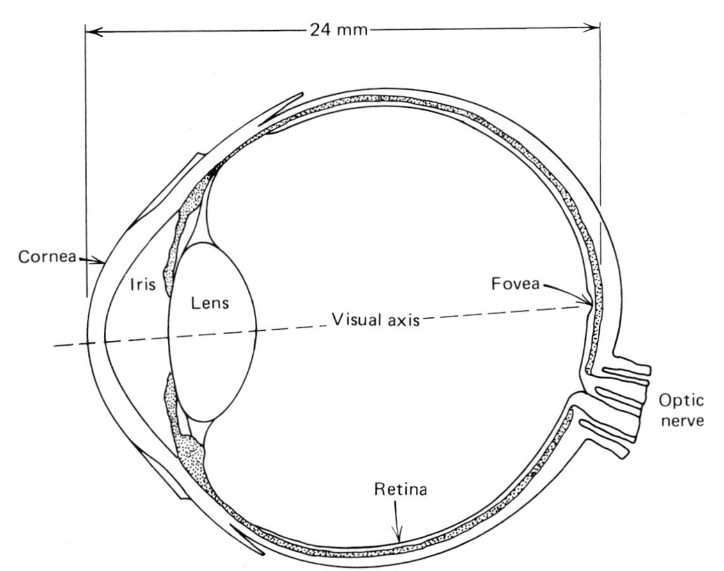

Figure 7.1 Cross section of the actual human eye. (Adapted from Cornsweet, *Visual Perception,* Academic Press, 1970)

responsible for the major portion (approximately 70%) of the refraction of light that occurs as light passes through the eye to the retina. The "fine tuning" of the eye, or adjustments that are made to focus correctly on the retina images of objects at various distances, is done by the lens of the eye.

The lens is unique in that it has the ability to change its curvature and thus its focal length. This is accomplished by the ciliary muscles that form a ring around the lens. Under normal conditions, with the eye relaxed, these muscles keep the front of the lens fairly flat and permit the lens to focus nearly parallel rays on the retina. If a near object is to be viewed by the eye, as is necessary for reading, then the focal length of the lens must be decreased. This is done by the ciliary muscles that contract and move forward and thus permit the front surface of the lens to assume a more curved shape. A shorter radius of curvature will permit the closer object to be focused on the retina. The process by which the lens adjusts so the eye can focus light coming from a wide range of distances is known as accommodation.

Between the cornea and the lens is a waterlike fluid known as the aqueous humor, which also acts as a refracting material. Likewise the jellylike vitreous humor between the lens and the retina also serves as a refractive medium. These different refracting media complicate the process of calculating in detail the position and size of any image that is formed. However a simplified model, which assumes that all the refraction takes place at the cornea rather than in the various media, makes calculations much easier. The simplified model is referred to as the *reduced* eye.

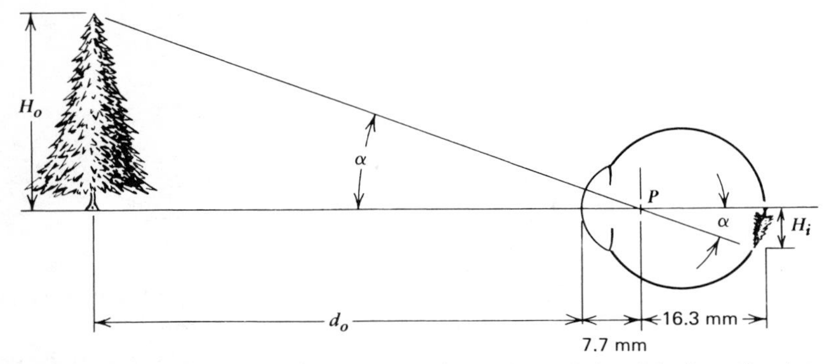

Figure 7.2 The reduced eye with pertinent dimensions. Point *P* is the effective center of curvature of the cornea. That is, rays striking the cornea perpendicularly will pass through point *P*. (Adapted from Cornsweet, *Visual Perception,* Academic Press, 1970)

If we assume that the index of refraction of the material of the eyeball is 1.47 (rather than the actual 1.336), the reduced eye will have the same dimensions as the actual eye; now, however, our calculations will be much more conveniently carried out. The properties of the retinal image calculated in this way are very close to those found for the actual eye.

Figure 7.2 shows the reduced eye, and how it can be used to find the image size corresponding to a given object. In the diagram some rays of light from the object are incident on the cornea. These rays strike the cornea along normals and hence do not deviate as they enter. Such rays will pass through the center of curvature of the cornea and form an inverted image on the retina as illustrated. From the similar triangles of the diagram we observe

$$\frac{\text{Height of the image}}{\text{Height of the object}} = \frac{16.3 \text{ mm}}{d_o + 7.7 \text{ mm}}$$

or

$$\frac{H_i}{H_o} = \frac{16.3 \text{ mm}}{d_o + 7.7 \text{ mm}}$$

When compared to the distance the object is from the eye (d_o), the 7.7 mm is very small and can be neglected for most purposes. Likewise, it is convenient to round off the 16.3 to 16 (since eyeballs do have variation in size). Thus we find to a good approximation that

$$\frac{H_i}{H_o} = \frac{16 \text{ mm}}{d_o}$$

We can now calculate the size of the image formed on the retina of an object a distance d_o from the eye. For example, a person 1.8 meters tall (about 6 feet) viewed at a distance of 15 meters would form an image of

$$\frac{H_i}{H_o} = \frac{16 \text{ mm}}{d_o}$$

$$\frac{H_i}{1.8 \text{ m}} = \frac{16 \text{ mm}}{15 \text{ m}}$$

$$H_i = \frac{16 \text{ mm} \times 1.8 \text{ m}}{15 \text{ m}} = 1.92 \text{ mm}$$

Thus the image would be 1.92 mm high.

Let us use this method to investigate another very important property of the eye — its resolving power or the limit of its ability to see the fine detail of the object being observed; sometimes called the *visual acuity* of the eye. From our studies of color printing (Chapter 2) we know that multicolored dots placed side-by-side and viewed from a distance blend together to give a solid color. In a similar manner the black and white photographs printed in the daily newspapers consist of thousands of tiny dots that are blended by the eye to give continuous looking pictures.

The ability of the eye to distinguish distance between two closely spaced adjacent points was investigated toward the latter end of the seventeenth century by Robert Hooke. He pointed out that it was necessary for two points to have an *angular* separation on the order of one minute of arc, that is, one sixtieth of a degree, if they are to be distinguished from each other. This is equivalent to being able to resolve at one mile objects that are one and one-half feet apart. Or on a smaller scale, objects that are about 10 inches (or 25 cm) from the eye can be seen as distinct individual points if they are not closer than 0.1 mm from each other. Let us use the method developed above to calculate the separation of the retinal images of two just barely resolved dots located at a distance of 10 inches from the eye.

We take as the height of the object the distance between the points, which is 0.1 mm. Since the object is 250 mm from the eye (d_o), we find the height of the image on the retina is

$$\frac{H_i}{H_o} = \frac{16 \text{ mm}}{d_o}$$

$$H_i = \frac{16 \text{ mm} \times H_o}{d_o} = \frac{16 \text{ mm} \times 0.1 \text{ mm}}{250 \text{ mm}}$$

$$H_i = 0.0064 \text{ mm} = 6.4 \times 10^{-6}\text{m}$$

Thus the extremes of the image, or points to be resolved, must be separated by this amount.

The question as to why this spacing will enable points to be resolved or seen separately can be understood through an examination of the cone size on the retina. On the retina we find that the cones have an average diameter of approximately 5×10^{-6}m. It is clear that in order to see the images of two

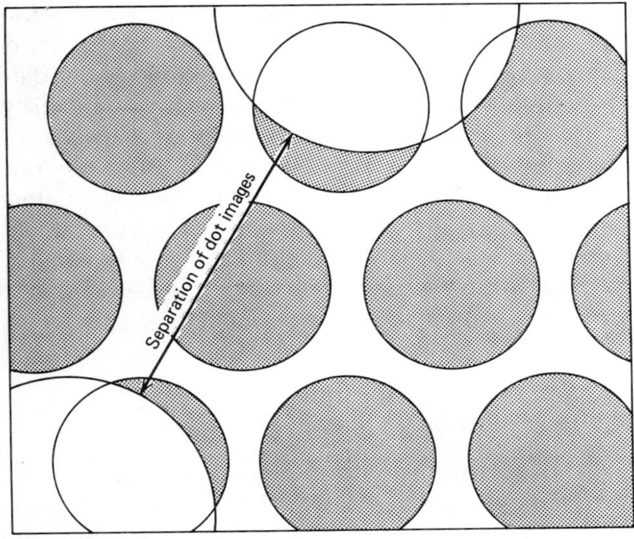

Figure 7.3 The separation of the images of two dots on the retina. The two larger light circles are the dot images. The smaller dark circles are the cones on the retina. In order for the dots to appear as two distinct images, the dot images must be separated by at least one *unstimulated* cone.

closely spaced dots as distinct, the images must be separated by at least one unstimulated cone (Figure 7.3). The calculated distance of 6.4×10^{-6}m agrees well with the average cone diameter of 5×10^{-6}m.

The photopic or cone vision of the eye is much more acute than that of the scotopic or rod vision. Again the anatomy of the retina indicates that the density of rods in the retina is much less than the density of cones in the fovea (that is, the rods are farther apart than the cones). In addition each cone in the fovea is usually connected to the brain by a single fiber of the optic nerve, while in contrast, large clusters of rods are connected by single optic nerve fibers. Thus, to be resolved as separate points in the rod area of the retina, the images of the two points must be separated not just by a rod, but by a clump of rods, greatly reducing the acuity of the scotopic vision.

7.2 ADAPTATION

The human eye has the capacity to function over a range of intensity of light that varies by a factor of 100,000. One of the mechanisms that makes this possible is the variable pupil or opening in front of the lens. The size of this opening is controlled by the colored iris, which automatically dilates or contracts as the quantity of light entering the eye increases or decreases. The pupil varies from a diameter of approximately 1.5 mm in bright light to a diameter of about 6 mm under the darkest conditions. These changes in diameter of the pupil under varying light conditions, known as adaptation, give rise to the ability to vary the intensity of light entering the eye by only a factor of 16 (since $A_{lens} \propto d^2$). Although this is most beneficial in aiding the eye to see under different lighting conditions, it would be insufficient were it not for the adaptability of the rods and cones themselves, and also for the different sensitivities of the rods as opposed to the cones. As discussed in Chapter 4 the cones operate only in relatively bright light while the noncolor-sensing rods can detect light under much less intensity. But, in addition, the cones (or the rods, depending on the light level) are able to adjust their sensitivity to light in a few minutes. This basically is a biochemical response to the prevailing light level. All of these factors together account for the tremendous range of intensity over which the eye can see.

7.3 DEFECTS OF VISION

The eye is capable of focusing light from objects at a variety of distances. The extremes of the distance range over which distinct vision is possible are known as the far point and the near point of the eye. For the normal eye the far point is at infinity. An object at that position would emit rays of light which, when reaching the eye, would essentially be parallel. Provided the eye is relaxed and has no visual defects, parallel rays will be brought to a focus on the retina.

If the eyeball is abnormally long in comparison with the radius of curvature of the cornea, the parallel light rays from an object at infinity will focus in front of the retina. If such a condition exists, the eye is said to be *myopic* or *nearsighted.* The accommodation ability of the eye is of no value in this situation since it can only shorten the curvature of the lens, which would cause the image to be even more out of focus, that is, the image would form further from the retina. As the object being viewed is brought closer to the relaxed eye, the incoming rays eventually will focus on the retina giving a clear image. This is because as an object is brought progressively nearer to a convex lens the image moves back away from the lens (refer to previous chapter). The distance at which focus first occurs is the far point of the myopic eye. Myopia (nearsightness) is corrected by placing a diverging or concave lens in front of the eye. Incoming parallel rays from a distant object then emerge from the lens and strike the eye as if they had come from its far point, and the relaxed eye is therefore able to focus them onto the retina (Figure 7.4).

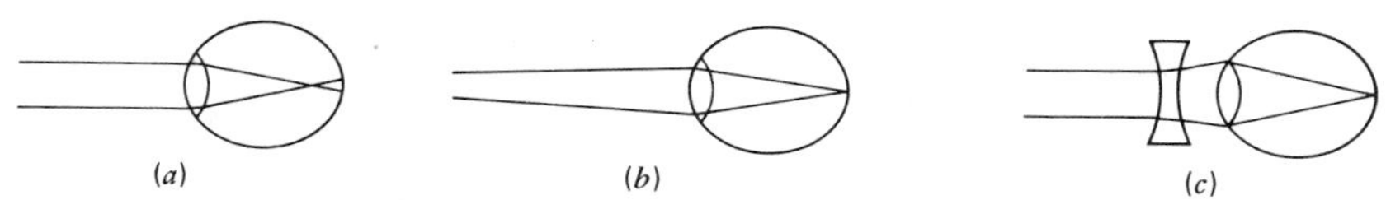

Figure 7.4 (*a*) The uncorrected nearsighted eye as it views a distant object. (*b*) The far point for the uncorrected nearsighted eye; note that the rays diverge as they enter the eye. (*c*) The corrected nearsighted eye; note that parallel rays are caused by the lens to diverge *as if* they were coming from the uncorrected far point.

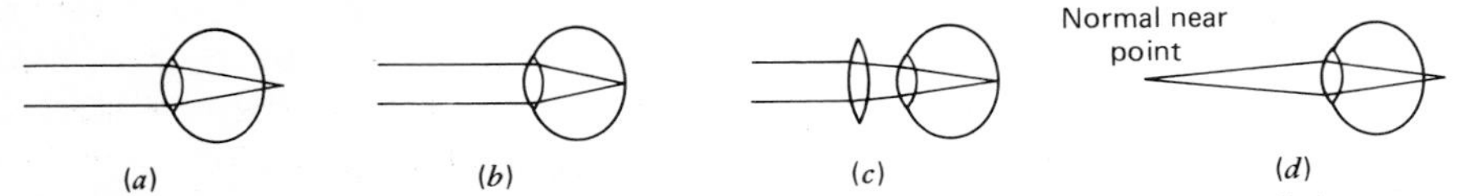

Figure 7.5 (*a*) The *relaxed* uncorrected farsighted eye; note the parallel rays are brought to focus behind the retina. (*b*) Because of the eye's ability to accommodate, the ciliary muscles will shorten the focal length of the lens so that parallel rays are focused on the retina. (*c*) The use of a convex lens allows the *relaxed* farsighted eye to focus parallel rays on the retina. (*d*) Without correction, the farsighted eye cannot focus on an object at the *normal* near point (~25 cm) even with the eye fully accommodated.

As an object is moved closer to the eye, the focal length of the eye is decreased by accommodation so that the image continues to form on the retina. That is, in the equation $d_o d_i = f^2$, if we want to keep d_i the *same* while *reducing* d_o, we must *reduce* f. This is how the eye maintains focus for objects at different distances, by adjusting f. The lens cannot comfortably focus images on the retina of objects that lie closer than a certain distance. That is, the focal length of the lens can be reduced only so much. The point at which the eye encounters this difficulty in focusing the object is called the *near point* of the eye and is about 25 cm (10 inches) in a young adult with normal vision. Before adulthood, when the lens still retains greater elasticity, the near point of the eye can be closer than this.

If the eyeball is abnormally short for the refracting system of the eye, the eye is referred to as hyperopic or farsighted. Incoming parallel rays passing through the *relaxed* lens system do not have ample path length to focus before striking the retina. However, the accommodating power of the lens will allow focusing to occur for distant objects. On the other hand, nearby objects cannot be brought into focus. Stated another way, the near point of the eye has increased beyond (perhaps a great deal beyond) the normal 25 cm distance; thus close objects cannot be focused by the hyperopic eye without assistance. To overcome this problem a converging or convex lens is used (Figure 7.5).

As people get older the lens of the eye loses some elasticity and the ciliary muscles also lose some of their strength. Because of these changes the eye's power of accommodation decreases with age and the near point of the eye moves further from the eye. This condition is called *presbyopia* or "aging sight," and poses the same problem for people who have normal vision, corrected myopia, or corrected hyperopia. Because of the loss of accommodating ability, the person can no longer focus on nearby objects. Of course myopic persons could simply remove their glasses to focus on nearby objects. But then they couldn't focus on distant objects. The general solution to this problem is bifocal lenses. The upper portion of each lens is ground to allow the person to focus on distant objects, while the lower half of each lens is shaped to bring nearby objects into focus. The two lenses (upper and lower) give the eye two focal lengths to replace the lost accommodating ability. The solution is not perfect but it works fairly well.

Another common vision problem is astigmatism, which refers to a defect of the eye in which the surface of the cornea is not spherical, but is more sharply curved in one plane than another. Because of this unequal curvature a collection of horizontal and vertical lines cannot be focused simultaneously. To correct for this problem a lens with a similar nonspherical, or cylindrical shape is used. This lens is oriented so that the extra curvature in one direction compensates for the lack of curvature in the corresponding direction of the cornea. Care must be taken to orient this lens in the proper position in the glasses; if this is not done the problem will be compounded instead of corrected. Astigmatisms are often found in conjunction with either myopia or hyperopia. In such cases the lenses to correct the problem are combinations of the lenses used for the simple condition. Thus, a nearsighted person with an astigmatism would need a concave toroidal lens.

CHROMATIC ABERRATION

In Chapter 6 we studied Snell's law and pointed out that the refraction of light was dependent on its wavelength. As the frequency of light increases, its refractive index becomes larger, causing more refraction of the shorter wavelengths. Consequently, when an image is formed through a lens all colors do not focus at the same distance; this gives rise to chromatic aberration. An image with chromatic aberration will appear to have a colored fringe around it.

The human eye, like other refracting devices, is subject to a small amount of chromatic aberration. Several strategies have been developed by the eye to minimize this difficulty. One of the means by which this problem is reduced is found in the yellow lens of the eye. This lens not only acts as a refracting device but also serves as a filter, cutting off the high frequency waves (violet end of the spectrum) at about 400 nm. Wavelengths smaller than this would stimulate the rods and cones in the retina, further distorting the image. Indeed, if one has the lens of the eye removed, as is done in a cataract operation, it is found that the eye then has excellent vision in the ultraviolet region of the spectrum. Thus, the lens limits the aberration from the shorter wavelengths. It may be noted here that as one ages the lens becomes a deeper yellow and further limits the older person's viewing of blues and violets.

A second factor that helps correct for the chromatic aberration is a yellow-colored pigment that acts as a filter over the fovea and the region of the retina just around it. This area is called the macula lutea. The pigment in the yellow patch absorbs light in the violet and blue zones of the spectrum just where absorption by the lens falls to very low values. Hence those waves of the spectrum that would add most to additional color aberration are removed before they strike the color sensitive cones.

Another important phenomenon that reduces the effect of chromatic aberration in the eye is the so-called achromatic response. This can be best demonstrated by taking a small piece of colored paper and moving it progressively farther from the eye. There comes a time when, although the shape of the paper is still discernible, the color can no longer be determined. That is, when the image of an object on the retina gets too small, color no longer registers. The tiny colored fringes that result from small amounts of chromatic aberration thus produce no color distortion.

In addition to the above-mentioned methods of color correction, when the eye is viewing an object under low illumination, the peak of the eye's sensitivity occurs at 500 nm. At higher light levels, vision shifts to the cones, which have a peak sensitivity at approximately 562 nm. This shift in sensitivity toward the red further reduces chromatic aberration.

Hence we see that the human eye, unable to correct for its color aberration by other means, eliminates those sections of the spectrum that would result in the largest aberrations. The yellow lens limits transmission to those waves longer than 400 nm, the macula lutea eliminates most of the blues and violets, the achromatic response reduces the effect of chromatic aberration, and the shift from rod to cones displaces vision in bright light toward the red. By these methods the eye avoids the negative effect of the chromatic aberration produced by the cornea and lens.

SPHERICAL ABERRATION

Another common aberration of lenses is spherical aberration. The eye, as is the case with other simple lenses bounded by spherical surfaces, cannot bring the rays that strike the marginal portion of the lens to focus in the same position as rays striking near its center. The results of this nonfocusing of the outer rays with the center ones is that the image of a point tends to form a small blur circle.

The eye is amazingly well corrected for this problem by two mechanisms. The first device is the actual shape of the cornea, which tends to have a flatter curvature at its edges than at its center. This corrects in part for the tendency of a spherical lens to refract more strongly at its edges.

In addition to the aspheric curvature of the cornea, the lens actually is denser at its center than at its margin. It therefore refracts light more strongly at its center than at the edges of the lens. Together with the flattened cornea, this lens of varying optical density corrects the eye astonishingly well for spherical aberration.

7.4 THE CAMERA

Cameras are available in different varieties, cost ranges, and sizes. In this section we shall consider only the principles of a 35 mm variable shutter, adjustable focus type of camera (Figure 7.6). Instamatic type cameras of fixed focus operate on the same principles but are not as versatile as the 35 mm variety.

The most obvious feature of a camera is its lens system. The lens system, which refracts incoming light, contains several different sections of glass to com-

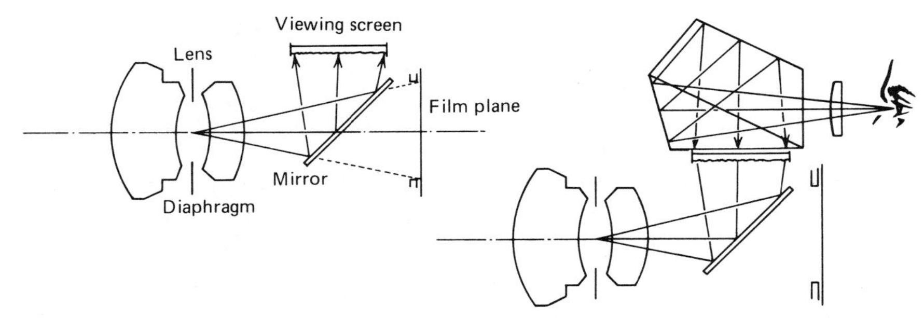

Figure 7.6 Typical single lens reflex (SLR) camera. When a picture is actually taken, the mirror snaps out of the way allowing light to reach the film. Just in front of the film plane is a shutter (not shown) that controls the exposure time. The purpose of the inclined mirror is to reflect the image onto the frosted glass viewing screen so that the user can view the scene. Since this image would be inverted and reversed left to right, a special internal reflection prism is used to reinvert the image. The eyepiece is added to magnify the image.

bat various different aberrations that would occur otherwise. It is convenient to treat this optical system as a single refracting device much in the same way that we work with the reduced eye instead of the individual refracting elements of the actual eye.

For most hand-held cameras, the lens system has a fixed focal length. This means that focusing in the camera cannot be done in the same way it is done in the eye. A camera is focused on a particular object when the image of that object falls on the film in the camera. In the previous chapter we developed a simple lens equation that allowed us to compute the location of an image formed by a convex lens. That equation was

$$d_o d_i = f^2$$

Strictly speaking, this equation must be used with some care when it is applied to thick lenses or to combinations of lenses. In these cases, f is not the distance from the center of the lens to each focal point. Rather, f must be measured from a principal plane of the lens. Two such planes (one for each focal point) lie inside the lens and do not coincide with either the center or the surfaces of the lens. Thus when we say, in what follows, that an image is located a certain distance from the lens, we mean that it is located that distance from a principal plane of the lens. Readers who are interested in a more complete discussion of this point should consult a physics or optics textbook. With this warning in mind, let us return to the above equation and algebraically rearrange it into the more immediately useful form:

$$d_i = \frac{f^2}{d_0}$$

Review Chapter 6 for the definitions of d_o and d_i.

Thus we see that objects at different distances from the camera (different values of d_o) will produce images at different distances from the focal point of the lens (different values of d_i). In order to focus a camera we must be able to adjust the distance between the film and the lens so that the image falls exactly on the film. Exactly how far from the lens should the film be located? First let us consider an object very far from the camera. In this case d_o can be taken as essentially infinite. Thus

$$d_i = \frac{f^2}{d_o} = \frac{f^2}{\infty} = 0$$

This means that the image is located exactly at the focal point (or, more properly, on the plane) of the lens. Thus in this case the film must lie behind the lens at a distance equal to the focal length of the lens. Camera lenses are labeled with their focal lengths. For example, there are 20 mm, 50 mm, and 100 mm lenses. In a camera with a 50 mm focal length lens, distant objects will be in focus only if the film is located 50 mm behind the lens.

For objects that are not so distant we see from our lens equation that d_i is no longer zero. As d_o is reduced (objects brought nearer), d_i increases correspondingly. Thus as objects are brought nearer to the camera, the film must be moved farther from the lens in order to maintain focus (Figure 7.7). There is a practical mechanical limit to how far the film can be moved from a lens. For example, for the typical all-purpose 50 mm lens, the distance between the lens and the film can only be adjusted over a range of 7.5 mm. That is, d_i can be varied from zero to 7.5 mm. We can use our lens equation to compute the smallest value of d_o for which the camera can focus.

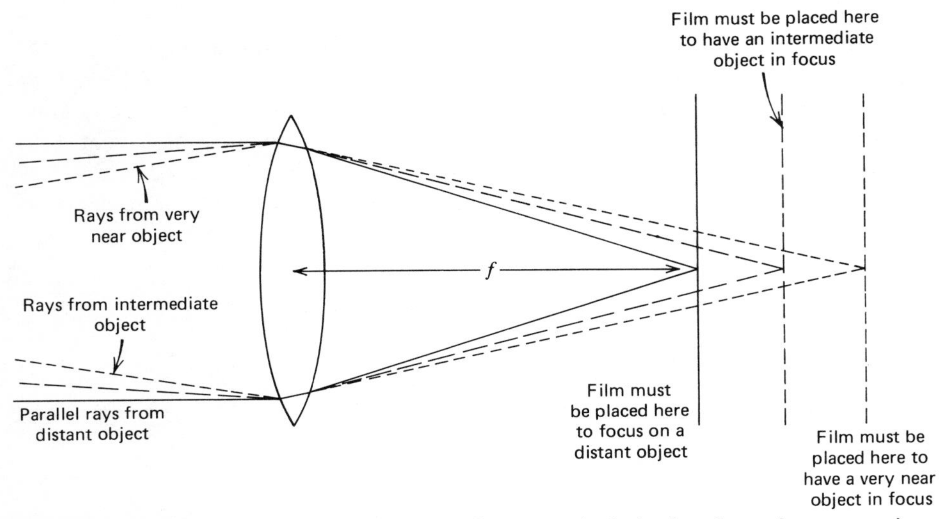

Figure 7.7 Film in a camera must be moved progressively farther from the camera lens so that progressively nearer objects are in focus on the film.

$$d_o d_i = f^2$$

or, upon algebraic rearrangement,

$$d_o = \frac{f^2}{d_i}$$

In our example, $f = 50$ mm, and we use the maximum value of d_i, that is, 7.5 mm. Thus

$$d_o = \frac{(50 \text{ mm})^2}{7.5 \text{ mm}} = 333 \text{ mm}$$

$$d_o = 33.3 \text{ cm}$$

Since d_o is measured from the focal point in front of the lens, this means that the nearest object that can be brought into focus will be a distance of 33.3 cm $+ f =$ 38.3 cm in front of the lens. Objects nearer than this cannot be brought into focus. Certain special lenses are constructed to allow for larger values of d_i. With these lenses it is possible to focus on extremely nearby objects.

Fixed focus cameras, such as instamatics, are constructed such that objects at about 8 to 10 ft form images that are in sharp focus on the film. Objects more distant than 10 feet will result in images that are slightly but insignificantly out of focus on the camera film. At distances closer than 8 ft the image falls significantly behind the film and is thus out of focus. Consequently this type of camera cannot be used if one desires to photograph things closer than 8 to 10 feet.

TELEPHOTO LENSES

As seen in the last section, the focal length of the lens used in a camera determines the distance that the film must be separated from the lens. The focal length also limits the size of the image that will occur on the film for any given object at a fixed distance. If we consider an object of height 2 meters at a distance of 10 meters from the camera, then the image height on the film can be found by using the magnification equation developed in Chapter 6. Thus we have

$$\text{Magnification} = \frac{H_i}{H_o} = \frac{f}{d_o}$$

or

$$H_i = H_o \frac{f}{d_o} = 2 \not{\text{m}} \cdot \frac{50 \text{ mm}^*}{10 \not{\text{m}}}$$

$$\text{Thus} \qquad H_i = 10 \text{ mm}$$

*Ten meters is used for the object distance even though the object distance is actually 10 m − 50 mm, which is not significantly different from 10 m.

If you desire a larger image of the object being photographed you can move closer to the object. This will decrease the object distance (d_o) and increase the image size. Often it is not possible or prudent to move closer to an object you would like to photograph, for example, taking a picture of a charging bull. In such cases the photographer can still get a large image if he has access to a telephoto lens—that is, a lens with a focal length longer than the standard 50 mm lens normally found in a camera. For example, suppose we compare the image height of an object taken with a 50 mm lens at a distance of 10 m to the image height of the same object taken with a 150 mm lens. We know

$$\frac{H_i}{H_o} = \frac{f}{d_o}$$

If the object's height is 1 meter then we find for the 50 mm lens

$$H_i = H_o \frac{f}{d_o} = 1 \text{ m} \times \frac{50 \text{ mm}}{10 \text{ m}} = 5 \text{ mm}$$

Likewise for the 150 mm lens

$$H_i = H_o \frac{f}{d_o} = 1 \text{ m} \times \frac{150 \text{ mm}}{10 \text{ m}} = 15 \text{ mm}$$

Thus we see that the image height when using the longer focal length lens is three times as large as with the 50 mm lens. The longer the focal length of the lens the larger the image of the object (Figure 7.8).

Increasing the focal length of a lens has several consequences. First, light coming from the object will be distributed over a larger area, making the image less bright. This problem will be discussed shortly. The second consequence is that the total image on film covers a smaller area of the field of view. As the images of individual objects grow in size there necessarily must be fewer of them on the film. Stated another way, the field of view (the area being photographed) decreases as the image of individual objects becomes larger.

SENSITIVITY TO LIGHT

As we noted in Section 7.3 the eye has the ability to adapt to a large range of light conditions. Cameras, likewise, can be used in a variety of light conditions. Both the eye and the camera have variable apertures that control the amount of light entering them. When the intensity of the light is large, the size of the aperture needed is small. As the light intensity decreases, the size of the aperture must increase to maintain the irradiance on the film. If the intensity decreases beyond a certain point, the maximum aperture opening is insufficient to properly expose the film. In such cases the photographer has the option of changing the type of film being used to one which requires less light or energy for proper exposure. This change of sensitivity to light, by changing types or speeds of films, is analogous to the eye switching from photopic to scotopic

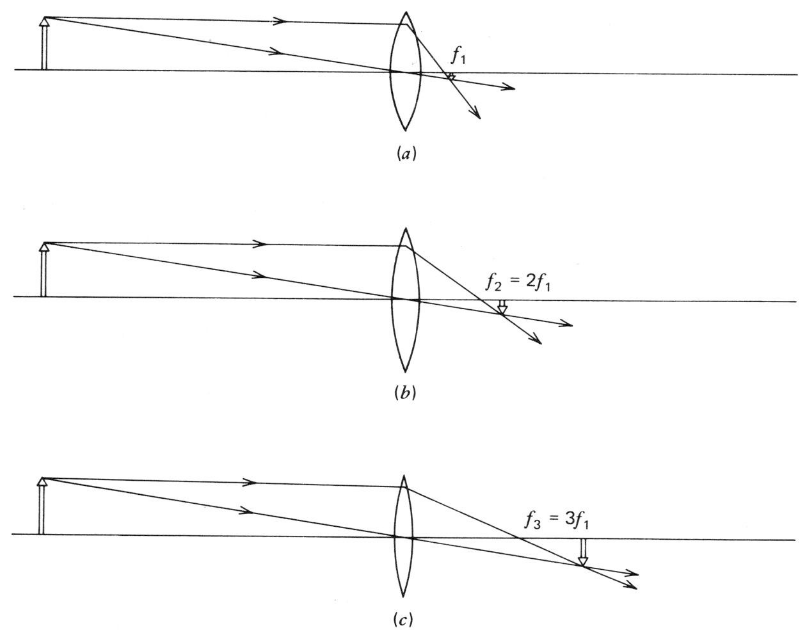

Figure 7.8 Relationship between image size and focal length. (*a*) Small image produced of an object located at a certain distance from the lens. (*b*) and (*c*) show the same object at the same distance, but with progressively longer focal length lenses that produce progressively larger images.

vision. Indeed, as one makes such a change in film, the grains of silver bromide in the film gelatin will be larger as the sensitivity to light becomes greater. These larger grains will have a higher chance of being struck by the fewer photons in less intense light. However, since the grains are larger there will be fewer of them and consequently the quality of the image produced will be poorer than if the grain size is smaller. This is similar to the eye's switching to the rods of the retina, which are more sensitive to light than the cones. Variations in the intensity of light can be adjusted for by changing the area of the aperture or the sensitivity of the light-responsive material that light strikes in either the eye or the camera. In both cases as the sensitivity increases the sharpness of the image is diminished.

INTENSITY OF THE CAMERA IMAGE; f-NUMBERS

An object being photographed is radiating or reflecting a fixed quantity of light. The brightness (as measured by irradiance) of the image produced on the camera film depends on two factors (in addition to the light intensity): the focal length of the lens being used and the area of the aperture.

Earlier we learned that the height of an image produced by a lens was directly proportional to the focal length of the lens. The width of the image being produced will also be directly proportional to the focal length of the lens. Suppose we had a square object that produced an image of 1 cm $\times$ 1 cm when photographed through a 50 mm lens. If we now used a lens of focal length 150 mm and photographed this same object from the same distance, the image produced would be 3 cm $\times$ 3 cm (Figure 7.8). Thus, the light that comes from the object would be distributed over an area that was 9 times as large, and the irradiance or brightness of the light on the film would 1/9 what it was when the 50 mm lens was used. Using a 200 mm lens would result in an image of the same object that was 16 times the area (4 cm $\times$ 4 cm) of the original image. The brightness of this image would therefore be 1/16 the value as when using the 50 mm lens. In general the brightness of the image is inversely proportional to the square of the focal length of the lens.

$$\text{Brightness} \propto \frac{1}{(\text{focal length})^2}$$

or

$$B \propto \frac{1}{f^2}$$

How bright the image is depends also on the size of the lens opening being used. If we use a lens with a small diameter opening, then the quantity of light that can pass through it will be smaller than the light that passes through a lens having a larger diameter opening. Since the aperture in a camera is circular, the area of an opening with a diameter of 1 cm would be

$$A = \tfrac{1}{4}\pi D^2 = \tfrac{1}{4}\pi(1\text{ cm})^2$$

$$A = \tfrac{1}{4}\pi\text{ cm}^2$$

If we used an aperture of 2 cm the area would be

$$A = \tfrac{1}{4}\pi D^2 = \tfrac{1}{4}\pi(2\text{ cm})^2$$

$$A = \pi\text{ cm}^2$$

Likewise a diameter of 4 cm would give an area of 4π cm^2.

Here we see that the areas of the apertures vary directly as the square of the diameter of the openings

$$A \propto D^2$$

Since the brightness of the image depends on the area of the opening, we find

$$\text{Brightness} \propto \text{area}$$

and

$$\text{Area} \propto \text{diameter}^2$$

so

$$\text{Brightness} \propto \text{diameter}^2$$

or

$$B \propto D^2$$

Combining this with our previous result gives

$$\text{Brightness} \propto \frac{\text{diameter}^2}{(\text{focal length})^2}$$

$$B \propto \frac{D^2}{f^2}$$

or

$$B \propto \left(\frac{D}{f}\right)^2$$

In most cameras, the focal length is fixed by the lens being used. Thus the brightness of the image on the film is controlled by adjusting the aperture diameter, D. Until recently this was done on most cameras by manually rotating the barrel containing the lens. On many newer cameras, photocells automatically sense the intensity of light entering the camera and then adjust the aperture to achieve a proper image intensity.

Since the image intensity is determined by the ratio of D to f, most cameras are calibrated directly in terms of this ratio, or more accurately, its inverse. This is done by defining a quantity called the f-number.

$$f\text{-number} = \frac{\text{focal length of a lens}}{\text{diameter of the lens opening}}$$

$$f\# = \frac{f}{D}$$

Substituting this in the expression for the brightness of an image, we find

$$B \propto \left(\frac{D}{f}\right)^2 \quad \text{or} \quad B \propto \left(\frac{1}{f\#}\right)^2$$

The f-number thus gives us a measure of the brightness of an image based on the diameter of the lens aperture and the lens focal length. Furthermore, when the f-number values are the same for two different lenses, the images they will produce will be equally bright even though they may not be the same size. For example, suppose we had a lens of focal length 50 mm. To what value would the aperture of the lens need to be adjusted if the f-number of the lens was 4? By definition we have

$$f\# = \frac{f}{D}$$

$$4 = \frac{50 \text{ mm}}{D} \quad \text{or} \quad D = \frac{50 \text{ mm}}{4}$$

$$D = 12.5 \text{ mm}$$

If we had a lens of focal length 200 mm and the diameter of the aperture was the same as the previous lens, then the *f*-number would be

$$f\# = \frac{f}{D} = \frac{200 \text{ mm}}{12.5 \text{ mm}}$$

$$f\# = 16$$

In these examples the diameter of the aperture and hence the area through which the light enters would be the same, but the brightness of the images produced would be different. The reason for the difference in brightness is that the focal length of the lenses are unequal. The second lens, having a focal length four times as large as the first, will produce an image whose height and width are each four times greater than that of the first lens. Thus the area of the second image is 16 times greater than that produced by the first lens and, consequently, the light that enters through the same size aperture will be only one-sixteenth as bright as with the first lens.

Note that in the previous example the higher *f*-number resulted in an image that was less bright than that produced by a lens having a smaller *f*-number. The range of *f*-numbers or *f*-stops that are usually found on a camera lens are:

1.4, 2.0, 2.8, 4.0, 5.6, 8, 11, 16

All cameras do not include all these values and some cameras will have *f*-stops between those given here. To change the camera lens from one stop to another, the aperture is adjusted by turning the barrel of the lens. This causes the iris of the aperture to become larger or smaller changing the ratio f/D. It should be obvious that as the diameter of the aperture becomes larger, the *f*-number will become smaller and a brighter image will be produced. Changing a lens' setting to a higher *f*-number will result in a smaller aperture and a fainter image.

A question that often occurs is, "why are lenses calibrated with the particular *f*-stops listed above? Why not numbers like 2,3,4,5, etc?" To answer this question let us examine the relationship between the brightness of an image and the *f*-number, that is,

$$B \propto \frac{1}{(f\#)^2}$$

If we take the given series of *f*-numbers, square them, and then find $1/(f\#)^2$, we have:

$f\#$	1.4	2.0	2.8	4.0	5.6	8.0	11	16
$(f\#)^2$	2	4	8	16	32	64	121	256
$1/(f\#)^2$	1/2	1/4	1/8	1/16	1/32	1/64	1/121	1/256

Thus we see that between any two adjacent *f*-stops the brightness, which varies as $1/(f\#)^2$, will just change by a factor of 2.

EXPOSURE TIMES; FILM SPEED

Taking a photograph can be compared to filling a bucket with water. If a large stream of water is flowing into a bucket, it takes a small time to fill the bucket. However, if the stream of water is small, the time to fill the bucket increases. In a similar manner, to expose a piece of film properly a fixed amount of energy in the form of light must strike the film. If the light intensity is very high, then the exposure time will be short. If the light is dim, the time of exposure must be longer.

Adjustable cameras have exposure times ranging from one or two seconds to approximately 1/1000 of a second. Depending on the intensity of the light and the nature of the object being photographed, the time of exposure is adjusted to give the best picture. Whatever time is chosen, the product of that time and the brightness of the image must be a specific constant value for a given type of film to give a correct exposure. That is,

$$B \times t = \text{constant} = \text{total amount of light striking film for proper exposure}$$

Different types of films require different amounts of light to expose them properly. Films are rated by an assigned number, called film speed, which specifies amount of light or energy necessary to produce a good picture or negative. This American Standards Association (ASA) number is higher for films that require a small amount of exposure as compared to films needing more light. Values vary from as high as ASA 3000 (a very fast film) to as low as ASA 20 for a very fine grain slow film.

For any given film we recognize that the product of the brightness of exposure multiplied by the period of exposure must be constant; that is,

$$B \times t = \text{constant (determined by the film speed)}$$

or, since $B \propto 1/(f\#)^2$, we can say $t/(f\#)^2 = \text{constant}$ for proper exposure of a given film.

Suppose we found that a proper exposure of a film could be achieved by taking a picture at 1/50 second with an *f*-stop of 8. If *under the same light conditions,* we wished to change the exposure time to 1/200 of a second in order to "freeze the action," what $f\#$ should we choose? We know for the original case

$$\frac{t}{(f\#)^2} = \text{constant} = \frac{1/50}{(8)^2}$$

Therefore the new setting must result in the same value. That is,

$$\frac{1/200}{(f\#)^2} = \frac{1/50}{(8)^2}$$

Solving for $(f\#)^2$, we find

$$(f\#)^2 = \frac{8^2 \times 1/200}{1/50} = 64 \times \frac{1}{4} = 16$$

Therefore,

$$f\# = \sqrt{16} = 4$$

Knowing the relationship between the different *f*-numbers, we could also have arrived at this value by the following logic: the time of exposure is 1/4 (1/200 compare to 1/50) as great in the second case, therefore, the brightness must be 4 times greater. Lowering the *f*-number two stops doubles the image brightness twice, making it 4 times brighter. Thus the correct value is *f*/4.

Most hand-held cameras today (except for inexpensive instamatic types) have provisions for varying both the *f*# (the aperture) and the exposure time. There are three common ways this is done, all involving the use of light sensitive photocells. In each case, the film speed must be set using the dial provided. This essentially tells the built-in light meter what amount of light is required for proper film exposure. In the first type of system the user simply looks through the viewer at the scene to be photographed and partially depresses the shutter release. This activates the light meter, which then indicates to the user whether the picture will be over or underexposed. The user then adjusts either the aperture (*f*#) or exposure time (or both) until the meter indicates that a proper exposure will occur.

In the second kind of system, the user presets the exposure time and again partially depresses the shutter release. The built-in light system then automatically adjusts the aperture (*f*#) for a proper exposure, and informs the viewer of which *f*# has been selected by means of a small red dot that appears next to the appropriate *f*# in the viewer. If the user is not satisfied with this *f*#, he or she may change the exposure time, which would result in a different *f*# being selected since

$$\frac{t}{f\#^2} = \text{constant}$$

for any particular film. The third type of system is just like the second, except that the *f*# is preset and the metering system then adjusts the exposure time. The user is then informed of which exposure time has been selected by means of a small red dot that appears next to the appropriate exposure time in the viewer.

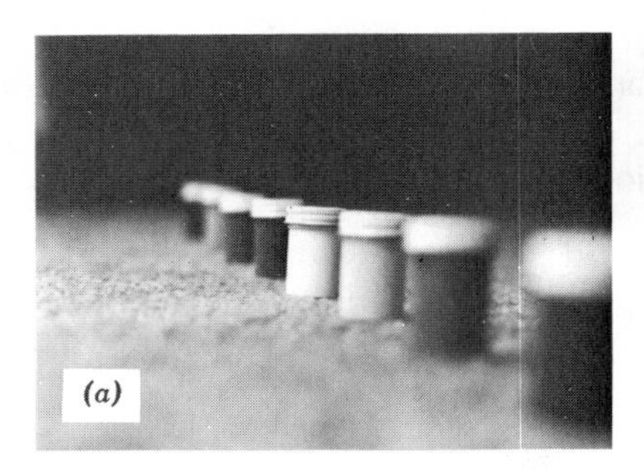

Figure 7.9 The reduction in depth of field that occurs *for a given lens* as the $f\#$ is decreased.

DEPTH OF FIELD

We have seen that a camera can be accurately focused on any object that is more than a foot from the camera. Suppose, as a specific example, we focus our camera on an object about 2 feet away. Obviously, objects closer or farther away cannot also be exactly in focus. Figure 7.9*a* illustrates these circumstances. Notice that for objects progressively further away or closer than 2 feet, the focus becomes progressively worse. There is a somewhat arbitrary range of distances centered on 2 feet where objects are in fairly good focus. Outside this range the focus is obviously bad. The depth or extent of the range in which objects are in fairly good focus is called the depth of field. It is interesting to note that the depth of field can be influenced greatly by changing the lens opening or $f\#$. Figures 7.9*b* and 7.9*c* are views taken with the *same lens* of the same scene as Figure 7.9*a*, except that Figure 7.9*a* was taken at $f/2$ while Figure 7.9*b* was taken at $f/5.6$ and Figure 7.9*c* was taken at $f/16$. Notice that as the $f\#$ is increased, the depth of field likewise increases. Figure 7.10 illustrates schematically why changing the lens opening (and thus the $f\#$) also changes the depth of field. The practicing photographer must take the depth of field into consideration when setting the $f\#$ for a given lens. Since the $f\#$ also controls the intensity of the image and thus the exposure time, it is apparent that compromises must often be made. In most cases, the photographer must juggle the exposure time, the $f\#$, and the film speed to get the best result. In particular:

Increasing exposure time — Increases light and reduces the ability to stop motion

Decreasing $f\#$ — Increases light and reduces depth of field

Increasing film speed — Reduces light requirements and increases image graininess

CHROMATIC ABERRATION

As stated earlier, all single lenses made of one material refract short, high frequency light waves more than they will refract those waves of longer wavelengths. This phenomenon, which is most strikingly demonstrated with a prism, is known as dispersion. Another way of looking at this is to say that the index of refraction of a particular material varies slightly with wavelength. For some materials the effect is quite large, and these materials are said to have high dispersive power. With respect to lenses, the problem is that for a given lens dispersion results in the violet and blue light from an object having a shorter focus

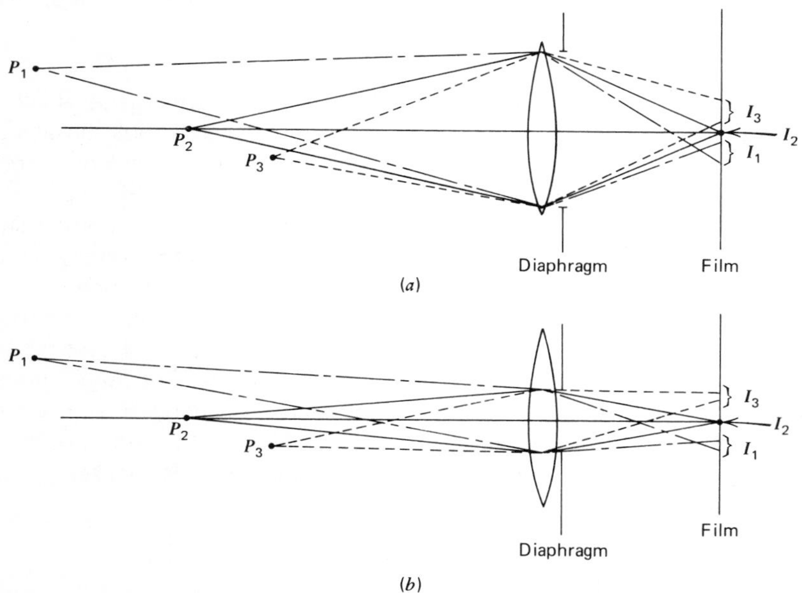

Figure 7.10 (*a*) Focus of three points attained at different distances from the lens. Film is positioned so that point P_2 is in focus. Light from P_3 has not yet converged completely when it reaches the film. Thus a blur circle is produced — the image is out of focus. Light from P_1 has already passed through its image point and is again diverging when it reaches the film. Again, a blur circle is produced. (*b*) When the lens opening is reduced by constricting the diaphragm, the blurred circles I_1 and I_3 become greatly reduced in size. On the film this produces images P_1 and P_3, which are in better focus.

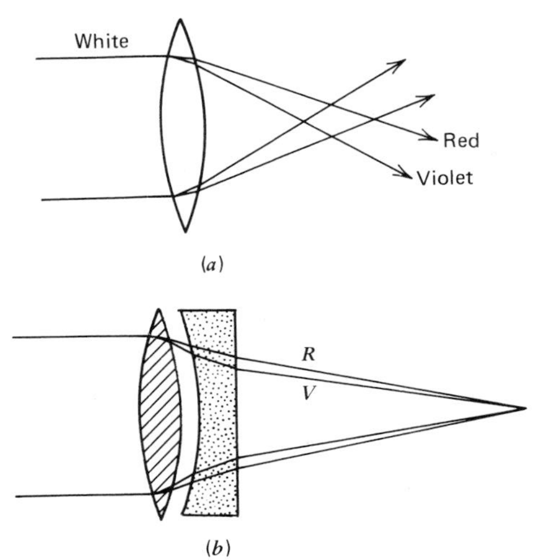

Figure 7.11 (*a*) Chromatic aberration (prism effect) associated with a typical lens. (*b*) An achromatic lens combination – the second lens is made from a different kind of glass than the first lens. The second lens reverses the color dispersion of the first lens, yielding a nearly distortion-free image.

than the red light from the same object. Consequently, the image of light from a white point is a blurred circle of colored fringes, which is most unsatisfactory for use in a camera. To combat this problem, nearly all camera lenses are color corrected by combining two lenses of different materials.

Such lenses, known as achromatic lenses, usually consist of a converging lens made of a material of low dispersive power combined with a diverging lens of high dispersive power. The combination of the two lenses produces a converging lens that will result in the extremities of the visible spectrum (red and violet) being focused at the same point. In Figure 7.11 the diagram shows white light being dispersed as it is refracted through the first segment of the combination. The second portion of the lens, being constructed of material having a higher dispersive power causes the blue to bend back toward the red, thus reversing the dispersive effect of the first lens. The combination thus focuses the rays at the single focal point of the compound lens.

SPHERICAL ABERRATION

Most lenses that are constructed for use in cameras produce good images when the light is incident near the principal axis of the lens. If the light strikes the edges of the lens, then it tends to form an image that will focus at a slightly different location. The eye corrects for this by the variable curvature of the cornea and nonuniform density of the lens. Lens makers are unable to match the ability of the eye in this fashion.

As with chromatic aberration, the lens maker usually adjusts for spherical aberration by combining more than one curved surface. This technique complicates the lens production but is necessary if a large lens aperture is desired. If only a small aperture is needed, it is not necessary to correct for the spherical aberration. However, if low *f*-number settings are required for a lens, large apertures must be obtained and the correction must be made. Correcting for this aberration may mean another correction must be included for an additional aberration that could be introduced.

While it is never possible to eliminate all aberrations, great strides have been made in the production of fine quality camera lenses since the advent of computers that can aid in the necessary calculations. Finding the correct radii of curvature for surfaces of lenses of different indices of refraction and dispersive powers has been reduced from years of calculations to a calculating period of less than a day. Thus new and better lenses are currently available for less money than was paid for inferior lenses 20 years ago.

7.5 OTHER OPTICAL DEVICES

THE MAGNIFIER

A simple magnifier is nothing more than the convex lens we studied in the last chapter used to view an object placed between the focal point and the lens. We examined how such positioning of an object would result in an image that was enlarged and virtual, that is, unable to be projected on a screen (see Section 6.5). While the size of the image divided by the size of the object gives us the magnification of the image (H_i/H_o = magnification) it really doesn't convey to us how much larger the image would appear to our eyes as we viewed it. The reason for this is that the *apparent* size of the image depends not only on the image size (H_i), but also the image location. To determine how many times larger the image of an object viewed through a simple magnifier appears than the image of the object viewed by the unaided eye, it is more convenient to change our basic lens equation into a new form.

The alternate form of the lens equation can be found by defining two new quantities, p and q, to replace the quantities d_o and d_i. Specifically (refer to Figure 7.12*a*)

$$p = f + d_o = \text{distance from object to lens}$$

$$q = f + d_i = \text{distance from image to lens}$$

Our basic lens equation is $d_o d_i = f^2$. But from the definitions of p and q we see that

$$d_o = p - f$$

$$d_i = q - f$$

Thus $d_o d_i = (p - f) \times (q - f) = f^2$. This can be rewritten algebraically into the form

$$\frac{1}{p} + \frac{1}{q} = \frac{1}{f}$$

which is the new form of the lens equation.

In terms of this new equation we can see in Figure 7.12*b* the relative positions of the object and image when a convex lens is used as a magnifier. Note that the position of the image is at $-q$, the negative sign indicating a virtual image on the same side of the lens as the object. Objects look larger when seen through a lens in this way, but the magnification as given by image size divided by object size does not describe this "enlarging" effect. The larger the image, the farther away it is; therefore the effect of greater magnification is counteracted by the increased distance. To properly describe the enlargement of the image we need to consider the magnifying power of the lens. The magnifying power is defined as the ratio of the *angular* size of the object viewed through the lens compared to the angular size when seen without the lens. In Figure

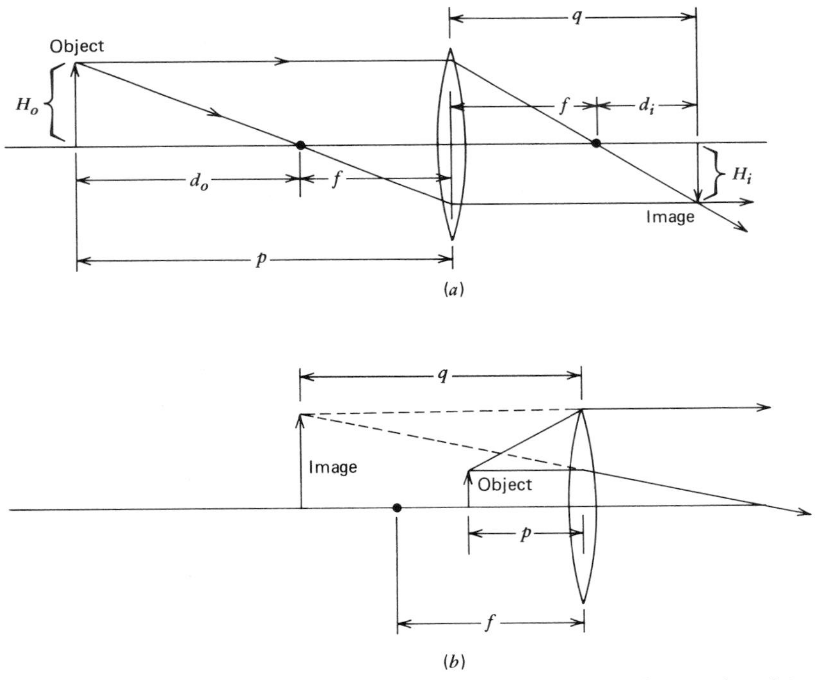

Figure 7.12 (*a*) The new quantities p and q that can be used to locate the object and image, respectively. (*b*) When the image is located on the same side of the lens as the object, q is taken to be negative. Likewise, if the use of the equation $1/p + 1/q = 1/f$ yields a negative value for q, the image is to be placed on the same side of the lens as the object.

7.13*a* an object is viewed without a lens. To be focused the object must be at least 25 cm (the near point) from the eye. Thus its angular size (measured in radians)* is $\alpha = H_o/25$ cm. In Figure 7.13*b* we note that if the image of the object is viewed through a lens its angular size is $\beta = H_o/p$. Hence the magnifying power is

$$MP = \text{magnifying power} = \frac{\beta}{\alpha} = \frac{H_o/p}{H_o/25\text{ cm}} = \frac{25\text{ cm}}{p}$$

Now considering this equation in terms of the alternate lens equation we have just written, we have

$$\frac{1}{p} + \frac{1}{q} = \frac{1}{f}$$

or

$$\frac{1}{p} = \frac{1}{f} - \frac{1}{q}$$

$$\begin{aligned}\text{Thus } MP &= \frac{25\text{ cm}}{p} \\ &= 25\text{ cm}\left(\frac{1}{p}\right) \\ &= 25\text{ cm}\left(\frac{1}{f} - \frac{1}{q}\right) \\ &= \frac{25\text{ cm}}{f} - \frac{25\text{ cm}}{q}\end{aligned}$$

The magnifying power does in fact depend on how far away the final image is. The distance q can range from the near point of the eye to minus infinity. Therefore the magnifying power of the lens can be from

$$MP = \frac{25\text{ cm}}{f} - \frac{25\text{ cm}}{\infty} = \frac{25}{f}$$

for

$$q = -\infty$$

to

$$MP = \frac{25\text{ cm}}{f} - \frac{25\text{ cm}}{(-25\text{ cm})} = \frac{25\text{ cm}}{f} + 1$$

for $q = -25$ cm, an object whose image is at the near point of the eye. Consider a lens of focal length 5 cm. Its maximum magnifying power would be

*Angle in radians is equal to the object height divided by the distance to the object. Also, the angle in radians = 1/57 times the angle in degrees (i.e., 2Σ radians = 360°).

$$MP = \frac{25 \text{ cm}}{5 \text{ cm}} + 1 = 6$$

for an object whose image was at 25 cm. In such a case the angular size of the object would appear 6 times larger than when it was viewed at 25 cm by the unaided eye. If the image of the object were at infinity the magnifying power would be (for $q = -\infty$)

$$MP = \frac{25 \text{ cm}}{f} - \frac{1}{q} = \frac{25 \text{ cm}}{f} = \frac{25 \text{ cm}}{5 \text{ cm}} = 5$$

This wide range in position of the image leads only to a difference of magnifying power of 1. Therefore, when a lens is rated in terms of magnifying power, the 1 is usually dropped and the magnifying power is usually calculated only in terms of 25 cm/f. The shorter the focal length of a lens the greater its magnifying power. A 5 cm focal length lens will thus have an *MP* of 5 while a 2.5 cm lens will have twice the magnifying power or 10. The magnifying power of a single lens is thus limited by the ability to use lenses of shorter and shorter focal lengths. To have a single lens with a magnifying power of 100 would require that it have a focal length of approximately 1 mm. Use of such a short focal length lens as a simple magnifier is virtually impossible.

THE COMPOUND MICROSCOPE

The compound microscope combines the concepts of magnification and magnifying power to produce an instrument enabling the user to produce magni-

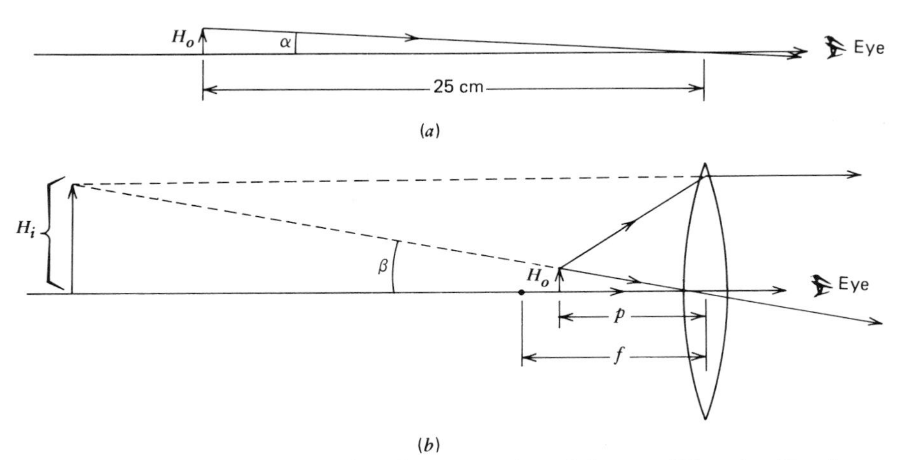

Figure 7.13 (*a*) An object viewed at the near point of the eye (25 cm) without a magnifier. The object's angular size (its apparent size to the eye) is given by the angle $\alpha = H_o/25$. (*b*) The same object viewed through a magnifier. Now the angular size of the image is given by the angle $\beta = H_o/P$.

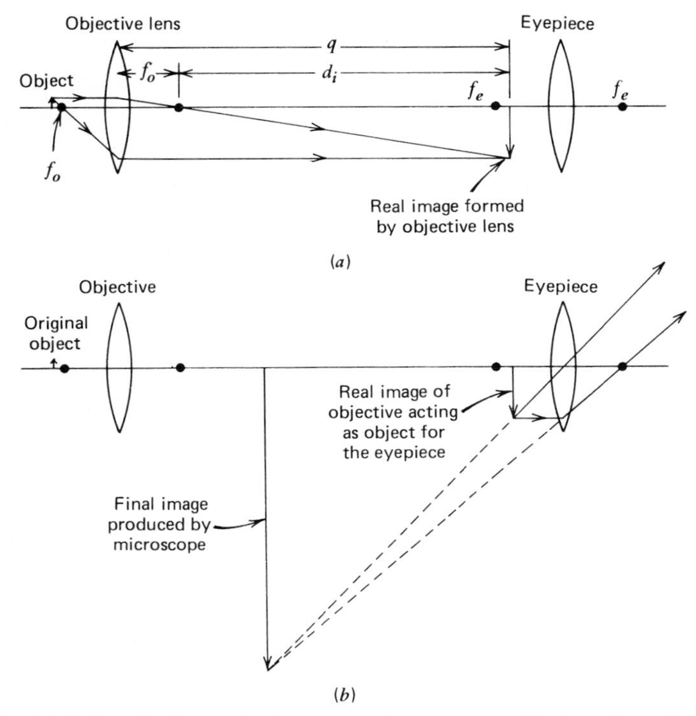

Figure 7.14 (*a*) Objective lens of microscope forms a magnified real inverted image just inside the focal point of the eyepiece. The focusing of the microscope amounts to moving the object slightly closer to or farther from f_o until the real image is in fact located just inside f_e. (*b*) The eyepiece uses the real image produced by the objective lens as an object. The eyepiece magnifies this "object" according to the simple magnifier formula.

fying powers of several hundred or even more. In principle, two lenses are combined so that the image of the object produced by the first lens serves as the object for the second lens.

A ray diagram showing how this process works is shown in Figure 7.14. In part (*a*) the *objective lens* (0) forms a real image of an object that is positioned just beyond the focal point of the lens. The image of the object will undergo magnification. The amount of this magnification is H_i/H_o, which we found before could be written as

$$M = \frac{H_i}{H_o} = \frac{d_i}{f} \quad \text{This is the same as } \frac{H_i}{H_o} = \frac{f}{d_o} \text{ since } d_i d_o = f^2.$$

But using

$$d_i = q - f_o \qquad f_o = \text{focal length of objective}$$

we get

$$M = \frac{H_i}{H_0} = \frac{q - f_o}{f_o} = \frac{q}{f_o} - \frac{f_o}{f_o}$$

or

$$M = \frac{q}{f_o} - 1$$

Most microscopes are designed so the image produced by the objective lens is focused at a distance of 16 cm from the lens; thus the magnification becomes

$$M = \left(\frac{q}{f_o} - 1\right) = \left(\frac{16\text{ cm}}{f_o} - 1\right)$$

This magnified *real* image now serves as the object for the ocular lens or eyepiece as shown in Figure 7.14*b*. The image is positioned just inside the focal point of the eyepiece lens so that a virtual image will be formed of the objective image. The eyepiece will act like a magnifying glass looking at an object. Its magnifying power to the microscope user will be $MP = (25/f_e + 1)$. The final image and its relationship to the initial object are shown in part (*b*) of the diagram. The total magnifying power achieved by the instrument is the magnification of the first lens times the magnifying power of the second lens, that is,

$$MP = M_{\text{obj}} \times MP_e$$

$$MP = \left(\frac{16\text{ cm}}{f_o} - 1\right) \times \left(\frac{25\text{ cm}}{f_e} + 1\right)$$

For example, suppose a microscope had an objective lens with a focal length of 1.6 cm and an eyepiece with a focal length of 2.5 cm. Its magnifying power would be

$$MP = \left(\frac{16\text{ cm}}{f_o} - 1\right) \times \left(\frac{25\text{ cm}}{f_e} + 1\right)$$

$$= \left(\frac{16}{1.6} - 1\right) \times \left(\frac{25}{2.5} + 1\right)$$

$$= (9)(11) = 99$$

The formula for the magnifying power is often written without the ones in both the parentheses since the numerical results are almost the same. Then $MP = (16\text{ cm}/f_o) \times (25\text{ cm}/f_e)$.

TELESCOPES

A telescope is a lens system used to examine distant objects. As in a microscope, two lenses are involved, with an eyepiece to enlarge the image produced by the objective lense. A telescope's objective lens, however, has a long focal length as compared to that of the microscope, which is very short.

As with the microscope the objective lens produces a real image of the object. However, since the object being viewed is very distant from the lens, the image that is produced is effectively located at the focal point of the objective lens and the image at this point is very small. This image is then viewed through the eyepiece or ocular as with the microscope. Here the object viewed is distant, and the size of the final image is generally very much smaller than the size of the original object. However, angular magnification is achieved. The extent of this angular magnification, or magnifying power, can be determined by comparing the angular size of the image as seen through the telescope with the angular size of the image that is seen with the unaided eye. Figure 7.15*a* shows an eye looking at an object of height H_o at a distance of p. The angular size of the image as seen by the eye is just

$$\alpha = \frac{H_o}{p}$$

Figure 7.15*b* illustrates the main features of an astronomical telescope. The first lens (the objective lens) has a focal length of f_1, and produces a real inverted image of height H_i at f_1. In this case, H_i is much smaller than H_o. But, if we use a short focal length lens (the eyepiece) to look at H_i, we can create a large

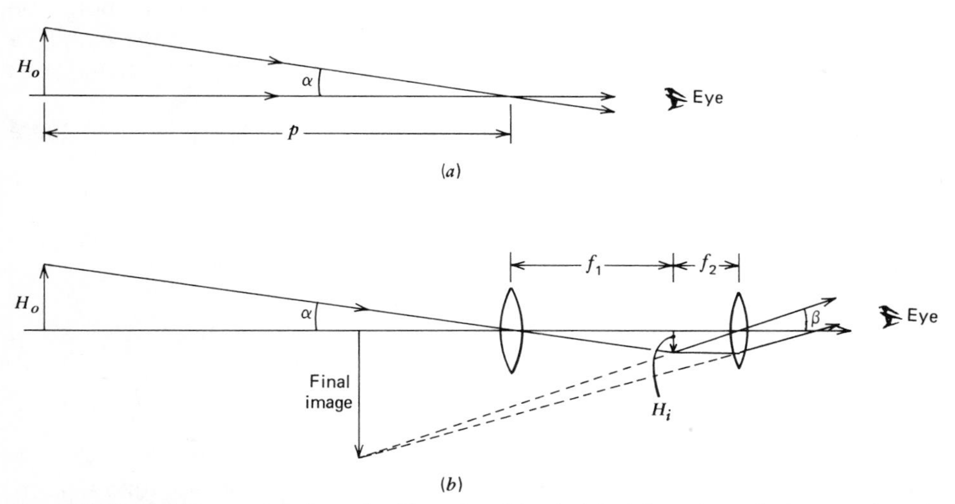

Figure 7.15 (*a*) The unaided eye looking at a distant object (imagine p to be very large). The angular size of the object is given by $\alpha = H_o/p$. (*b*) The same object at the same distance being viewed through an astronomical telescope. Now the angular size of the image is given by $\beta = H_i/f_2$. In (*b*) it is clear that α can also be written as $\alpha = H_i/f_1$. Thus the magnifying power of the telescope is $MP = \beta/\alpha = f_1/f_2$.

image. The eyepiece is placed so that H_i falls just inside the focal length of the eyepiece, f_2. Thus H_i acts just like an object of height H_i being observed by a magnifier of focal length f_2. The apparent angular size of H_i can be seen from Figure 7.15*b* to be equal to

$$\beta = \frac{H_i}{f_2}$$

This is the apparent angular size of the distant object H_o when seen through the astronomical telescope. The original unaided angular size of the object is α, which from Figure 7.15*a* is seen to be equal to H_o/p. But in Figure 7.15*b* it is also apparent that α must be given by

$$\alpha = \frac{H_i}{f_1}$$

The magnifying power of the telescope is just

$$MP = \frac{\beta}{\alpha} = \frac{(H_i/f_2)}{(H_i/f_1)} = \frac{f_1}{f_2}$$

$$MP = \frac{f_1}{f_2}$$

Thus the magnifying power of an astronomical telescope is just given by the ratio of the objective focal length to the eyepiece focal length. It should be noted that the objective lens produces an inverted image. Since the eyepiece (acting as a simple magnifier) does not reinvert the image, the final image seen by the eye is inverted.

If a terrestrial telescope is desired (one that produces an upright image), there

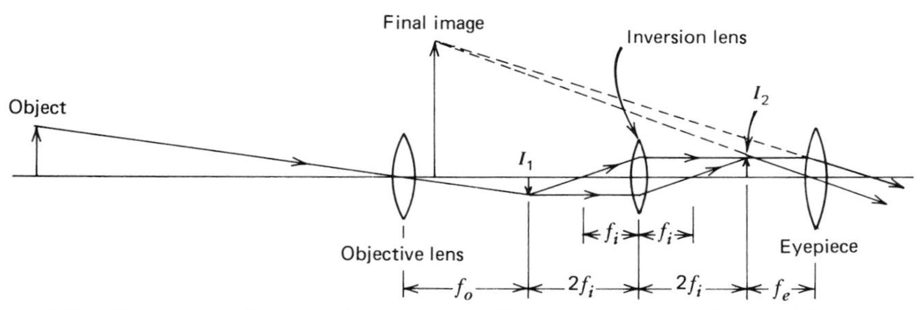

Figure 7.16 One type of terrestrial telescope. An inversion lens is inserted into an astronomical telescope. If no additional magnification is desired, the inversion lens is located so that the image produced by the objective falls a distance $2f_i$ from the inversion lens. The inversion lens then produces a new unmagnified upright image at a distance of $2f_i$ on the opposite side of the inversion lens. The eyepiece then magnifies this image exactly as in the case of the astronomical telescope. Note that the inclusion of the inversion lens increases the length of the telescope by an amount $4f_i$.

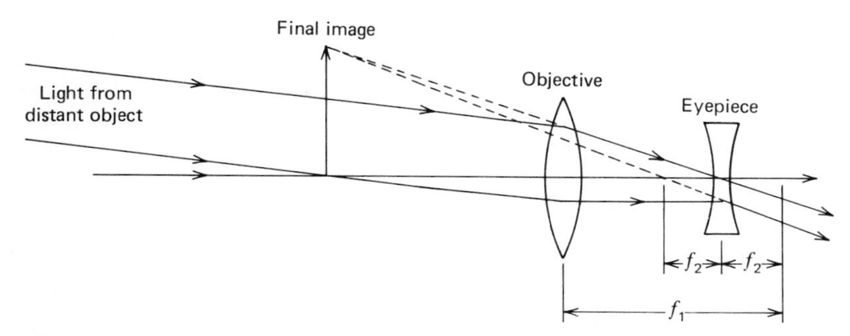

Figure 7.17 Galilean telescope. Converging light from the objective lens is intercepted by the concave eyepiece before a real image is formed. The eyepiece causes this light to diverge and produce an upright image as shown.

are several ways to make one. First, simply insert an additional lens into an astronomical telescope as in Figure 7.16. If the focal length of the additional lens is f_i, this procedure will increase the length of the telescope by at least a distance of $4f_i$ (this is shown in Figure 7.16, and also results in no additional magnification). An alternate way of producing a terrestrial telescope is to construct the so-called Galilean telescope. This telescope uses the standard convex objective lens of relatively long focal length f_1. The eyepiece, however, is a concave lens that is placed *inside* the focal length of the objective lens so that the rays of light from the objective are intercepted *before* they form an image. These rays are caused to diverge by the concave eyepiece so that an upright virtual image is formed. Figure 7.17 shows the details of a Galilean telescope. If the focal length of the eyepiece is $-f_1$ (minus because it is a concave lens), the magnifying power of the Galilean telescope is the same as that of an astronomical telescope with the same focal lengths.

To most people the magnifying power of a telescope is its most obvious characteristic. However, the large objective lens of most telescopes produces two other beneficial effects. The first of these is the greatly enhanced light gathering power of the telescope. The objective lens has a diameter many times larger than the pupil size of a typical eye. The amount of light collected and brought to focus by the objective lens is proportional to the *area* of the lens opening, as is also true for the eye. Thus a lens with a diameter of 10 cm will collect much more light than the unaided eye, which has a diameter of only about 0.5 cm. In fact, the 10 cm lens will collect more light by a factor of

$$\text{Increased light factor} = \frac{\frac{1}{4}\pi D_o^{\,2}}{\frac{1}{4}\pi D_{\text{eye}}^2} = \frac{D_o^{\,2}}{D_{\text{eye}}^2} = \frac{(10\text{ cm})^2}{(0.5\text{ cm})^2} = 400$$

On the other hand, if the 10 cm objective is coupled with an eyepiece such that the telescope's magnifying power is 5, this magnification will increase the height *and* width of all images on the retina by a factor of 5. The *area* of an image on the retina will thus be increased by a factor of 5^2 or 25. Such an increase will spread the light out and *reduce* its intensity by this same factor of 25. The overall increase in the intensity of the retinal image will thus be (in the case of a 10 cm objective)

$$\text{Overall increased light factor} = \frac{400}{25} = 16$$

This overall intensification of the image means that it is possible to see fainter objects with the telescope than with the unaided eye.

The second benefit of increasing the diameter of the objective lens is the increase in resolution this brings about. Resolution is the ability of an optical instrument to separate clearly the images of two closely spaced objects. If the *minimum* angular separation that can be resolved by an optical instrument is σ_{min} then the resolving power of the instrument is said to be

$$RP = \frac{1}{\sigma_{min}}$$

where σ_{min} is expressed in radians (1 radian = 57°). For example, suppose a telescope can just barely distinguish two objects that are 0.57° apart. We see that 0.57° = 0.01 radian. Thus in this example

$$RP = \frac{1}{0.01} = 100$$

Calculations show that for an astronomical telescope the resolving power is controlled directly by the diameter of the objective lens. In fact,

$$RP \propto D_o$$

Thus, for example, doubling the diameter of the objective lens will double the resolving power and will, thus, allow objects that are more closely spaced to be clearly resolved. If we combine the equation that defines resolving power with the fact that resolving power is proportional to lens diameter, then we can see that σ_{min} and D_o are related by

$$\sigma_{min} \propto \frac{1}{D_o}$$

THE REFLECTING TELESCOPE

Although terrestrial telescopes and small astronomical telescopes use pairs of lenses (Figures 7.15, 7.16, and 7.17) as the main elements of the instrument, large astronomical telescopes replace the objective lens by a concave mirror (review Section 6.3). There are three basic reasons for this. First, all lenses introduce chromatic aberration into the final image while mirrors avoid this problem.

Second, more light is lost in passing through an objective lens than in being reflected from an objective mirror. Finally, since light gathering power is so important in astronomy, it is important to have a large diameter objective element. Supporting a large diameter lens is very difficult since all glasses are somewhat plastic and tend to sag slightly under their own weight. It is much easier to support a large mirror.

One of the problems with a reflecting telescope is getting the light to a position where it can be observed. If the telescope is very large, it is actually possible to *climb inside* the telescope and place an eyepiece in the proper location for observing (Figure 7.18). The large telescope at Mount Palomar has a mirror diameter of 200 inches (over 16 feet), making it large enough for a person to climb into. If it is not possible or convenient to climb into the telescope itself, then it is necessary to get the light out of the telescope before an image is formed. Two common ways of doing this are illustrated in Figures 7.19*a* and 7.19*b*.

Most people imagine astronomers peering through their telescopes night after night. Actually, most observing is done on film. To convert an observational telescope into a photographic telescope one need only remove the eyepiece and place a piece of film at the focal point of the objective mirror (Figure 7.20). With the film in this position an exposure of several hours can be taken

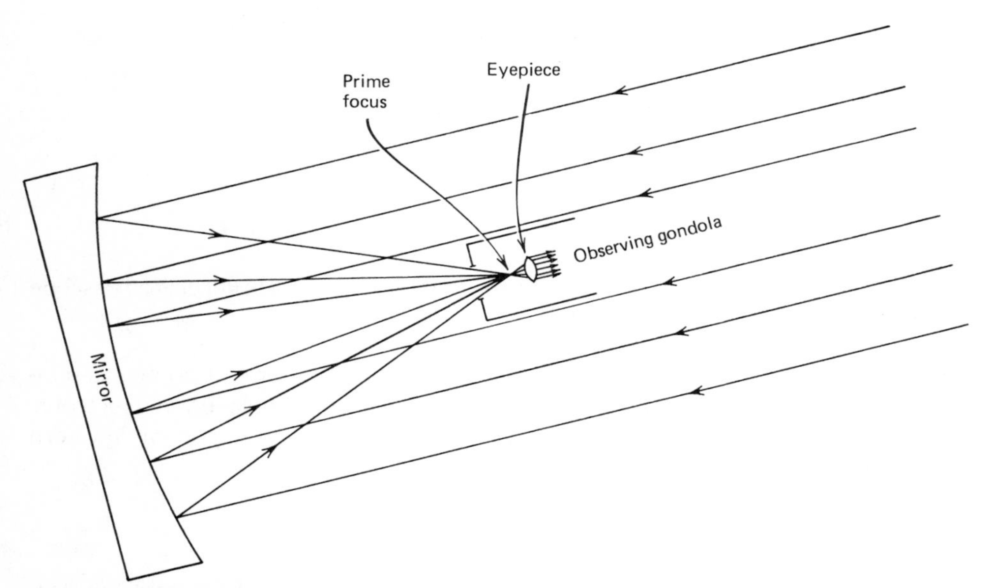

Figure 7.18 A large reflecting telescope. With a very large telescope, an observing gondola is actually constructed *inside* the telescope so that an eyepiece can be placed just outside the *prime focus*. The observer rides inside the gondola.

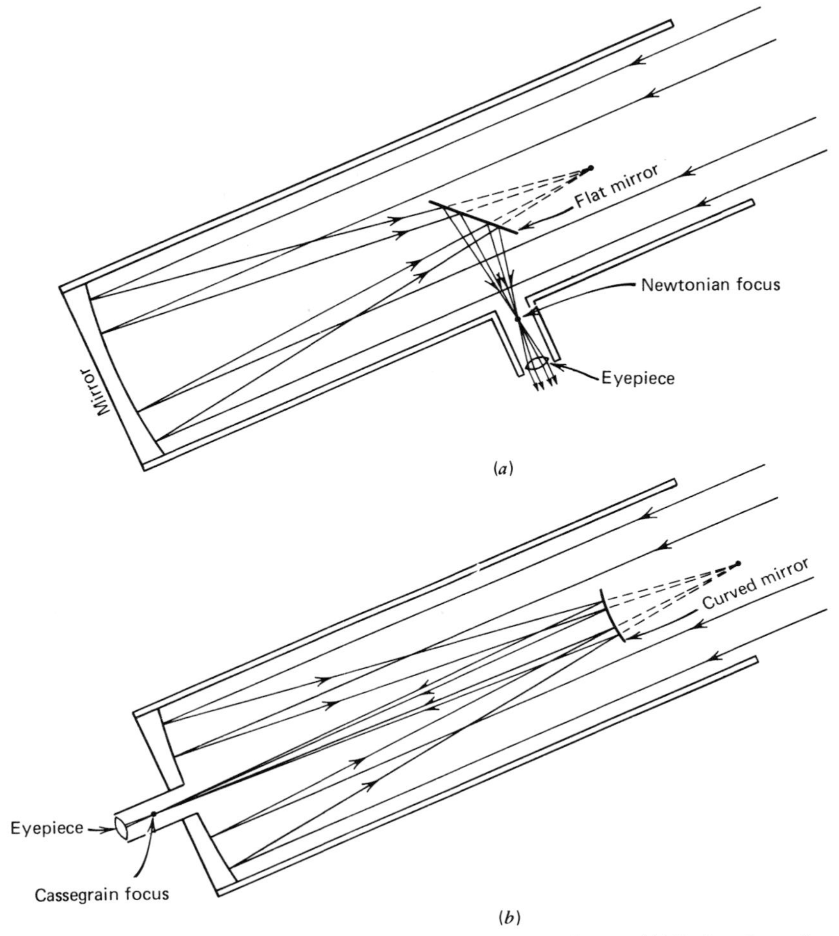

Figure 7.19 (*a*) Reflecting telescope using the Newtonian focus. (*b*) Reflecting telescope using the Cassegrainian focus.

if necessary. In this way the light of incredibly faint astronomical objects can be recorded on film. More recently, the use of film has been replaced by electronic image intensifiers that have greatly extended the sensitivity of modern telescopes.

THE SLIDE PROJECTOR

The final optical device that we shall discuss is the common slide projector. Figure 7.21 shows a schematic diagram of a simplified slide projector. Light from the bulb (both directly and by way of the mirror) goes to the condensing lens where it is slightly converged to illuminate the slide. The highly illuminated slide

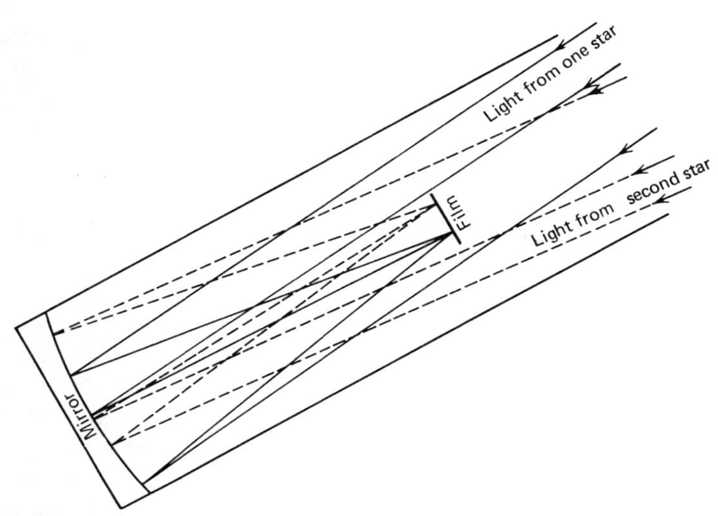

Figure 7.20 Reflecting telescope with film placed at the prime focus. Light coming from stars in different parts of the sky (the solid and dashed lines, respectively) will focus on different parts of the film, thus producing an accurate picture of the sky.

then acts as an object for the projection lens. If the slide is placed just outside the focal point of the projection lens, the image produced will be real, magnified several times, and inverted. The location of this image must be made to coincide exactly with the location of the screen on which the image is to be displayed. By varying the distance, *d*, between the slide and the projection lens, the location of the real image can be adjusted until it coincides with the screen. This is

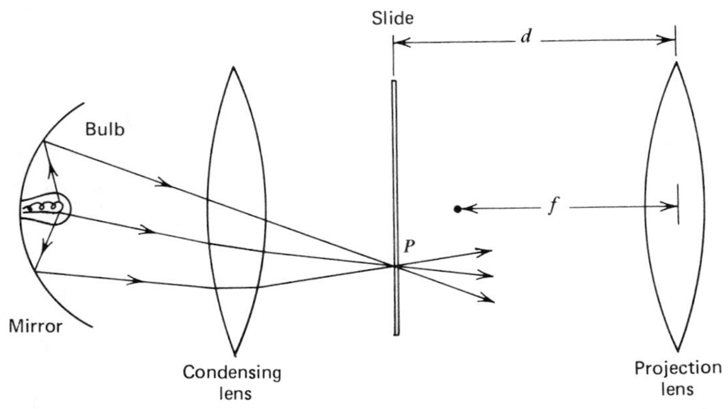

Figure 7.21 Schematic diagram of a simple slide projector. Light from the bulb follows many different paths to each point (such as *p*) on the slide. Light passing through each point on the slide is focused by the projection lens at a unique point on the screen. The slide in effect acts as an object for the projection lens. The exact location of the projected image is controlled by varying the distance *d*.

how a slide projector is focused. Since the real image is inverted, the slide itself must be placed into the projector in an inverted position so that the final image will appear upright.

PROBLEMS AND EXERCISES

1. The eye and the camera are each capable of focusing on objects at varying distances. Compare the method the eye uses to focus on these objects to the method used in a camera.
2. Under bright light conditions the pupil of the eye might have a diameter of 2 mm. At dusk it dilates to 6 mm. How does the area of the pupil in the bright light compare to its size at dusk? What variation in the intensity of light does this allow you to make? If the change in the intensity of light is greater than this, how does the eye now adjust so your vision continues? What effect does this last change make in your vision?
3. Why do you seldom find a young person who wears glasses in need of bifocal lenses but find it quite commonly in older people.
4. Many older people have difficulties distinguishing between navy blue and black objects. Explain why this is true.
5. Compare and contrast the compound microscope and astronomical telescope.
6. Discuss why it is difficult if not impossible to read from a newspaper in twilight or moonlight.
7. A successful picture is taken using an exposure of 1/100 of a second and a camera lens setting of *f*/8. The photographer wishes to repeat the shot, with his lens "stopped down" to *f*/16. What should the exposure time be?
8. While taking pictures you can make adjustments in your aperture openings to high or low *f*-number, your shutter speed to very fast (1/500 sec) or very slow speeds (1 or 2 sec), and your film from a high ASA number to a low ASA number. Give reasons why a photographer would desire to make such changes, that is, what advantages would there be to each condition listed?
9. You are provided with two lenses of focal length 10 cm and 1 cm. If you construct a telescope using these lenses, what magnification might it achieve? Sketch this telescope showing the location of the lenses, object, and image.
10. A photographer uses a camera with 50 mm focal length lens to photograph a distant object. He then uses a 150 mm focal length lens to photograph the same object. How will the height of the object compare on the two resulting photographs? How do the areas compare?

8 WAVE OPTICS

In the preceding two chapters we discussed the laws of geometrical optics and then used these laws to analyze the performance of several optical devices and systems. Geometrical optics is, of course, a simplification of the true situation. As we saw in Chapter 1, light does, in fact, exhibit wavelike properties that become apparent under certain circumstances. An understanding of optical devices that operate under such circumstances requires that we take the wave-like properties of light into account. Let us begin our discussion of wave optics with a more detailed look at Young's two slit interference experiment, which we considered briefly in Chapter 1.

8.1 YOUNG'S TWO SLIT EXPERIMENT

In Section 1.4 we described how waves of equal wavelength can interfere with each other. For example, two waves that are "in phase" with each other (such as shown in Figure 8.1*a*) will reinforce and are said to interfere constructively. On the other hand, two waves that are completely "out of phase" (Figure 8.1*b*) will cancel each other (assuming they are of equal strength or amplitude), and are said to interfere destructively. Of course, two waves need not be completely in phase or completely out of phase. In these intermediate cases the resulting interference will be between the extremes shown in Figure 8.1.

Young's two slit experiment is designed to show the interference between two light "waves" of equal wavelength. Figure 8.2 illustrates a simple modern version of Young's experiment. An opaque screen is prepared with two narrow parallel slits cut in it. Generally the widths of the slits are equal to each other and are slightly less than the wavelength of visible light. The separation between the two slits is usually a few times the wavelength of visible light and thus the slits appear very closely spaced. Incident light of a pure color or wavelength illuminates the slits from one side of the screen. In setting up Young's experiment, it must be arranged that the incident light reaches both slits in phase. The light then diffracts through both slits and passes into the region on the other side

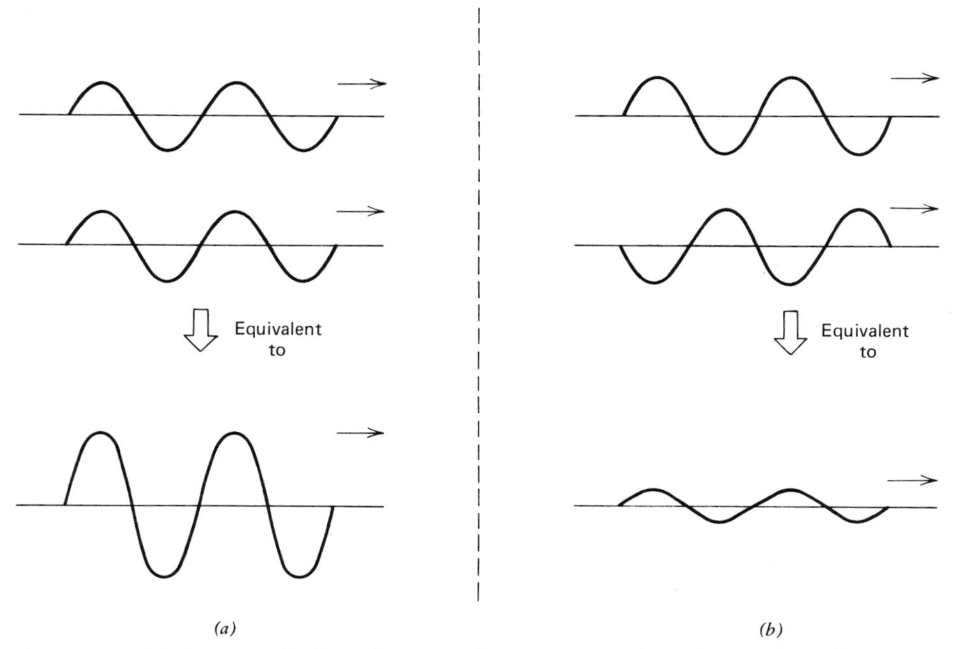

Figure 8.1 (*a*) Constructive interference of two waves of nearly equal amplitude; (*b*) destructive interference of two waves of nearly equal amplitude (the wavelengths are identical in both cases).

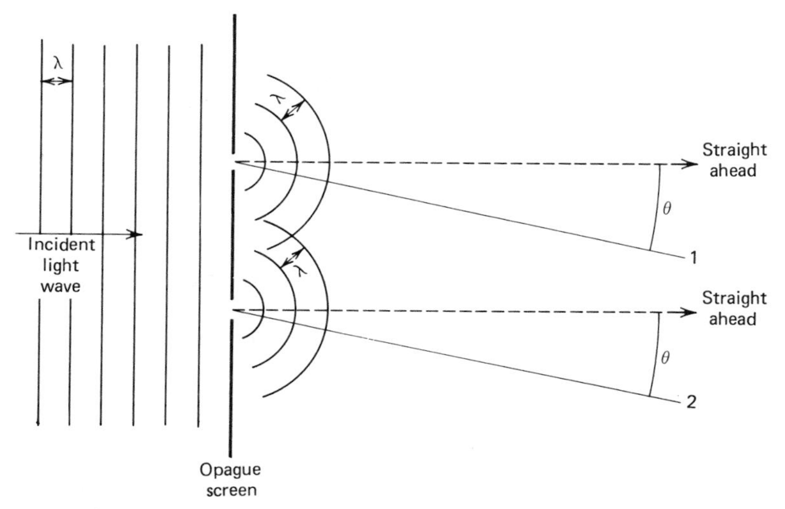

Figure 8.2 Schematic of a two slit demonstration of the interference of light waves.

of the screen. Because the slits are so narrow, the light that diffracts through the slits spreads out over a wide range of angles. If we define straight ahead as zero degrees, then the light leaving each slit can be described by an angle θ as shown in Figure 8.2. We shall generally be concerned with the light that leaves the slits at the *same angle* (i.e., in the same direction), as drawn specifically in Figure 8.2. Let us now examine what happens to light that leaves the two slits in the same direction (such as along lines 1 and 2 in Figure 8.2).

Figure 8.3*a* shows the geometry of Figure 8.2 in greater detail. First, let us note that lines 1 and 2 are parallel. Light traveling along these lines thus consists of parallel light. In order to determine whether this light is in phase or out of phase, we must bring it together at a common point. This can be done with a lens oriented perpendicular to the light traveling along lines 1 and 2, as shown in Figure 8.3*a*. Now let us recall that the light reaching slits *A* and *B* of Figure

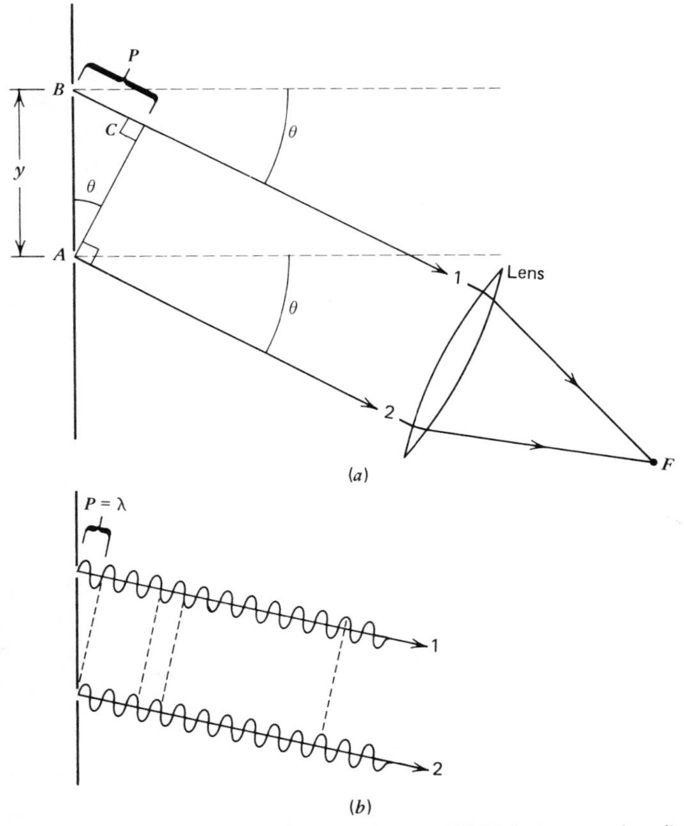

Figure 8.3 (*a*) Geometry of the two slit experiment. (*b*) Light leaves the slits in *just the right direction* so that $p = \lambda$, resulting in constructive interference. Note that the peaks and troughs of both 1 and 2 are exactly in phase.

8.3*a* is in phase as it reaches the slits. However, the light that passes through slit *B* must travel *farther* to reach the common point *F* than the light traveling along line 2. Thus light from slit *B* falls behind the light from slit *A*. How far behind? Clearly, from the diagram, this distance is the distance from *B* to *C*, which is labeled p. This distance p (the *path difference* between light along line 1 and line 2) will be different for light leaving the slits at different angles θ. We can calculate this specifically if we recall our trigonometry. Notice that triangle *ABC* is a right triangle with p as one leg and y (the distance between the two slits) as the hypotenuse. Line *AC* is perpendicular to both lines 1 and 2. Thus the angle labeled θ inside the triangle is the same angle θ that defines the direction of lines 1 and 2. Our trigonometry now lets us write

$$\sin \theta = \frac{p}{y}$$

or

$$p = y \sin \theta \qquad (1)$$

Equation (1) tells us exactly how much light traveling along line 1 lags behind that traveling along line 2. Of what use is this? Consider what must occur if we are to get constructive interference. Light traveling along line 1 must be exactly in phase with light traveling along line 2. That is, p must be zero, or p could be exactly one wavelength (see Figure 8.3*b*), or two wavelengths, or three, and so on. That is, constructive interference (what is called an *interference maximum*) occurs whenever

$$p = n\lambda \qquad n = 0,1,2,3, \text{etc.} \qquad (2)$$

combining this with eq. (1) gives

$$n\lambda = y \sin \theta$$

or

$$\sin \theta = \frac{n\lambda}{y} \qquad n = 0,1,2,3, \text{etc.} \qquad (3)$$

Equation (3) tells us how to compute all those directions for which constructive interference occurs. If we examine these directions experimentally using a lens as in Figure 8.3*a*, we will find a bright spot if we put a screen at point *F*.

It is worth pausing here to reexamine Figure 8.2. In drawing lines 1 and 2, we have arbitrarily drawn them below the straight ahead direction. They could just as easily have been drawn above the straight ahead direction at the same angle θ. That is, the light diffracts symmetrically both up and down. Thus there will be a series of interference maxima (and minima) symmetrically arranged on either side of the straight ahead direction. Equation (3) gives the position of these maxima on either side of straight ahead.

Now what about destructive interference? This occurs whenever p amounts

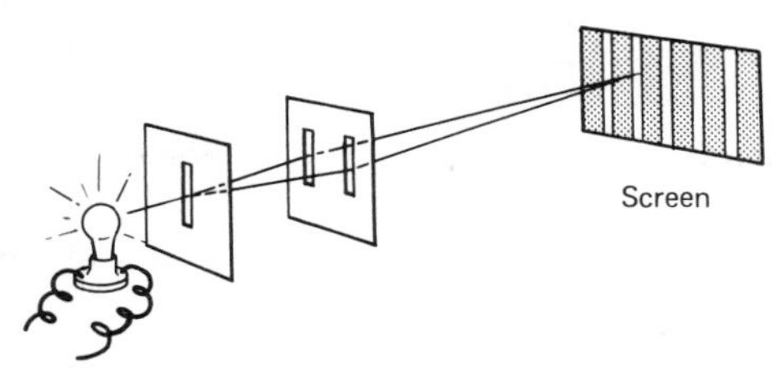

Figure 8.4 An interference pattern formed by passing light through two narrow slits. The first single slit is used to ensure that the light reaching each slit of the double slit pair is in phase with the other slit of the pair.

to exactly one-half of a wavelength, or one and one-half wavelengths, or $2\frac{1}{2}$, or $3\frac{1}{2}$, and so on. That is, destructive interference (what is called an *interference* minimum) occurs whenever

$$p = (n + \tfrac{1}{2})\lambda \qquad n = 0,1,2,3, \text{ etc.} \tag{4}$$

Combining this expression with eq. (1) gives

$$\left(n + \frac{1}{2}\right)\lambda = y \sin \theta$$

or

$$\sin \theta = \left(n + \frac{1}{2}\right)\frac{\lambda}{y} \qquad n = 0,1,2,3, \text{ etc.} \tag{5}$$

Equation (5) tells us how to compute all those directions for which destructive interference occurs. If we examined these directions with our lens, we would find a dark spot at point *F*. Figure 8.4 shows a typical two slit interference pattern (the lens is omitted for simplicity and is not, in fact, needed if the screen is far enough from the slits).

A close examination of an interference pattern such as that of Figure 8.4 shows that it consists of alternating bright and dark lines corresponding to where constructive or destructive interference occurs. These lines are not particularly sharp, but rather blend slowly into one another. The spacing between adjacent bright (or dark) lines is controlled by the distance between the two slits and by the particular wavelength of light used. In fact, from eq. (3) for constructive interference,

$$\sin \theta = n\left(\frac{\lambda}{y}\right)$$

Thus it is actually the ratio of λ to y that determines the spacing of the lines. Thus, for a given wavelength, placing the slits closer together (reducing y) will increase the value of $\sin \theta$ (and thus of θ also). The interference pattern will therefore spread out to larger angles. The opposite will occur if the slits are moved farther apart.

As a numerical example, suppose we have two slits that are separated by a distance

$$y = 2 \times 10^{-4} \text{ cm}$$

Suppose also that we illuminate the slits with light of wavelength

$$\lambda = 550 \text{ nm} = 0.55 \times 10^{-4} \text{ cm}$$

What shall we see? The pattern will resemble that of Figure 8.4, with the specific location of the bright and dark lines determined by the values of y and λ. The angular location of the bright lines (the interference maxima) can be computed from eq. (3);

$$\sin\theta = \frac{n\lambda}{y} \qquad n = 0,1,2,3, \text{ etc.}$$

For $n = 0$ this gives $\sin\theta = 0$. Thus $\theta = 0$, and there is a maximum straight in front of the pair of slits, the *central maximum.* For $n = 1$, eq. (3) gives

$$\sin\theta = \frac{(1)\lambda}{y} = \frac{0.55 \times 10^{-4} \text{ cm}}{2 \times 10^{-4} \text{ cm}}$$

$$\sin\theta = \frac{0.55}{2} = 0.275$$

A sine table shows that the angle with a sine of 0.275 is about 16°. Thus there is a bright line at $\theta = 16°$ on either side of the central maximum.

In our example we can continue to compute the locations of the various maxima. Thus:

$$\text{for } n = 2 \qquad \sin\theta = \frac{2\lambda}{y} = 0.55$$

$$\text{or } \theta = 33.4°$$

$$\text{for } n = 3 \qquad \sin\theta = \frac{3\lambda}{y} = 0.825$$

$$\text{or } \theta = 55.6°$$

$$\text{but for } n = 4 \qquad \sin\theta = \frac{4\lambda}{y} = 1.1$$

Since $\sin\theta$ must be no more than 1.0, there is no maximum for $n = 4$ or higher. Thus in our particular example there will be interference maxima on either side of the central maximum at $\theta = 16°$, 33.4°, and 55.6°. The location of the minima can be computed from eq. (5). If we changed the value of either λ or y, the location of the maxima and minima would obviously be affected. What is not so obvious is the effect of increasing the *number* of slits.

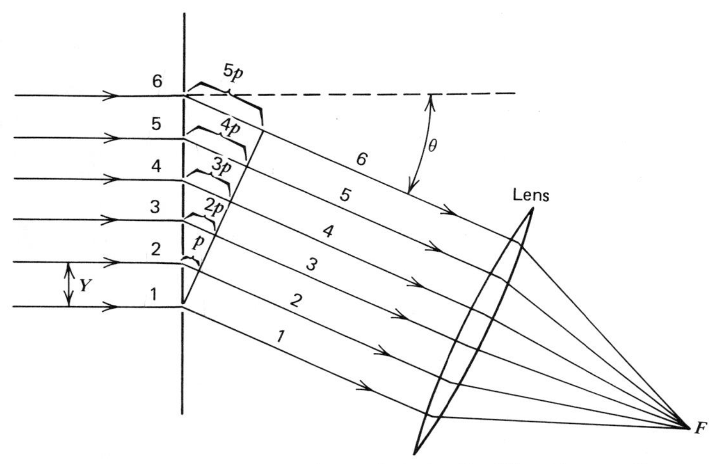

Figure 8.5 Light passing through a six slit diffraction grating.

8.2 THE DIFFRACTION GRATING

A series of many equally spaced narrow parallel slits is called a diffraction grating. Such a grating operates in the same basic fashion as the double slit except that we must now consider the interference of many more waves. As a concrete example let us consider a six slit grating. In Figure 8.5 light is shown exiting each slit in a specific direction given by the angle θ. Notice that each ray must travel a different distance from a slit to the point F. In fact, ray 2 must travel a distance p farther than ray 1, where p is given by eq. (1), which we previously developed for a double slit. That is,

$$p = y \sin \theta$$

Notice that ray 3 must travel a distance p farther than ray 2, and likewise for each succeeding ray. Now, constructive interference will occur when all six rays are in phase. This will be the case whenever

$$p = 0, \lambda, 2\lambda, 3\lambda, \text{etc.}$$

or

$$p = n\lambda \qquad n = 0,1,2,3, \text{etc.}$$

This leads to

$$n\lambda = \sin \theta$$

or

$$\sin \theta = \frac{n\lambda}{y}$$

This is exactly the same equation that we developed for the double slit. The conclusion is that increasing the number of slits *does not change* the location of the interference maxima. However, the intensity and the sharpness of the maxima will change. The intensity will increase substantially since many more waves are adding together constructively to produce each maximum. The fact that the maxima are much sharper can be understood from the following argument.

Suppose a grating had 1002 slits and we were considering the formation of a first order maximum (the $n = 1$ maximum). At the point of the first order maximum the difference in path length between adjacent slits would equal λ. Thus the ray coming from slit 1002 would differ in path length by 1001 λ from that of the ray coming from the first slit (Figure 8.6*a*). Now if we change the angle *very slightly* to θ' such that the path length of the ray from the top slit is increased by 1λ, becoming 1002λ longer than the ray from the beginning slit, the path length of the middle (501) slit under these conditions would become 0.5λ longer (Figure 8.6*b*). Thus ray 1002 and ray 501 would destructively interfere with each other, since they would differ in path length by 501.5λ (a half-integer number). Likewise ray 1001 and ray 500 would also cancel each other for the same reason. Similarly, each ray from the bottom half of the grating would cancel another from the top half of the grating where the path difference between the two was some half-integer number of wavelengths. This cancel-

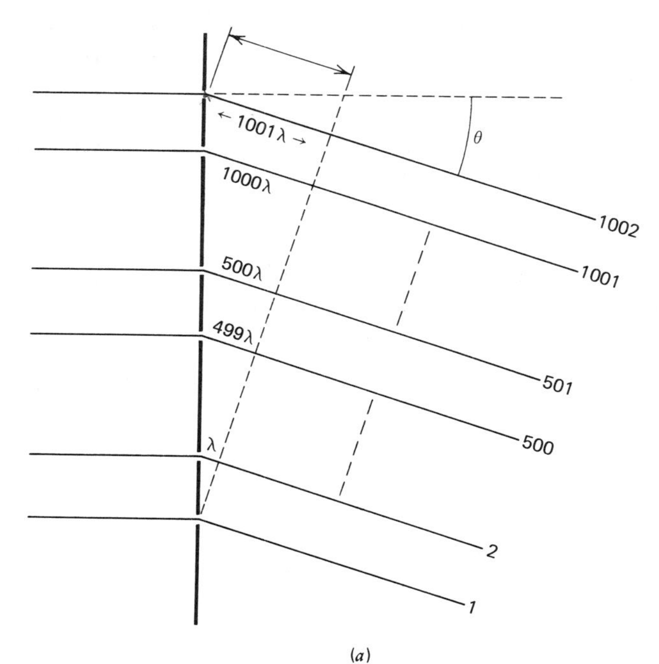

Figure 8.6*a* Grating of 1002 slits. All waves are in phase at angle θ.

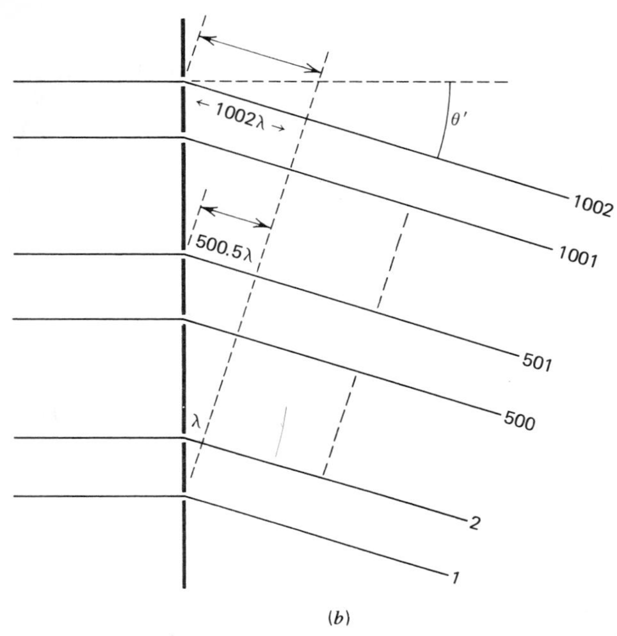

(b)

Figure 8.6b Grating of 1002 slits at slightly different angle θ. Rays cancel in pair to give minimum intensity.

lation at the slightly different angle gives rise to a much sharper interference pattern. That is, we go from a maximum to a minimum in a *very small* angular distance, much smaller than for the double slit. Since, as we have seen, the spacing of the maxima is unchanged as we increase the number of slits, the minima between adjacent maxima are very wide. That is, the maxima appear as sharp bright lines separated by wide dark regions. If the same variation in angle is considered for only two slits some cancellation will occur, but the reduction of the brightness or intensity will not be nearly as great. Figure 8.7 shows the qualitative difference between the interference pattern of a grating and a pair of very narrow slits. Interference gratings of several thousand slits can be produced, and the resulting interference pattern, for a particular wavelength, consists of maxima that are exceedingly sharp. In fact, a detailed calculation shows that the angular width of each maximum varies inversely with the number of slits. The location of these maxima is given by eq. (3).

If a light source consists of only a few wavelengths, these will each produce a series of sharp interference maxima or bright lines located according to eq. (3). By measuring the angles at which these maxima occur, and by knowing the slit spacing y, we can use eq. (3) to calculate the wavelengths of light in the source. Thus the diffraction grating is a very useful tool in the study of the spectral composition of light.

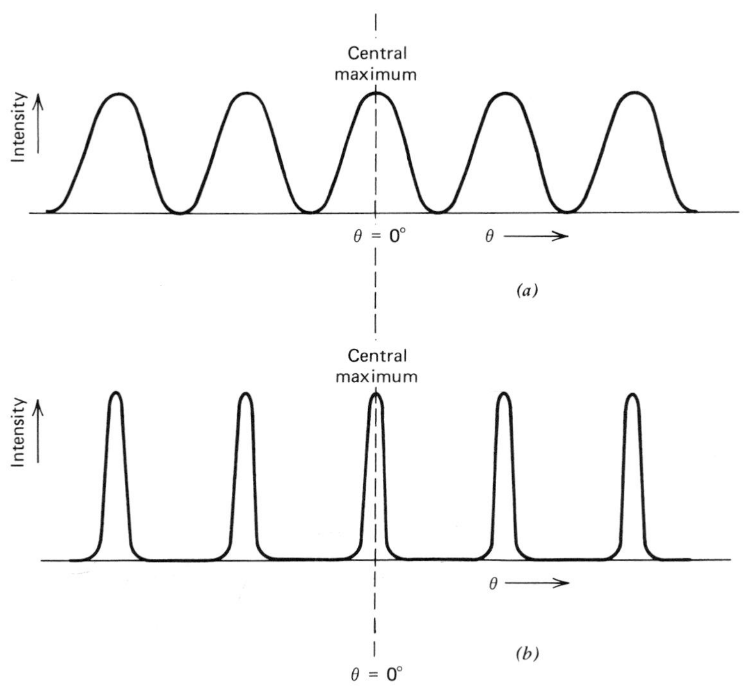

Figure 8.7 (*a*) Two slit interference pattern, (*b*) six slit interference pattern [same slit spacing as (*a*)].

8.3 SINGLE SLIT DIFFRACTION

Thus far in our discussion of wave interference we have been considering the effects of light passing through "narrow slits." The implication of a "narrow slit" as it has been used here is that it is narrow enough so that all the waves passing through it are in phase with one another from one side of the slit to the other. This is true if the width of the slit is small compared to the wavelength.

The question now to be considered is what happens to light if it is passed through a single slit that is not small in comparison to the wavelength of the light being considered. Examining the passage of light through this wider slit compels us to take into account the phase difference of rays moving through various parts of the opening. Figure 8.8 illustrates a "wide" slit, which in this case allows 10 rays of light (an arbitrarily chosen number) to pass through it. This type of slit will also produce an interference pattern. In this case, the interference pattern comes from interference between the component rays of a single broad wavefront. An interference pattern produced in this way is called a diffraction pattern.

Suppose, in Figure 8.8, we consider the pattern formed by rays leaving the slit at an angle θ such that $\lambda = W \sin \theta$ where W is the width of the slit. Under these conditions the tenth ray will be λ units longer than the first ray, making the sixth ray $\frac{1}{2}\lambda$ unit longer than the first ray. Since the sixth ray is $\frac{1}{2}\lambda$ longer than the first ray, the two rays will be out of phase and will destructively interfere with each other. Likewise the second ray will cancel the seventh, the third the eighth, etc. Thus the light passing through this wide slit will produce a point of minimum intensity at the angle θ. This cancellation of light that takes place is similar to the cancellation that occurs in a diffraction grating. The difference is that we now are considering a continuous group of many waves instead of the 1002 we considered in the discussion of the grating.

As we have seen, if the angle θ is chosen so that the path difference between rays from opposite sides of the slit is λ, we will have a position of minimum amplitude. Likewise, if the difference in path length between the extremes is 2λ, 3λ, or any integral number of λ's, a minimum will occur at an angle predicted by

$$n\lambda = W \sin\theta \qquad \text{for } n = 1,2,3,\text{etc.}$$

where n is the *order* of the minimum and W is the slit width.

Note that the above relationship does not give a minimum where $n = 0$ and $\sin\theta = 0$. This is because no path difference would exist at this position (directly in front of the slit) and all the rays would be in phase giving a central maximum. The diagrams of the diffraction patterns (Figure 8.9) clearly show this relationship. In general, most of the energy of the wave passing through the slit will lie

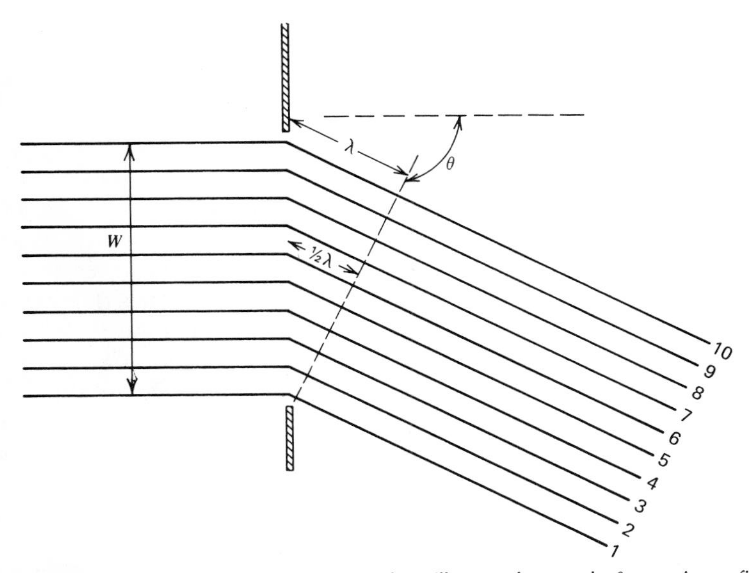

Figure 8.8 The light passing through a wide slit will cancel at angle θ, causing a first order minimum, provided $W = \lambda \sin\theta$. A second order minimum will occur at that value of θ for which $W = 2\lambda \sin\theta$, a third order minimum when $W = 3\lambda \sin\theta$, and so on.

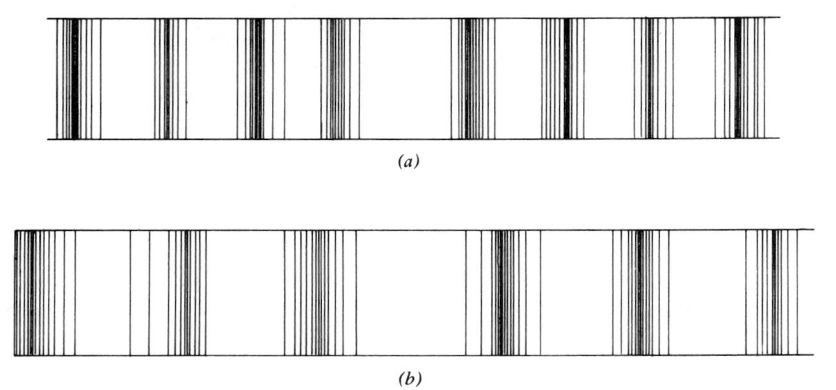

Figure 8.9 Diffraction patterns formed by coherent light passing through a narrow slit (*a*). If the slit is further narrowed the pattern spreads as in (*b*).

in the central bright band. As the slit width becomes greater the ratio $\lambda / W = \sin \theta$ will become smaller and the light will spread or diffract less. When the slit becomes wide in comparison to the wavelength ($\lambda / W \ll 1$) the diffraction is very small and is usually unnoticed. Since doors and windows are much larger than the wavelength of the light passing through, the light effectively travels in straight lines as it enters or leaves a room. These effects become noticeable only as the openings become small.

8.4 THE LASER

In Chapter 2 we examined the means by which an atom gives off light. We noted that, initially, some energy was absorbed, causing an electron to move to an excited state. The electron then deexcited, moving from the outer orbit to its normal lower orbit. In doing so it emitted the energy it had absorbed. This radiated energy of transition to a lower orbit was given off as a burst of light energy called a photon. Usually an electron will stay in an excited state for an extremely short time (about 10^{-8} seconds), but under certain conditions will be delayed in returning to the lower state. When the electron is delayed in returning to its ground state, it is said to be in a metastable state. The existence of such metastable states provides the basis for the operation of lasers.

The theory behind the laser was first developed in 1958, and the initial working model was constructed in 1960. Since then the news has been filled with items of the current uses and possible future development of this instrument. Its operation is described in the acronym naming it. *Laser* stands for *L*ight *A*mplification by *S*timulated *E*mission of *R*adiation. The first laser used a ruby crystal cut in the shape of a cylinder. One end of this cylinder was polished and silvered, making it totally reflecting. The other end was only partially silvered, thus making it able to reflect some light while permitting the rest of the light to pass through. A spiral flash tube was then placed around the ruby cylinder (Figure

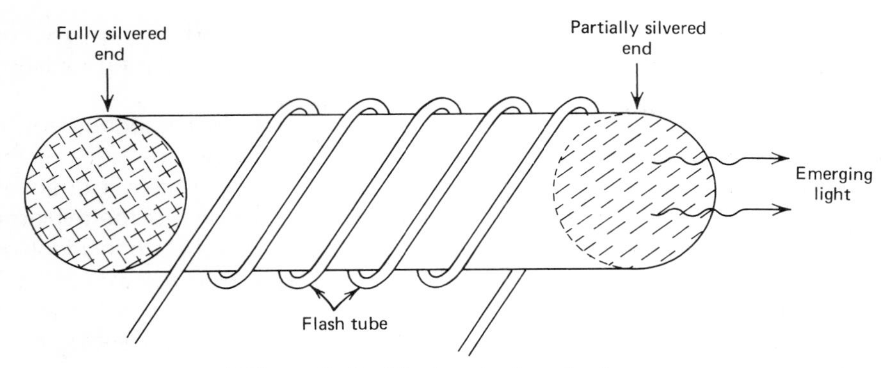

Figure 8.10 The heart of a ruby laser.

8.10) to act as a source of energy. The energy provided by the flash tube caused the electrons in the crystal to be raised to excited states, one of which is metastable.

The nature of ruby crystal is such that when the electrons remain in the metastable state they can be stimulated to emit their radiation in a particularly advantageous manner, giving the laser its unique properties. After a period of time one of the excited electrons will spontaneously drop to a lower orbit and emit a photon in the process. This photon will begin to travel through the ruby crystal until it encounters another atom. If this atom is in the ground state, the photon will probably be absorbed by the atom. However, if the atom that the photon encounters is in the same metastable excited state as was the atom that originally produced the photon, then this new atom will be stimulated to deexcite and emit a photon. The resulting photon is *in phase* and moves in the *same* direction as the photon that stimulated its emission (Figure 8.11). Since the new photon is in phase with its stimulator, the two photons will interfere constructively, and the result will be a wave of greater intensity. This wave will continue to move through the crystal, encountering more atoms as it goes. It will continue to grow by the process of stimulated emission if it encounters more atoms in the excited metastable state than in the ground state. This condition, of more atoms being in the excited rather than ground state, is called a *population inver-*

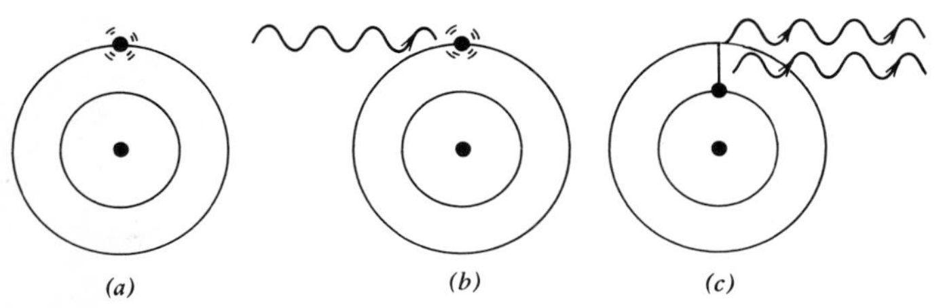

Figure 8.11 An excited atom in a metastable state (*a*) is stimulated by a passing photon (*b*) to give off its photon in phase with the stimulating photon (*c*).

sion since normally the ground state is more heavily populated than any excited state. The successful operation of a laser requires the creation of a population inversion.

The direction of the photon given off by the initial atom that is spontaneously deexcited is random. Thus the photon may pass out through the sides of the crystal. However, some spontaneously emitted photons will have a direction that is parallel to the axis of the crystal, and will move to one of the mirrored ends of the cylinder, striking it along a normal. If the photons are incident on the fully silvered end, they will be totally reflected parallel to the axis and will travel toward the other end of the ruby, stimulating deexitation of other atoms along the way. At the partially silvered end a portion of the incident light will pass through the mirror, while the rest is reflected, continuing the process until all atoms that were initially excited by the flash tube are deexcited. The laser light produced in this fashion has these unique properties:

- **a.** It is coherent – all the waves are in phase giving the light great intensity.
- **b.** The emerging light from the partially silvered end is extremely nondivergent, that is, it is unidirectional.

These characteristics make the laser an extremely useful source of light for a variety of applications, some of which will be discussed below.

This early pulsed type of laser has a major shortcoming in that it cannot give off a continuous beam light. Instead, its light comes in pulses, one for each flash of the flash tube.

Today the most often used type of lasers are gas lasers, which can be operated on a continuous basis. The most common of these lasers use a mixture of helium and neon gases to produce their laser action. The principal aspects of their operation are essentially the same as those of the ruby laser. A glass tube with mirrored ends, one completely silvered and one partially silvered, encloses the mixture of the gases. As with the fluorescent tubes studied in Chapter 2, a source of high voltage forces the electrons of the atoms to sweep back and forth. The collisions between the helium and neon atoms causes them to reach excited states. However, one of the excited states of helium is metastable and seldom is deexcited on its own. The most likely method by which it is deexcited

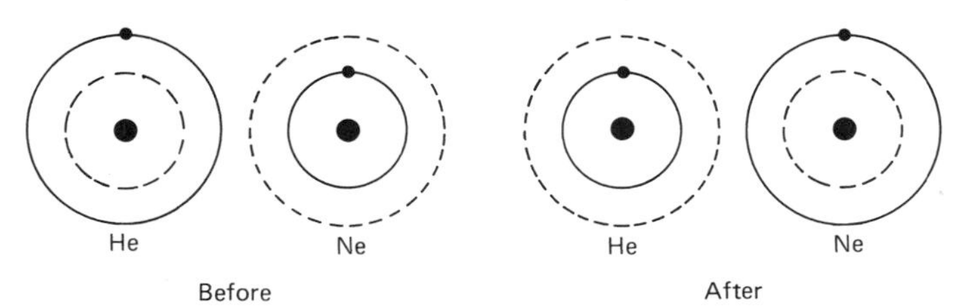

Figure 8.12 Before and after a collision between an excited helium atom and an unexcited neon atom.

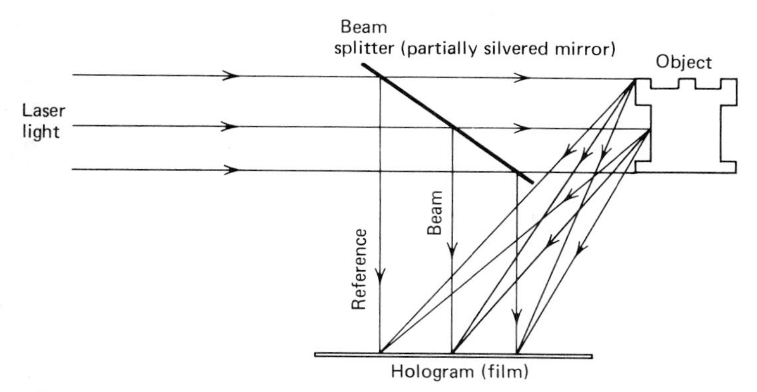

Figure 8.13 One geometric arrangement for the production of a hologram.

is through a collision with a neon atom. Such a collision results in the neon atom being raised to an excited state (Figure 8.12).

The consequence is that neon can be raised to an excited state in two ways: by collision with the surging electrons and by collision with excited helium atoms. The result is that a great many excited neon atoms can exist in the tube at any time. This permanent population inversion means that the laser can be continuously operated. This type laser has many more uses than the original type, one of which is the production of holograms.

THE HOLOGRAM

Conventional photography uses a lens to produce a two-dimensional image of a three-dimensional object. Holography is a method of recording on film a three-dimensional image of a three-dimensional object without the use of a lens. However, holograms must be made with a source of monochromatic light and usually must be viewed under a similar light. Understanding the method by which holograms are constructed gives a great deal of insight into how they are able to record the information they do.

Various geometric arrangements of light sources can be used to produce holograms. One simple arrangement is shown in Figure 8.13. In this type of arrangement a laser beam is initially split into two parts by shining it on a partially silvered mirror. After the light has been diverged by use of two concave lenses, one portion of the light, the reference beam, is allowed to shine directly onto the film. The second portion of the light illuminates the object whose image is to be recorded. Some of the light that strikes the object reflects back to the film. Every section of the film receives light from each illuminated part of the object. The light from the reference beam and from the object interfere with each other when they meet at the film. The result is a very complex interference pattern that is meaningless when viewed in normal light. To reconstruct the image stored in the interference pattern, light from a laser or other monochromatic source must be shown through the hologram. Shining the light produces a virtual image that can be viewed as shown in Figure 8.14. How such an image can be produced requires us to keep in mind that each part of the hologram film contains a complex pattern of light and dark areas that act like mini-diffraction gratings. Thus, monochromatic light passing through a *particular portion* of the hologram will be diffracted in a very definite direction. Calculations show that this direction turns out to be identical to the direction of the light reflected from the original object to this particular part of the hologram. Thus when the hologram is illuminated in its entirety, it reconstructs the entire original beam of light that was reflected from the object to the hologram. Since the hologram film has a definite width, different parts of the hologram "see" the object from different directions or perspectives. The reconstructed beam shown in Figure 8.14 preserves these perspectives, which can be seen by moving the head from side to side, thus giving the image its three-dimensional quality.

An alternate method of observing the image results in real images being projected on a screen as in Figure 8.15. These images are actually first order diffraction patterns of the information stored on the film. Again, different views of the original object can be obtained by moving the undiverged laser beam to different points on the film. Each different point will contain information about the illuminated portion of the object, but each point will give that information from a different perspective. Strange as it might seem, the hologram could be cut into many pieces and each individual piece could reproduce an image of the original object! Of course, with progressively smaller pieces the range of "views" of the original object is progressively reduced.

The method described here for producing a hologram would only enable approximately 50% of the surface of an object to be recorded on film. Alternate arrangements of light can be used to produce holograms giving a full 360° view

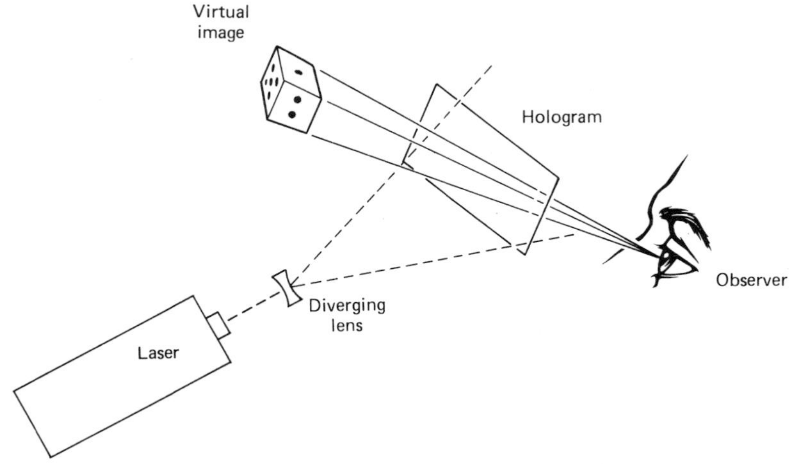

Figure 8.14 Viewing the virtual image of a hologram.

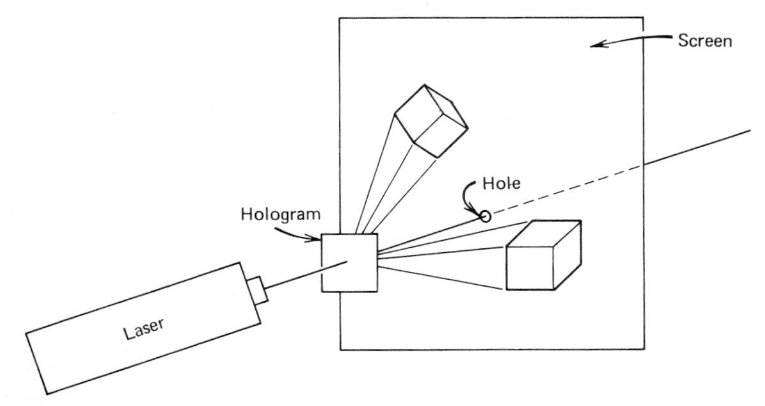

Figure 8.15 Viewing the real images of a hologram by projecting the undiverged laser beam directly through the hologram. The interference pattern of the hologram form first order diffraction patterns, which are the images.

of an object as the hologram is viewed from different positions. In the future, by the use of these techniques, possibilities exist for the production of films and television that are truly three-dimensional.

OTHER LASER USES

The fact that the light from the laser is so unidirectional can also be put to good use. Whenever problems requiring highly precise straight line adjustment arise, nothing can surpass the laser in producing a reference line whether it be to guide the driller of a tunnel under a river or to align airplane wings in construction. This nondivergence also enables the laser to be used to measure accurately movements of the earth by monitoring light reflected from strategically placed mirrors.

The total energy output of a laser can never exceed the energy input. However, since most of the energy output is channeled into the narrow beam, which can be further focused, it is very concentrated and hence extremely powerful. Consequently it is capable of being used as a cutting tool for extremely hard materials as well as an instrument for precisely cutting patterns of dresses and shirts.

There are also applications in the medical field. One of the most dramatic is the use of a laser for correcting a detached portion of the retina of the eye. In this procedure the patient, whose retina has become damaged, is moved into a position so the detached portion of the retina assumes its proper position. Light from a laser beam is then focused by the eye on the edges of the detached section. This intense beam then burns a very small spot on the retina. As the burn heals, it bonds the retina to the back of the eye, thus preventing further detachment of the retina (Figure 8.16). Other surgical procedures in hard to

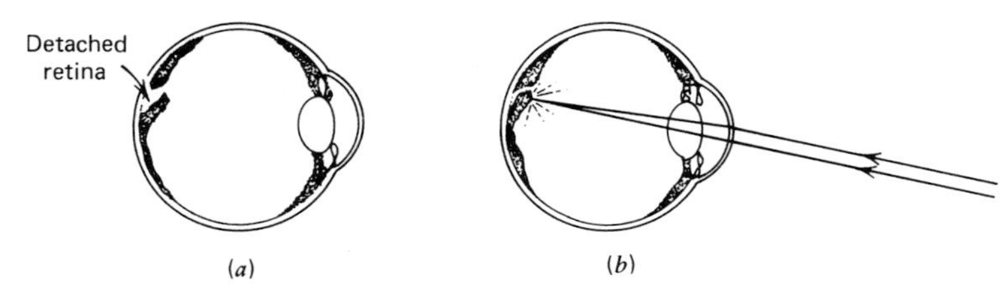

Figure 8.16 A laser can be used to treat a detached retina (*a*). Once the detached retina has been manipulated to its proper position, it is scarred by the laser beam (*b*). After healing the retina will detach no further.

reach parts of the body can also be done advantageously with a laser. For example, nonmalignant nodules, which sometimes appear on vocal cords, can be removed by focusing a laser, via series of mirrors, onto the nodule and burning it off. This procedure is much easier for the patient than other surgical alternatives.

Laser technology may also be combined with fiber optics for some types of internal treatment. For example, gastric bleeding, which must be stopped, can sometimes be treated in this way. A fiber optic gastroscope is inserted, and the cause of the trouble found. Laser light of high intensity can then be directed onto the bleeding area cauterizing it.

When working with lasers, great care must be taken never to permit an undiverged beam to shine into the eye because of its potential to damage that organ. Few people would be foolish enough to look directly into a laser beam, but you must also be alert to avoid reflected beams from shiny polished objects.

8.5 POLARIZATION

Another wave phenomenon that was briefly discussed in Chapter 1 was the polarization of light. Light can be thought to consist of an enormous number of photons of different frequencies or wavelengths. Each photon might be pictured as a mini-electromagnetic wave that vibrates transverse to its direction of travel. A given electromagnetic wave (or photon) from a single atom will, since it is a transverse wave, vibrate in a definite plane like the rope illustrated in Figure 1.5. Viewed "end on" the transverse vibration might look like a single line oriented in whatever crosswise direction the wave happened to be vibrating. The light from most light sources does not consist of a single wave, but of an uncountably large number of waves coming from many atoms and molecules. Each of these waves will be oscillating transverse to the direction in which the light is traveling, but the orientations of these transverse vibrations will be random when viewed "end on" as in Figure 8.17.

Polarization of light refers to having the electrical portion of the light waves moving in a single direction rather than in random directions. Figure 8.17 shows a head-on view of an approaching light beam in which there is no polarization

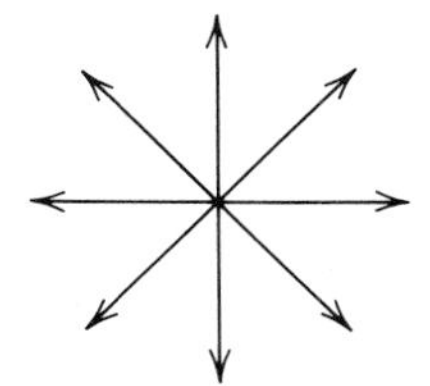

Figure 8.17 Head-on view of unpolarized light.

of the light. Certain types of materials, some of them naturally occurring, will polarize these randomly aligned light waves. In Figure 8.18 the light has passed through a polarizing material and now only waves of a selected direction are present. Common polarizing materials consist of many tiny crystals of a quinine compound oriented with their axes parallel to each other and spread out on a transparent base. The individual crystals act like one large crystal and have the unusual property of transmitting only light whose vibrations are properly oriented to the crystals. Light vibrating in other directions is absorbed by this polarizer. Thus when randomly oriented waves of light are directed through the polarizing material only the light that vibrates in the direction determined by the crystals will pass unattenuated. Waves vibrating in other directions will have only the component of their vibration that is in the chosen direction transmitted by the material. On the average, approximately 50% of an unpolarized light beam will pass through a polarizer.

Rotating a single piece of polarizing material in a beam will not result in a reduction of the intensity of the light at any given orientation of the polarizer. However, if a second piece of polarizing material is placed in the beam and held stationary as the other is rotated, then a position will be found where the intensity of the transmitted light will be nearly zero. This will occur when the polarized light passing through the first polarized filter is perpendicular to the direction of polarization of the second sheet. Under these conditions the light is almost 100% blocked by the combination of the two polarizers (Figures 8.19*a* and *b*).

Polarization is most commonly used in the construction of sunglasses. Without any other effect, polarized material would cause light passing through a pair of polarized lenses to be reduced by 50% (Figure 8.20). However, the polarizing characteristic gives polaroid glasses an added advantage over nonpolarized lenses. When light is reflected from a surface it becomes partially polarized so

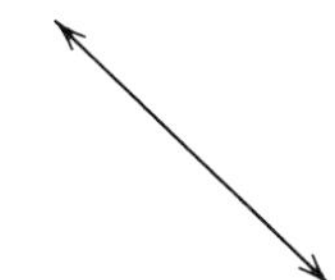

Figure 8.18 Head-on view of polarized light.

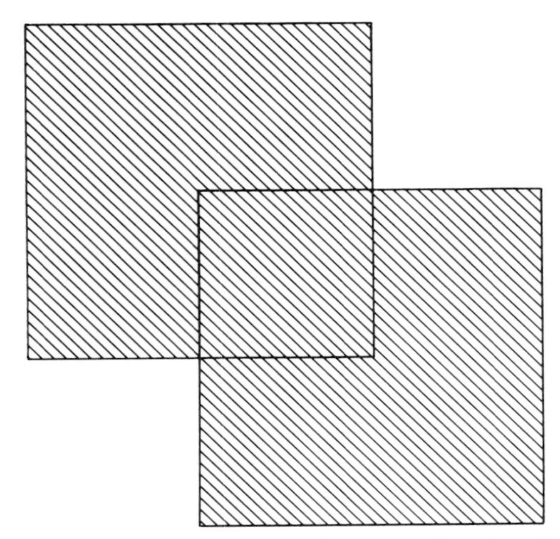

Figure 8.19*a* With the polarizers oriented in the same direction about 50% of the total light passes through them.

that only vibrations in the same direction as the plane of the reflecting surface are unattenuated. All other vibrations are reduced in amplitude. Thus, the light reflecting from a plane surface (such as a lake) can be reduced by orienting the polarized lenses so that their plane of polarization is perpendicular to the plane that is reflecting the light. This ability to reduce glare is a particularly attractive feature of polarized sunglasses. Glare is light reflected from a surface so intensely that it interferes with your ability to see objects in the same direction

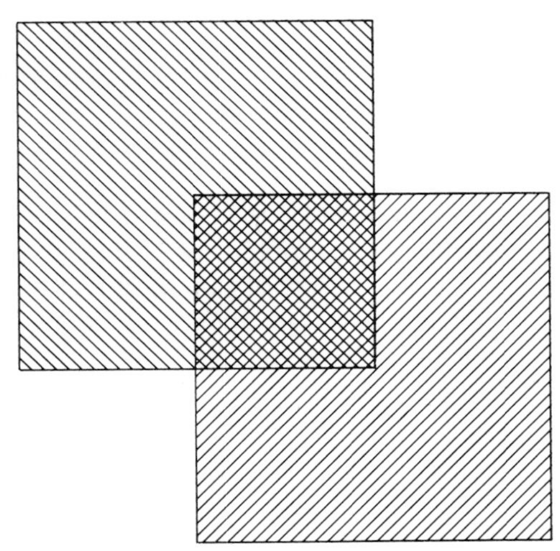

Figure 8.19*b* Crossing the polarizers (orienting them of 90 degrees to each other) blocks the light in their common area.

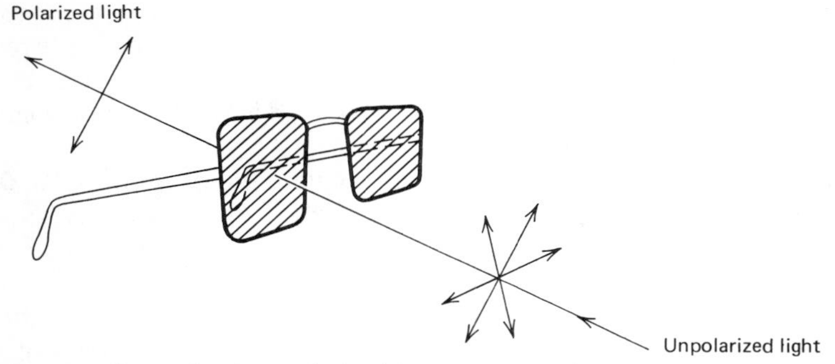

Figure 8.20 Sunglasses having polarized lenses pass only components of light in one direction.

from which the glare is coming. Thus, the ability to orient the polarized lens material in sunglasses so as to block reflected glare is quite valuable. A commercial advertisement for a brand of sunglasses shows how light reflected from the hood of a car can reduce your ability to see a person along a road. With the use of polarized lenses this reflected light can be greatly reduced while the unpolarized light from the pedestrian remains strong (Figure 8.21). The same principle is used in photography to reduce unwanted reflections.

The sky receives its color by reflecting light and, thus, light from the sky is naturally polarized. Unpolarized light from the sun is reflected by the air and particulate matter. This reflection results in different portions of the sky being

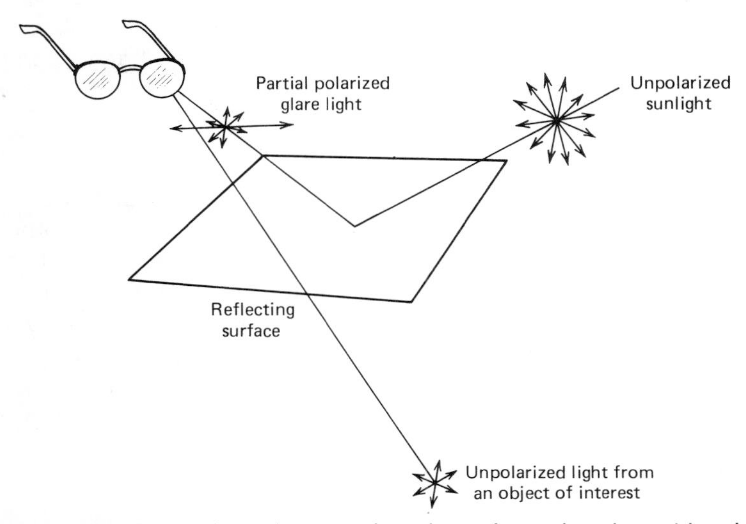

Figure 8.21 Light reflected from the smooth surface of a car hood provides glare to the driver. By properly orienting the polarizing lenses the majority of the glare can be removed.

more highly polarized than others. The effect is most pronounced if the observer looks through a polarizing sheet at right angles to the sun's rays — for instance, if he looks straight up at sunset or sunrise. Photographers aware of this fact often use polarizing filters to darken the sky when photographing landscape scenes.

Certain substances have the property of rotating the plane of polarization of a beam of polarized light. This property is called optical activity. Numerous materials exhibit this characteristic, which can be used in various types of analysis. For example, it can test for the concentration of sugar in a juice or syrup. By measuring the amount of rotation that occurs when polarized light passes through a standard size sample, we can determine the sugar content of a liquid. Vineyards use this quick procedure to determine the optimum time for picking their grapes.

Glass and some transparent plastics become optically active when subjected to mechanical stress. Therefore, manufacturers developing new parts often build models of such materials. These models, placed between polarizing sheets, are subjected to various forces. Stress patterns become visible as the plane of polarization in the stressed regions is rotated. Using such analysis, engineers can thus design parts that are less likely to break.

The optical activity of materials can also be used to generate displays of color. Cellophane, folded or crumpled and placed between two polarizing sheets, will exhibit various colors depending on the thickness of the material. Each color of light rotates at a different rate just as it refracts at a different rate. Consequently, white light, when polarized and passed through the cellophane, can be separated into different colors because each color rotates a different amount on passing through the optically active material. When the second polarizer is then used to view the transmitted light, only one color of light at any given position will be vibrating in the proper direction to pass through it.

IN CONCLUSION

In this chapter we have discussed a few devices that rely for their operation on the wave nature of light. The interference of light waves with each other, especially, is found to be important in the understanding of many optical effects. In addition, the transverse wave nature of light, the fact that it can be polarized, is very important in the operation of many optical devices and in the analysis of various materials. The extent to which light from a remote source (such as a star or nebula, for example) is polarized can yield important information about the conditions under which the light was produced. All in all, the wave properties of light play a central role in modern optical science.

PROBLEMS AND EXERCISES

1. You suspect that a pair of sunglasses advertised as having polarized lenses are really just tinted glass. How might you check if this suspicion is true?

2. Radio waves diffract substantially around buildings while light waves, which are also electromagnetic, do not. Why?
3. Can longitudinal waves undergo diffraction and interference similar to transverse waves? Can they be polarized?
4. When an interference pattern is formed, what becomes of the energy of the light waves whose destructive interference leads to the dark lines in an interference pattern?
5. When you are using sunlight, interference fringes in the double slit experiment are colored. Why aren't they simply white like the sunlight?
6. Light shines through a grating that has 5000 lines/cm. If the second order maximum on the interference pattern produced is found at an angle of 30°, what is the wavelength of the light?
7. A laser shines light of wavelength 633 nm through a slit of width 0.1 mm. How far from the central maximum will the second minimum be found on a screen 5 meters from the slit? (NOTE: This problem requires some additional knowledge of trigonometry.)

9 LIGHT AND COLOR IN NATURE

My heart leaps up when I behold
A rainbow in the sky:
So was it when my life began;
So is it when I am a man;
So be it when I grow old
or let me die!
W. WORDSWORTH

This famous poem by Wordsworth reveals the wonder and delight most of us experience when we encounter a spectacular color phenomenon in nature. In the course of our daily lives we are provided with countless wonderful examples of beautiful and entrancing optical displays. Rainbows are perhaps the most impressive of these, but there are many others as well. Why is the sky blue, or a sunset red? What produces the beautiful colors in oil drops on water or in soap bubbles? What causes a mirage? Why do distant mountains look bluish-gray? What produces a halo around the sun or the moon? All of these phenomena can be explained using the principles of light and color we have developed in the previous chapters. Rather than detracting from the wonder of these natural delights, an understanding of their origin should make them all the more appealing and significant.

9.1 RAINBOWS AND HALOS

RAINBOWS

Color plate 12 is a photograph of a rather vivid double rainbow, a truly spectacular natural phenomenon. Let us examine this photograph closely. First, the lower rainbow is clearly much brighter than the upper one. The lower rainbow is called the primary, and the upper rainbow is called the secondary. Further examination of these two rainbows reveals that the colors they contain are reversed in their order of appearance. The primary rainbow begins with blue or violet on its lower side, while the secondary rainbow begins with red. Notice also that below the primary bow lies a series of green and pink bands that are called the supernumerary arcs. These arcs are not always visible but, if you look for them, you should see them fairly often. Finally, notice that the area of sky between the primary and secondary rainbows is decidedly darker than the rest of the sky. This dark region is called Alexander's dark band. A precise theoretical explanation of all these aspects of rainbows lies beyond the scope of this discussion. But we can easily understand in a qualitative way how these features arise.

Let us begin by examining the geometry of a double rainbow more closely. Figure 9.1 shows parallel rays of sunlight incident on a region of air containing water droplets. The observer, looking into the sky, will always see the two rainbows at the same angular location *with respect to the direction of sunlight.* That is, with the sun at the observer's back, the primary rainbow will be seen at 42° above the line passing from the sun through the observer, while the secondary rainbow will be 50° above this line. This fixed angular location of the two rainbows relative to the direction of sunlight means that when the sun is high the rainbow will be lower in the sky, while when the sun is quite low the rainbow will appear higher. If the sun is more than 42° above the horizon, no primary rainbow can be observed from ground level. Another way of looking at the geometry of the rainbow is to say that the light that makes up the primary rainbow has been deflected from its original direction through an angle of 138° by the water droplets, while the light of the secondary rainbow has been deflected through an angle of 130°. To understand the origin of the locations of the primary and secondary bows we need only consider what happens when sunlight strikes a single spherical drop of water.

Figure 9.2 shows a ray of sunlight striking a spherical water droplet. Because the droplet is spherical, the fate of the ray of light is determined by a single

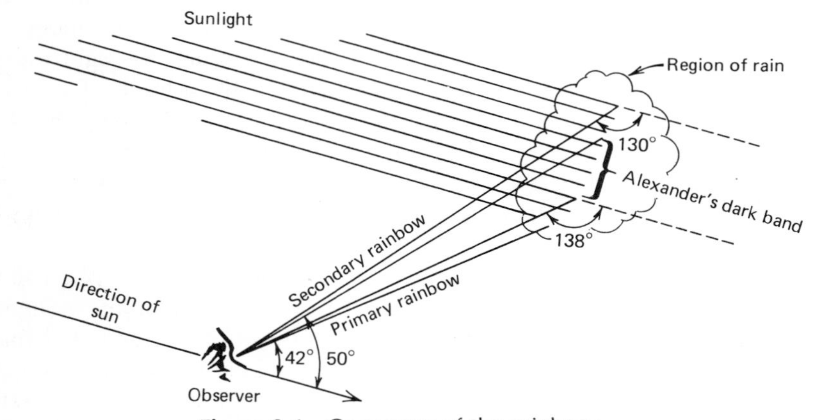

Figure 9.1 Geometry of the rainbow.

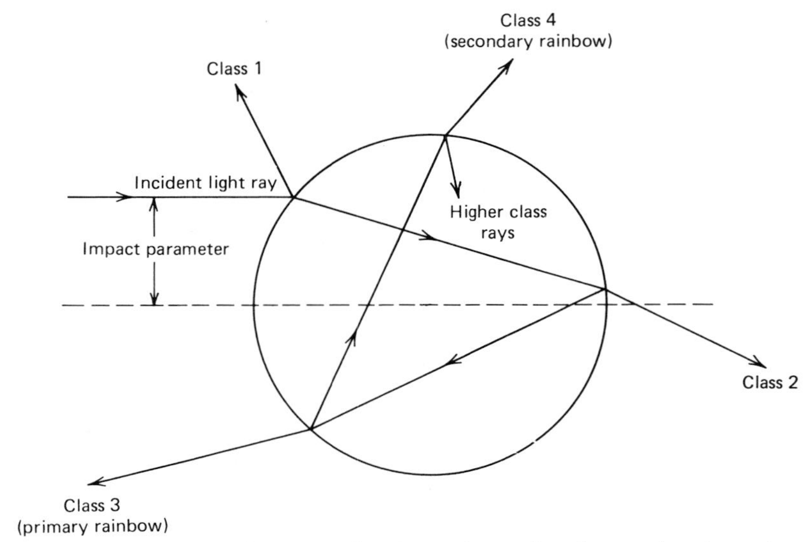

Figure 9.2 Geometry of a water droplet and associated rays of various classes.

factor, the impact parameter. This parameter is the displacement between the incident ray and the axis of the droplet. As Figure 9.2 shows, the light ray will strike the droplet at a definite location, and will be partly reflected and partly refracted. The reflected portion of the ray is called *class 1* ray. The refracted portion of the ray continues through the droplet until it reaches the droplet's back surface. Here the ray is also partly reflected (back into the droplet) and partly refracted (passing out of the droplet). The portion of the ray that is refracted and passes out of the droplet is called a *class 2* ray. The part of the ray internally reflected continues on until it meets the surface of the droplet again, where part will be reflected and part will be refracted. The refracted part will leave the droplet, and is called a *class 3* ray. Class 3 rays produce the primary rainbow. The part of the ray that is internally reflected continues on and gives rise to a whole series of rays, each produced when the light meets the surface of the droplet and partly reflects and partly refracts. For example, whereas class 3 rays suffer one internal reflection before escaping the droplet, class 4 rays suffer two internal reflections, class 5 rays suffer three internal reflections and so on. The secondary rainbow is produced by class 4 rays. In principle there exist a whole series of rainbows corresponding to class 5 and higher rays but, in practice, these rainbows are too faint to be observed.

Figure 9.2 shows the fate of a single ray of light entering a spherical water droplet. Actually, many parallel rays enter the droplet at all points along its surface. A careful examination of the paths of all these rays shows that many of the class 3 rays are deflected through an angle of nearly 138°. Thus a concentration of light is focused at this angle and, therefore, that is where the rainbow appears (Figure 9.3). Light entering the droplet at precisely the right location to exit exactly at 138° is called the *rainbow ray,* and the angle of 138° is referred to as the *critical angle* for class 3 rays that form the rainbow.

To understand the origin of the colors, we need only recall that light of different wavelengths is refracted by slightly different amounts as it passes from one transparent medium to another. This dispersion effect is what causes a prism to produce the spectrum of colors from white light. In the case of the water droplet, blue light will be refracted through a slightly greater angle than red light. Thus the critical angle for the blue end of the spectrum will be slightly greater than for the red end. This will cause the blue band in the primary rainbow to be seen below the red band (see Figure 9.4).

The secondary rainbow is produced by class 4 rays in a manner similar to the way that class 3 rays produce the primary rainbow. There is a slight difference, however. For zero impact parameter, a class 4 ray reflects back and forth along the droplet axis *twice* so that the ray ends up going out the far side of the droplet at *zero* deflection angle. For larger parameters, the deflection angle increases to a *maximum* value of 130°. Therefore the class 4 rays are concentrated at a deflection angle of 130°, and this is where the secondary rainbow is located. As with the primary rainbow, the effect of dispersion is to cause each wavelength of light to have a slightly different critical deflection angle. Now, however, because there have been two internal reflections within the droplet for the class 4 rays that form the secondary rainbow, the red band will be seen at the bottom of the rainbow and the blue band will be at the top (Figure 9.5).

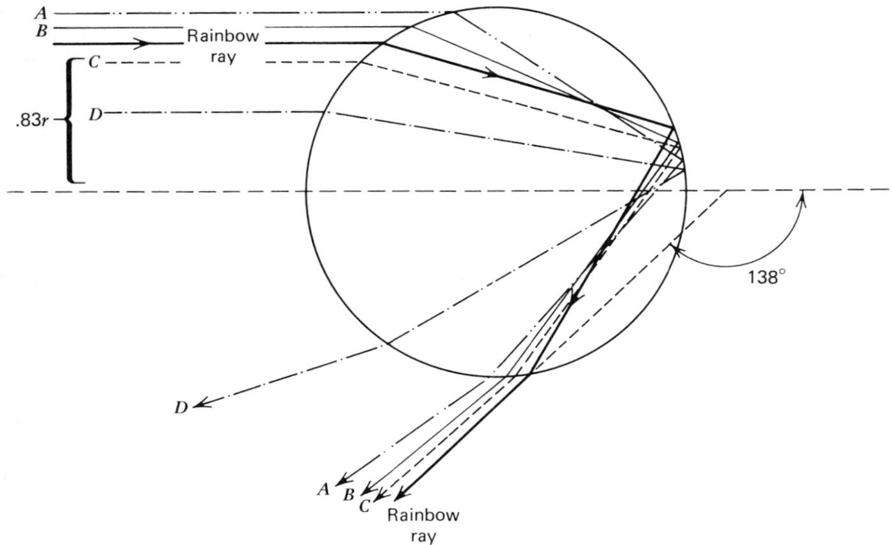

Figure 9.3 Effect of different impact parameters on the angle of deflection. Note that rays with slightly greater (*A* and *B*) or smaller (*C*) impact parameters than the rainbow ray will all be deflected through an angle greater than 138°.

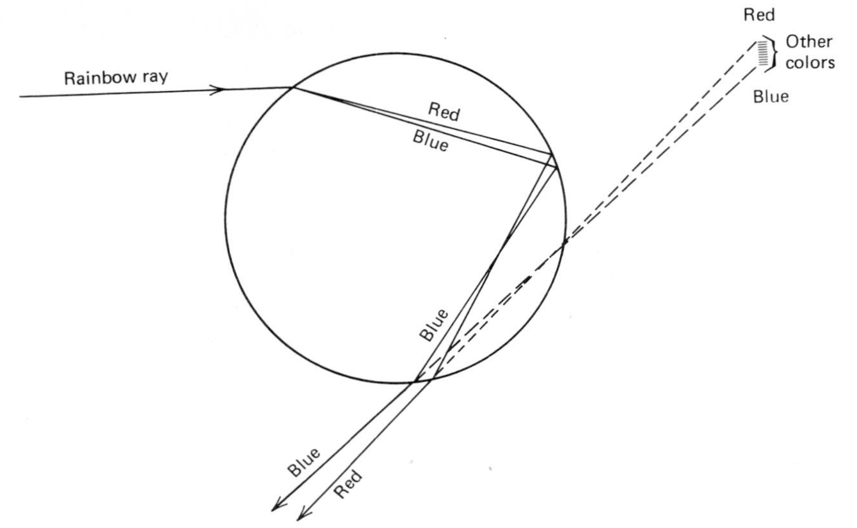

Figure 9.4 Effect of dispersion on the order of colors of the primary rainbow. Since red light is refracted less than blue, the red light of the rainbow appears higher in the sky.

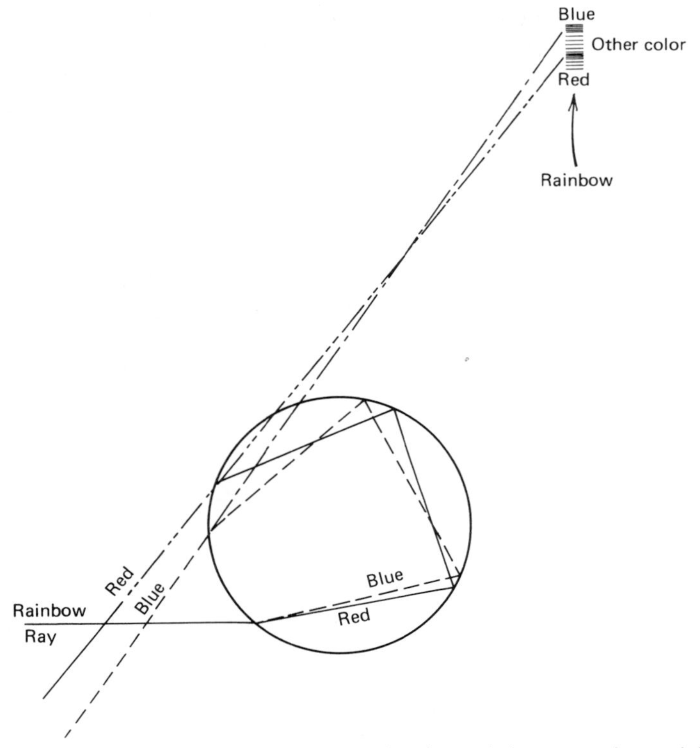

Figure 9.5 Effect of dispersion on the order of colors of the secondary rainbow. Note in this case it is blue that suffers the smallest angle of deflection and thus appears higher in the sky.

For the primary bow, no class 3 rays are deflected through an angle *less* than 138°. For the secondary bow, no class 4 rays are deflected through an angle *greater* than 130°. Thus between 130° and 138°, no class 3 or class 4 rays will be observed at all. But this is just the region *between* the two rainbows. Thus this region will look quite dark compared to the rest of the sky (Alexander's dark band).

The only feature of the rainbow we have not dealt with is the existence of the supernumerary arcs. These arcs are the result of interference effects. To understand this effect, look at Figure 9.6. The dark ray (the so-called rainbow ray) is at the critical impact parameter, and is deflected at the critical minimum angle that produces the primary rainbow. Ray 1 has a slightly *smaller* impact parameter than the rainbow ray, but exits the droplet at a slightly *larger* deflection angle. Ray 2 has a slightly *larger* impact parameter than the rainbow ray, but it too exits the droplet at the *same* deflection angle as Ray 1. Thus these two rays exit the droplet *parallel* to each other. As the two rays originally approached the droplet (position *A* in Figure 9.6), they were in phase. However, by the time the two rays are again traveling parallel to each other (position *B* in Figure 9.6) they will have traveled along different paths. Thus, depending on the wavelength of the light, the exit angle, and the size of the droplet, the two rays may be either in phase or out of phase at position *B* in Figure 9.6. If the rays are in phase, *constructive* interference occurs and a bright supernumerary arc is

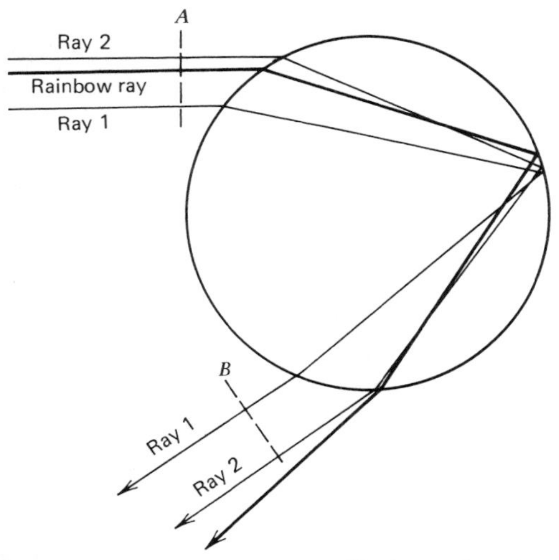

Figure 9.6 The origin of the supernumerary arcs. Both ray 1 and ray 2 leave the droplet at the *same* exit angle. Thus, to the eye, these rays appear to originate from the same part of the sky. The two rays originally in phase at position *A* have traveled different paths in reaching position *B*. Thus either constructive or destructive interference may occur at this exit angle depending on the wavelengths involved and other geometrical factors.

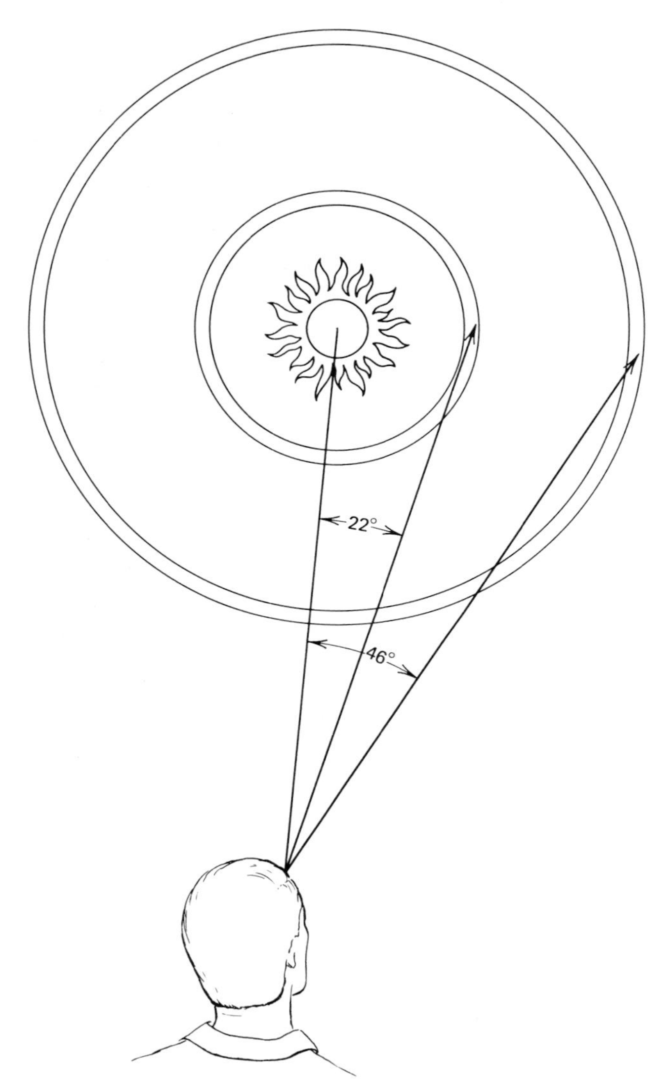

Figure 9.7 The appearance of the small (22°) and large (46°) halos that often appear around the sun or moon.

seen. If the rays are out of phase, *destructive* interference occurs and a dark supernumerary arc is seen. The true situation is a bit more complicated than this simple analysis makes it appear. The white light striking the water droplets contains light of all visible wavelengths. A complete description of the supernumerary arcs must take all these wavelengths into account. In essence what occurs is that where a bright arc is seen, most of the wavelengths in the middle of the visible spectrum are undergoing constructive interference (which gives the arc

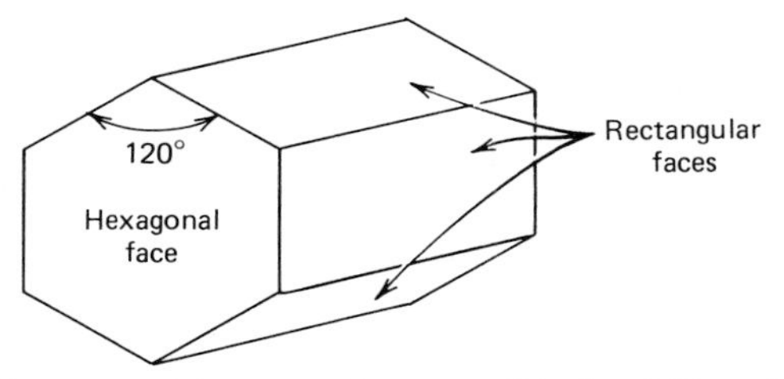

Figure 9.8 Typical hexagonal ice crystal responsible for 22° and 46° halos around the sun and moon.

a slight greenish appearance). The dark arcs occur for angles where the wavelengths in the middle of the visible spectrum suffer destructive interference. (Since the blue and red ends of the spectrum will not suffer complete destructive interference, these arcs will appear magenta.)

Notice how many basic optical effects have been used in our discussion of rainbows. Reflection, refraction, dispersion, and interference all played a direct role, not to mention the visual variation of color with wavelength. This is typical of the complexity of many optical effects in nature.

HALOS

It often happens that a change in weather from fair to foul is signaled by the arrival of high wispy cirro-stratus clouds. These clouds do not hide the sun, but appear to draw a light feathery veil across the sky. During such times, the sun will be surrounded by one or perhaps two halos (Figure 9.7). Or, if the clouds arrive at night, it will be the moon that is adorned by the circles of light. These halos are formed in a manner quite similar to rainbows, except that it is small ice crystals rather than water droplets that are responsible. The ice crystals that produce the most common halos are shaped like tiny hexagonal prisms (Figure 9.8). These ice crystals are oriented in all directions within the high clouds, but only a few of these crystals will be oriented in a direction that is effective in producing a halo. Consider a line joining the sun with the observer's eye. Now imagine a circle drawn so that its central axis lies along this line. (Figure 9.9 shows

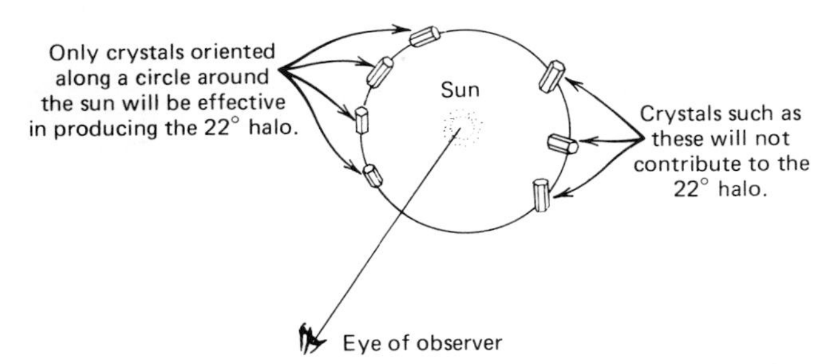

Figure 9.9 Orientation of hexagonal crystals that produce the 22° halo.

such a circle.) In order for an ice particle to refract sunlight to the observer's eye, the crystal must be oriented with its long axis along the circumference of such a circle. With the crystal so oriented, it may refract light through a number of different angles depending on the orientation of its rectangular faces (see Figure 9.10). But, as with the raindrop, something interesting occurs. A close examination of the effect of different orientations of the rectangular faces shows that there exists a minimum angle of deflection at 22°. This deflection is produced when the incident sunlight strikes a rectangular face at an angle of 41° (Figure 9.11). For ice crystals that are oriented so that the sunlight strikes one of their rectangular faces at an angle slightly smaller or larger than 41°, the deflection angle will still be very close to 22°. Thus a concentration of light occurs at this viewing angle, and a halo is produced.

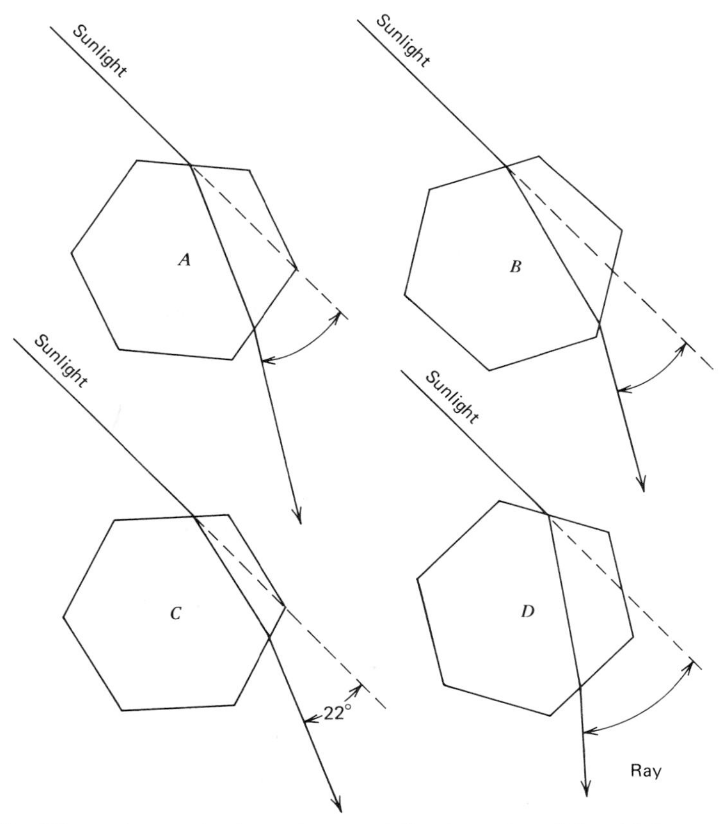

Figure 9.10 Effect of crystal orientation on deflection of sunlight. All four crystals are struck by the rays of the sun from the same direction. Crystal *C* is oriented in such a way that it deflects the sun's rays through the smallest angle.

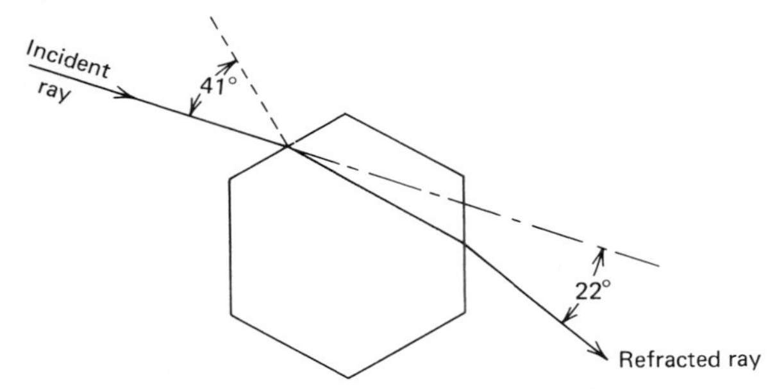

Figure 9.11 Geometry for minimum deflection of ray incident on rectangular face. Rays incident at angle *other than* 41° will be refracted through an angle larger than 22°.

As with the rainbow, the refraction that produces the halo affects blue light more than red light. Thus blue light will suffer a slightly larger minimum deflection than red light, leading to the red appearance of the inside of the halo. Also, since no light is deflected through angles of less than 22°, the area inside the halo appears darker than the rest of the sky.

In addition to the halo at 22°, a second larger halo will frequently be seen at 46°. This occurs because there is a second way that the ice crystals can refract light. Instead of having light enter one rectangular face of a crystal and leave by another rectangular face (as in Figures 9.10 and 9.12), it is possible for light to enter the crystal at a rectangular face and leave by a hexagonal one (or vice versa; see Figure 9.12). This new possibility involves two faces that meet at a 90° angle; thus the angle of minimum deflection is 46°, which is the angle observed for the large halo (Figure 9.13).

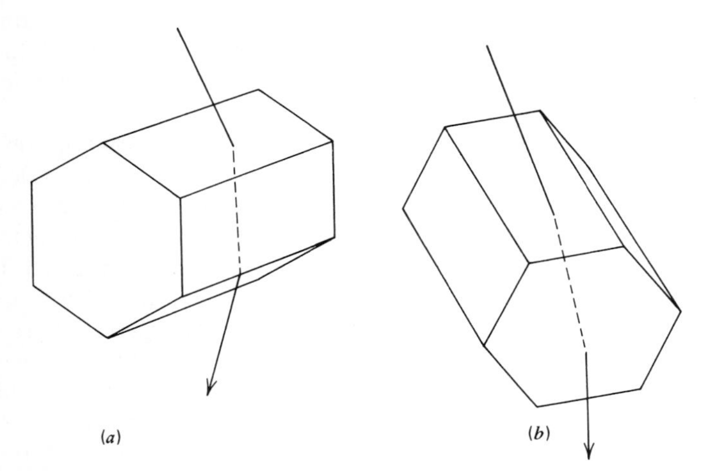

Figure 9.12 Two basic types of refraction for hexagonal crystals. Configuration (*a*) produces the 22° halo while configuration (*b*) produces the 46° halo.

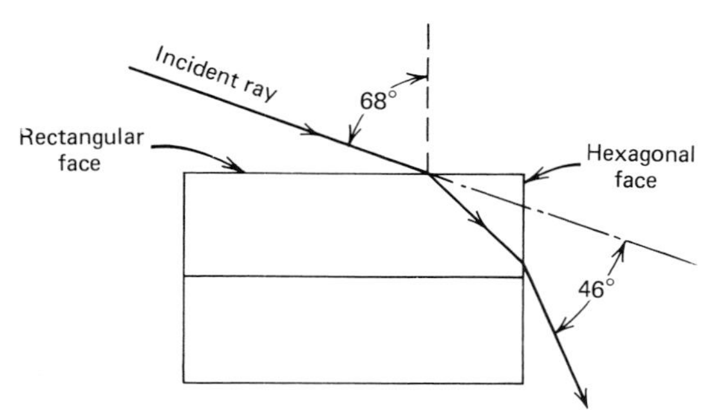

Figure 9.13 Geometry for minimum deflection of ray incident in such a way that it enters a hexagonal crystal via a rectangular face and exits via a hexagonal face. In this case the minimum deflection angle is 46°.

9.2 INTERFERENCE PHENOMENA

OIL SPOTS AND SOAP BUBBLES

Most of us have seen the colors produced by a thin layer of oil floating on the surface of water or by light reflecting from a soap bubble. The origin of these colors can be most easily understood by considering a thin film of one material (such as oil) sandwiched between two other materials (such as air and water, or air and air for a soap bubble). Before continuing with our analysis, however, we need to consider one important point.

When light encounters an interface between two different transparent media, some of the light will be reflected from the interface and some will be refracted into the second medium. If the second medium has a *higher* index of refraction than the first, then the reflected light will suffer a 180° phase change as it is reflected. This is equivalent to the light having traveled an extra distance equal to one half of its wavelength (see Figure 9.14). With this in mind let us now consider a thin film of oil floating on water.

Figure 9.15 shows a typical thin film situation. In this illustration only one incident ray is depicted to simplify the analysis. Of course in the real situation many rays will strike the oil. As the figure shows, part of the ray striking the oil will be specularly reflected and part will be refracted into the oil. The refracted ray will itself be partially reflected at the oil-water interface and will return to the upper surface of the oil. Here, part of the ray will pass out of the oil into the air in a direction *parallel to* the original reflected ray (these last two rays are labeled 1 and 2 in Figure 9.14). Notice that the light making up ray 2 has traveled farther (through the oil and back again) than ray 1, but ray 1 has suffered a 180° phase change. Since these two rays were originally in phase (in fact they were part of the same incident ray), they may now be out of phase because of the different distances they have traveled before coming back together again and because of the phase change. Thus, depending on the original angle of incidence, the oil film thickness, the index of refraction of the oil, and the wavelength of light, ray

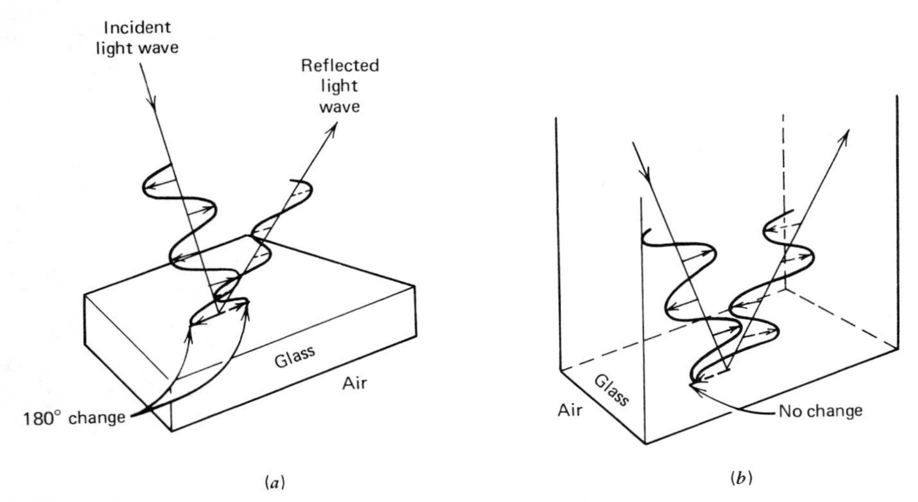

Figure 9.14 (*a*) 180° phase change occurring when light reflects from a more dense medium. (*b*) No phase change when light reflects from less dense medium (in this case the air below the glass).

1 and ray 2 may interfere with each other constructively or destructively. Normally, the incident light contains all visible wavelengths. Some of these wavelengths will interfere with each other destructively and thus be removed from the light reflected from the oil spot. Other wavelengths will be such that constructive interference will occur, and these wavelengths will reflect especially well from the oil. Thus the reflected light from each part of the oil spot will

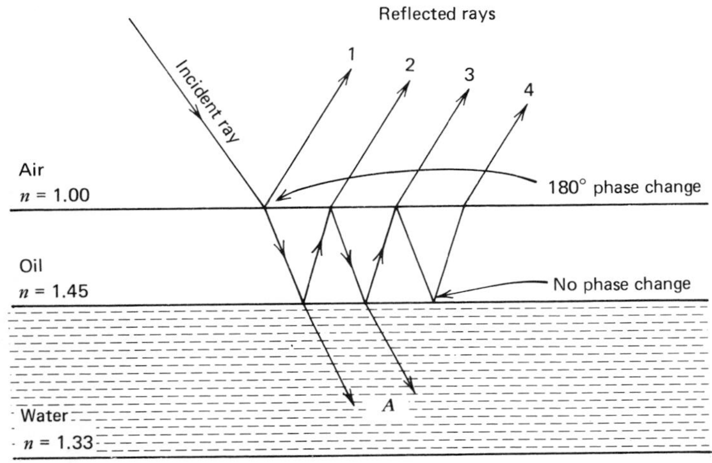

Figure 9.15 Analysis of light reflected from oil floating on water. Notice that this reflected light is actually composed of several rays (1, 2, 3, etc.) that interfere constructively or destructively with each other depending on a number of factors.

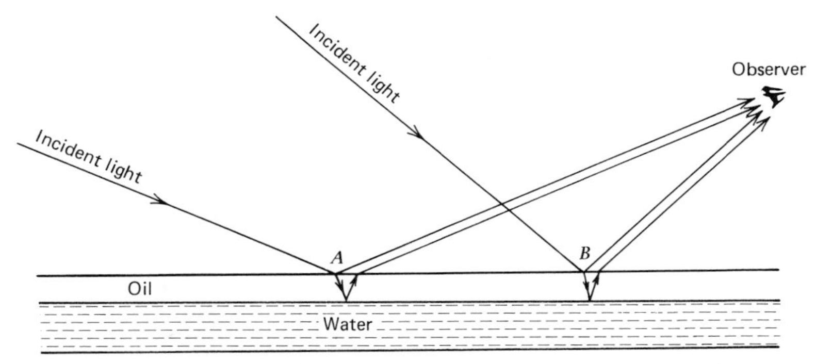

Figure 9.16 An observer will see different parts of an oil slick (*A* and *B*, for instance) as having different colors. This is because the viewing angle is different for each part of the oil. Thus the rays reflected from near point *A* will interfere constructively with each other at different wavelengths than will those reflected from point *B*.

appear in different colors because for each region of the spot the viewing angle is different (Figure 9.16). Changing the viewing angle will change the difference in the distances traveled by ray 1 and ray 2 and, thus, change the colors that interfere constructively and destructively. If we could view the oil spot from below the water, we would see the color of each region of the spot changed to the complementary color. This is because colors that reflect well off the spot will obviously not be transmitted; likewise colors that do not reflect well will be transmitted.

Soap bubbles produce colors in the same fashion as do oil spots. The only difference is that the thin film composing a soap bubble has air on both sides. Thus, as seen in Figure 9.17, reflections always take place at an air-water inter-

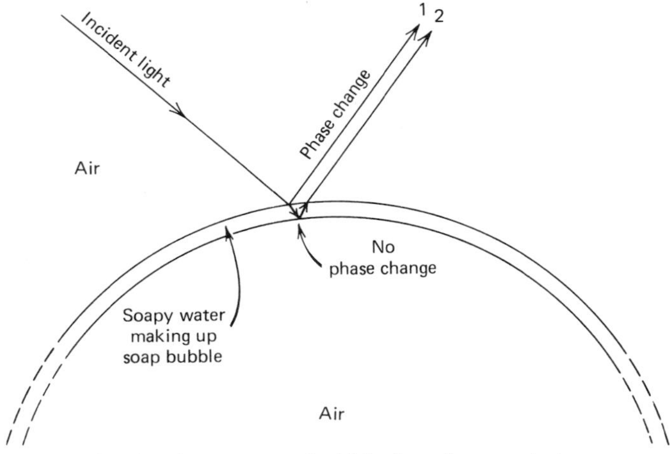

Figure 9.17 Light reflecting from a soap bubble. Interference between rays 1 and 2 will produce a variety of colors. Note that air is on both sides of the soap film.

face. An interesting phenomenon occurs with soap films as they are made progressively thinner. Remember that a 180° phase change occurs for the light that reflects from the outer air-water interface, but not for the light reflecting from the inner air-water interface. Thus as the film gets very thin, rays 1 and 2 in Figure 9.17 become more nearly out of phase regardless of wavelength. For very thin films all colors suffer destructive interference and the film will look dark when viewed by reflected light. If viewed from the other side, however, it will look clear and colorless.

COATING ON GLASS

Interference due to thin films can be used to advantage in producing nonreflecting glass. Figure 9.18 shows a thin film of transparent MgF_2 on a piece of glass. Since MgF_2 has an index of refraction intermediate between that of air and glass, reflections from both the air–MgF_2 and MgF_2–glass interfaces involve the same 180° phase change. Thus the extent to which interference occurs between ray 1 and ray 2 will be determined primarily by the film thickness and the wavelength of light. What is usually done is to use a film thickness such that light of 550 nm suffers destructive interference. Thus the brightest part of the visible spectrum will not reflect from the coated glass. The effect is to create a nearly nonreflecting glass surface.

IRRIDESCENCE

A number of animals exhibit a particularly striking coloration usually called irridescence. This is actually an interference effect of one kind or another. Certain

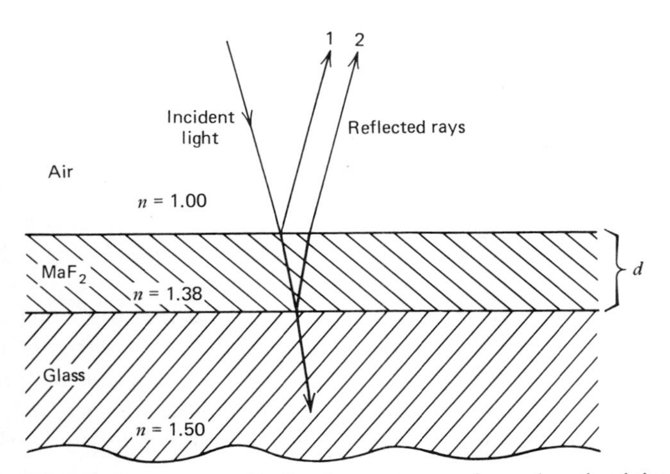

Figure 9.18 MgF_2 coating on glass. If d is chosen to equal one fourth of the wavelength of 550 nm light in MgF_2 (a wavelength of 550 nm in air becomes 550 nm/1.38 = nm in MgF_2), then destructive interference will occur for this light upon reflection at head-on incidence with the MgF_2 coating.

creatures such as fish and lizards have scales with a thin film coating. This thin film produces colors in the same way as do oil spots and soap films. On the other hand, certain birds and insects appear irridescent for another reason. The feathers of some birds consist of many closely spaced parallel reflecting grooves. These grooves act just like a reflection diffraction grating. As we saw in Chapter 8, this type of grating is highly selective in the colors it reflects in a given direction. Thus, birds with such feathers appear to reflect light of different colors from different parts of their bodies. But as the viewing angle is changed (because either you or the bird moves), the colors also change just as with a diffraction grating. Some insects produce their colors in this fashion. In fact, frequently the actual outer surface of the insect is virtually colorless, because the color comes almost totally from interference effects.

9.3 SCATTERING EFFECTS

RAYLEIGH SCATTERING

Have you ever noticed how smoke just as it leaves a cigarette or steam just as it comes out of a kettle tends to look bluish? On the other hand, if you look *through* the smoke or steam at a white wall, the smoke or steam tends to look somewhat yellow. This interesting phenomenon is caused by the way small particles of matter scatter light. If the particles of matter are substantially smaller than the wavelength of light that strikes them, then the efficency of scattering is given by the Rayleigh formula

$$e_s \propto \frac{1}{\lambda^4} \qquad (1)$$

That is, small particles of matter cause individual photons to deflect in new directions. It is important to keep in mind that no light is lost in this process.

Equation (1) shows that short wavelength light will be scattered much more than light of longer wavelengths. This means that blue light is scattered much more efficiently than red light. Thus from a beam of white light, a great deal more of the blue light will be scattered in other directions than will the red light. Consider the beam of white light shown in Figure 9.19. This light is shown encountering a small cloud of smoke (or perhaps steam). As the light proceeds through the cloud, the shorter wavelengths will be more frequently scattered in new directions and, thus, will no longer be a part of the original beam. An observer on the far side of the cloud will see a light beam with much of the blue end of the spectrum removed. The beam will thus look yellow. On the other hand, an observer viewing the cloud from off to one side will see a great deal of the scattered blue light from the original beam. Thus the cloud will look bluish in appearance.

The preferential scattering of shorter wavelengths (primarily violet and blue light) is responsible for a number of common natural color effects in addition to

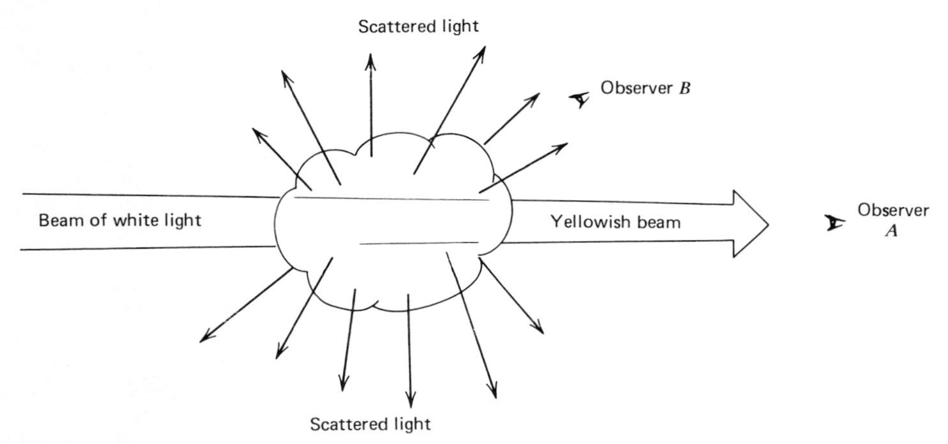

Figure 9.19 White light passing through smoke or steam. Observer *A* will see the original beam with much of the blue light removed. Observer *B* will see the scattered blue light.

that produced by smoke. For example, the blue sky, red sunsets, and the apparent blue-gray color of distant mountains all result from short wavelength (Rayleigh) scattering. Let us consider the blue color of the sky first.

The molecules of the air act as very tiny scattering centers. Light coming from the sun is initially a very balanced white light. However, the atmosphere is several miles thick so that as the light propagates through the air a substantial amount of the blue light is scattered in all directions. Thus the direct sunlight reaching our eyes (or illuminating any object directly) will have a definite yellow appearance. However, if our eyes are directed away from the sun to some other part of the sky we will see the blue light which has been scattered by all the portions of the atmosphere along our line of sight (Figure 9.20).

When the sun is very high in the sky it appears much whiter than when it is lower. The reason is quite simple, as a glance at Figure 9.21 will confirm. As the sun gets progressively lower in the sky, the sunlight must traverse an ever longer path of air in order to reach us. This longer path will result in more blue light being scattered out of the sunlight leaving the remaining light more yellow. As the sun nears the horizon, the path the sunlight must follow becomes so long that most of the violet, blue, and green parts of the spectrum have been scattered. The remaining light gives the sun a very red appearance, thus producing red sunsets.

The scattering of blue light by the air has another effect besides that of producing the blue color of the sky. Everyone has noticed that distant features of the landscape appear more bluish and less distinct than nearby features. This is primarily a result of the scattering of blue light by the intervening atmosphere (Figure 9.22). There are two opposing effects to be considered. First, light from a distant object will lose some of its short wavelengths because of scattering.

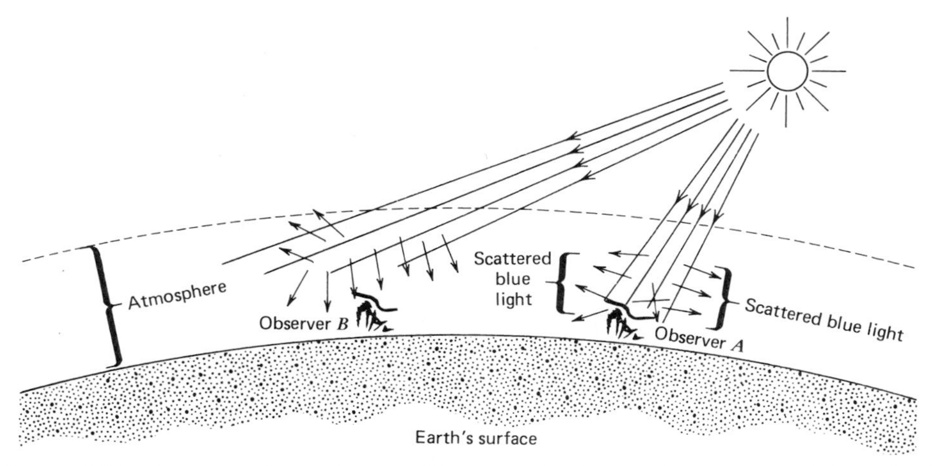

Figure 9.20 Atmospheric scattering. Observer A will see the sun as slightly yellow because the blue light has been scattered by the atmosphere. Observer B will see the scattered blue light and thus perceive the sky as blue.

But sunlight falling on the air between the observer and the object will result in a substantial amount of blue light being scattered in the direction of the observer, just as if it had come from the distant object itself. Usually this second effect predominates, and distant objects appear bluer than similar nearby objects. Because the added blue light reduces the contrast between adjacent portions of a distant object, the sharpness of the image will be reduced as well. There is one common example of a case where the loss of blue light from the distant object actually predominates over the effect of scattered sunlight from the intervening air. This occurs when the distant object is a cloud directly illuminated by sunlight. In this case the object is so brightly illuminated and so highly

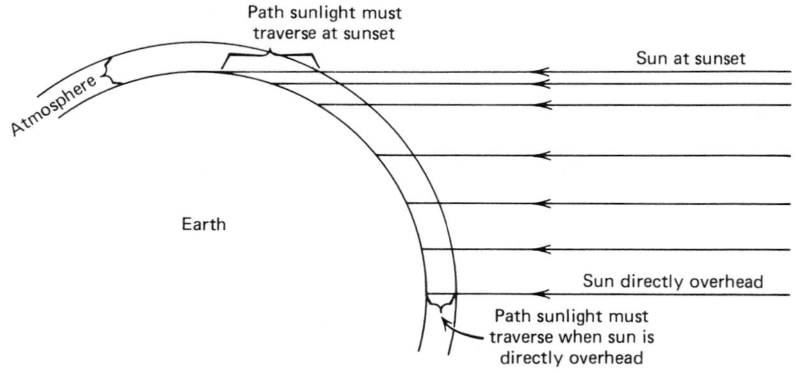

Figure 9.21 Sunlight must pass through a greater amount of atmosphere at sunset or sunrise. This causes the red colors that are observed — most of the shorter wavelengths are scattered by the atmosphere.

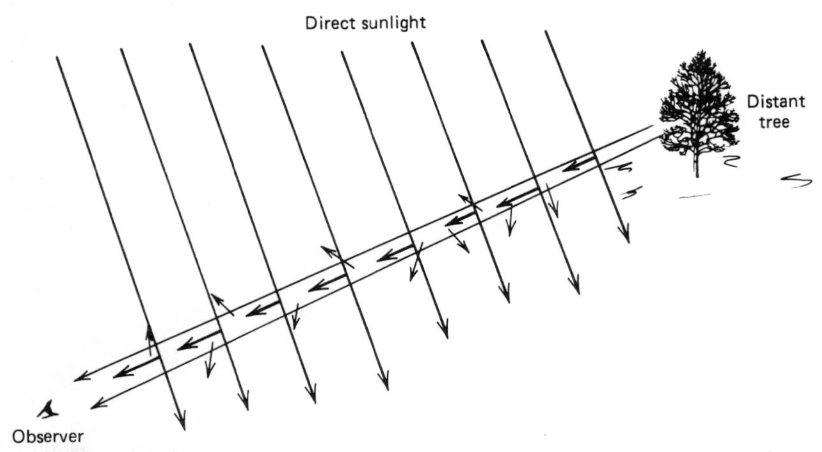

Figure 9.22 Light from the distant tree loses some of its blue component on the way to the observer because of atmospheric scattering. However, sunlight incident on the atmosphere between the tree and the observer results in blue light being scattered from the sunlight directly toward the observer. The net effect is to increase the apparent blueness of the tree.

reflecting that distant clouds actually take on a yellow appearance, an effect identical to that which causes red sunsets.

SCATTERING BY LARGE PARTICLES

We have seen that very small particles of matter are able to scatter light in all directions, with blue light being more effectively scattered than red. When the particles become larger, however, the nature of the scattering changes substantially. This is especially true for particles that are significantly larger than the wavelength of visible light. In this case, most of the scattering is in the forward direction. That is, the light is deflected by only a small angle each time it is scattered. In addition, all wavelengths are affected more or less equally.

The most common example of this type of scattering is seen in observing clouds or mists. In a heavy mist or fog, for example, objects are seen in their true colors (because the scattering effects all colors equally), but the images are quite indistinct because of the loss of light due to scattering. The predominantly forward nature of the scattering produces some eerie shadow effects as well. Consider sunlight shining through a fog and illuminating a tree as shown in Figure 9.23. Suppose an observer walks along line *AB*, a path that traverses the shadow of the tree. As long as the observer is outside the shadow the fog will seem well illuminated, almost self-luminous. Light scattered by the water droplets of the fog will seem to come from every direction. As the shadow is entered, however, a dramatic reduction will occur in the luminosity of the fog. The fog in the tree's shadow will not be illuminated by sunlight and will thus

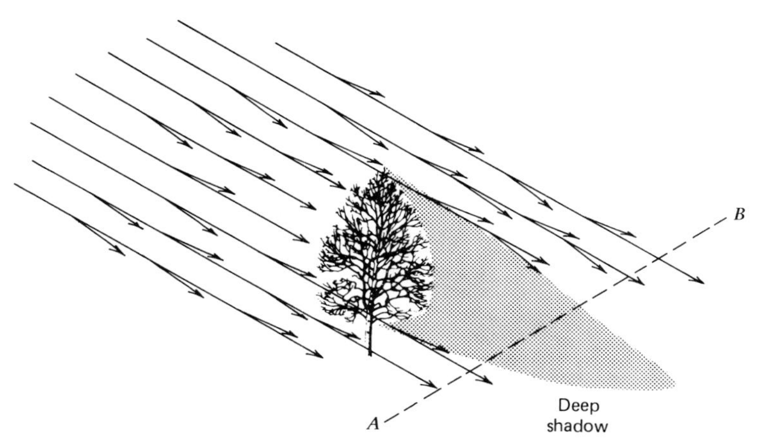

Figure 9.23 Dense shadow cast by tree during a heavy fog. Observer walking from *A* to *B* will perceive shadow as almost tunnel-like. Most light scattered by fog particles will be scattered nearly forward and thus miss the shadow.

seem very dark. It is as if the observer has entered a tunnel. This same effect occurs even more dramatically at night when a streetlight cuts through a dense fog to illuminate any number of odd-shaped objects and produce shadows everywhere.

Smoke and steam provide very interesting examples of both small particle (Rayleigh) and large particle scattering. For instance, as steam leaves a tea kettle the particles are quite small, and the steam looks somewhat blue. Several inches from the kettle the steam has condensed into much larger droplets. The resulting larger particle scattering makes the vapor look white. The same thing can be observed by following the smoke produced from a bonfire.

9.4 MIRAGES

One of the more common sights on a hot summer day is the appearance of shimmering "pools" that seem to hang just above the surface of paved roadways. Such pools are really examples of an inferior mirage, and occur whenever the air adjacent to a flat surface becomes heated several degrees above the temperature of the air away from the surface. The main principle at work here

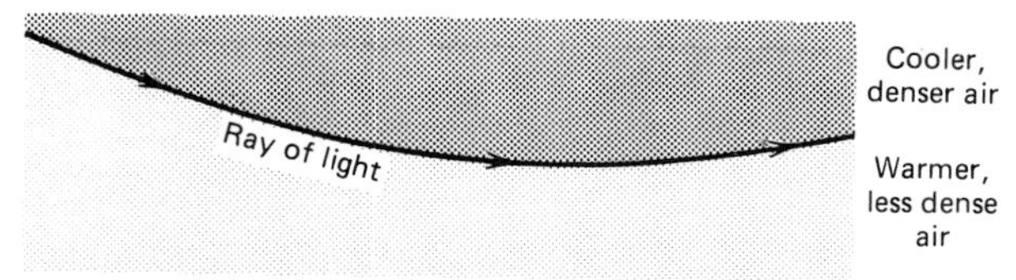

Figure 9.24 Effect of a temperature gradient on the path of light in air. Note that light will curve back toward the cooler, denser air.

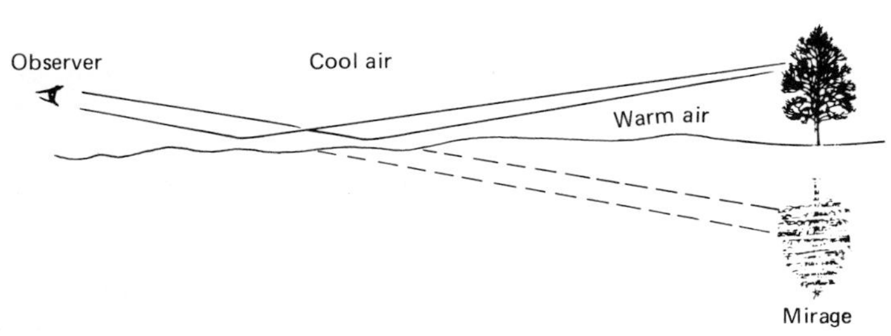

Figure 9.25 Inferior mirage. A layer of warm air near the ground causes the light from the tree to be bent upward (toward the cooler air) producing a mirrorlike effect.

is the fact that the index of refraction of air increases as the density of air increases. Hot air has a lower density, and thus a smaller index of refraction, than cool air. Figure 9.24 shows the effect of nonuniform air temperature on the path of light traveling through the air. As the diagram illustrates, light will tend to bend toward the cooler air. In practice, this bending often occurs when relatively large temperature differences (a few degrees Celsius) occur within a fraction of an inch of a surface. Thus the bending is almost mirrorlike (Figure 9.25). The mirage produced in Figure 9.25 is the most common type observed, and is called an inferior mirage because the mirage is observed below the actual object. In the case of the "pools" seen on a hot highway, the original object is the blue sky and the mirage is the "reflection" of the sky from the roadway.

On rare occasions the atmosphere may have an abnormal temperature distribution, the so-called temperature inversion. For example, warm air over a cold body of water may result in a layer of cold air near the water with warmer air above it. If a sufficient temperature gradient exists, a superior mirage may be formed. Such a mirage is illustrated in Figure 9.26. It should be emphasized that both the inferior and superior mirages are formed by the same basic mechanism, that of a relatively strong temperature gradient in the air. The two different names are simply derived from our viewing perspective, whether the original object or the mirage is on top.

9.5 THE AURORA

Where do the northern lights come from? Most of us have seen photographs of these shimmering lights that frequently dance across the polar skies. Unfortunately, few people live where the aurora is clearly visible, since the most spec-

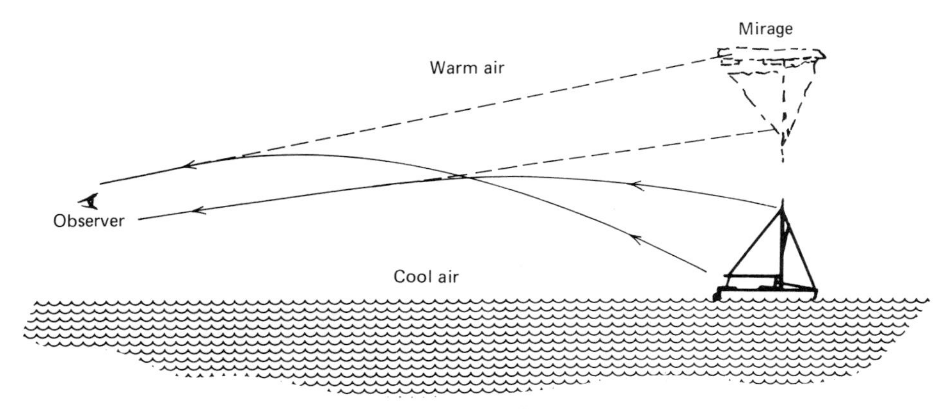

Figure 9.26 Superior mirage. Occasionally a temperature inversion will occur leaving cool air at the surface and warm air above. Light from the sailboat will bend downward (toward the cooler air) producing a mirage that seems to hang above the original object.

tacular displays (in the northern hemisphere) are confined to the extreme northern latitudes. The origin of this strange and beautiful phenomenon lies in the interaction of the solar wind, the earth's magnetic field, and the atmosphere.

The solar wind is a stream of high velocity particles, primarily electrons and protons, which is continuously given off by the sun in every direction. This wind, whose strength varies with solar surface activity, does not penetrate directly to the surface of the earth because of the presence of the earth's magnetic field (Figure 9.27). The action of a magnetic field on a moving charged particle is very interesting. Figure 9.28 is a sketch of a magnet with the magnetic field lines drawn in. Also shown are two small positively charged particles approaching the magnet. Particle *A* will enter the region of the magnetic field in a direction

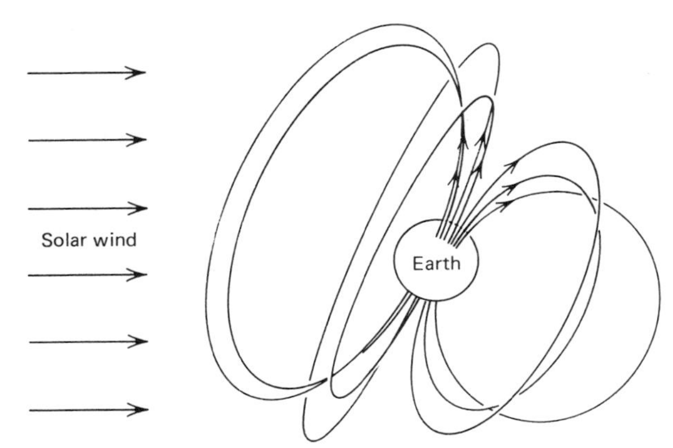

Figure 9.27 The Earth's magnetic field. The high energy particles of the solar wind are funneled to the earth's poles by the earth's magnetic field.

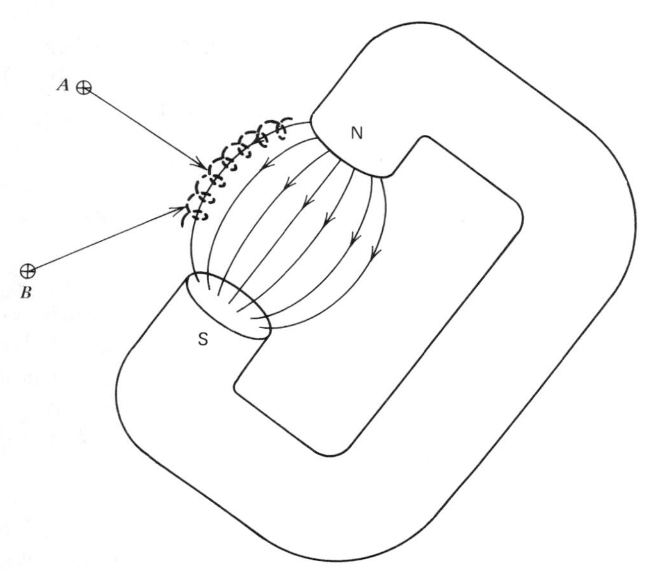

Figure 9.28 Magnetic field of a horseshoe magnet. Note particle B will spiral around field lines toward north pole of magnet, while particle A will simply orbit in a fixed plane.

perpendicular to the magnetic field lines. The laws of electrodynamics tell us that a magnetic field can only exert a force on a moving charged particle in a direction that is perpendicular to both the magnetic field lines and the direction in which the particle is moving. Thus particle A will be deflected sideways by the magnetic field, and will tend to move in a circular path that lies in a plane. Particle B is shown approaching the magnetic field in an oblique direction. The motion of particle B can best be visualized as consisting of a component perpendicular to the magnetic field and a component parallel to the magnetic field (Figure 9.29). The perpendicular velocity component causes the particle to be

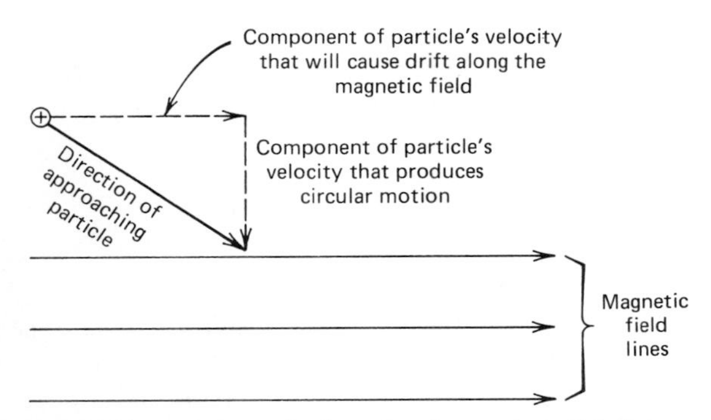

Figure 9.29 Decomposition of the velocity of a charged particle into a component parallel to the magnetic field and a component perpendicular to the magnetic field.

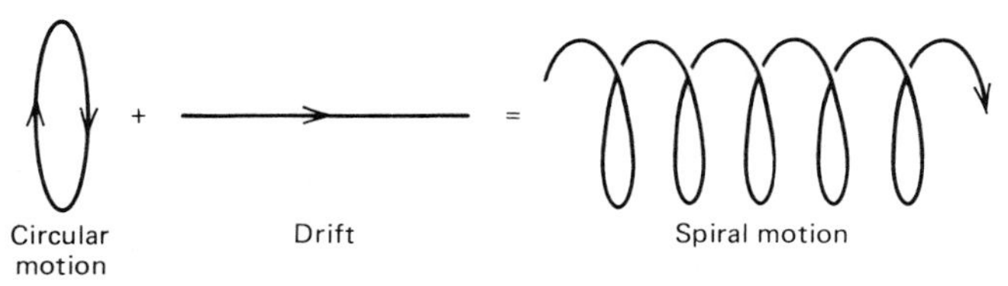

Figure 9.30 The combination of circular motion plus drift causes the motion of a charged particle in a magnetic field to be a spiral.

deflected sideways just as in the case of particle *A*. If no parallel velocity component were present, particle *B* would also move in a plane circular path. However, the parallel velocity component causes particle *B* to drift along the magnetic field in addition to moving in a circle. The combination of the circular motion and the drift causes the particle to follow a spiral path along the magnetic field lines (Figure 9.30). This is essentially what happens to the charged particles of the solar wind as they approach the earth. The particles become trapped by the magnetic field and begin to spiral around the field lines, following them to their source. In the northern hemisphere the magnetic field lines mostly originate in the vicinity of the geomagnetic pole in northwestern Greenland. Thus the high speed protons and electrons that comprise the solar wind tend to be focused on the regions surrounding the geomagnetic pole.

The focused stream of high speed charged particles begins to come into contact with the atmosphere at an altitude of a few hundred kilometers. Most of the auroral light is produced by the interaction of the incoming particles with nitrogen and oxygen atoms or molecules. In some cases the atoms or molecules are ionized by the incoming particles while in other cases they are simply excited. When the ionized or excited atoms or molecules return to their original condition, light consisting of a definite set of wavelengths is given off. For example, excited oxygen atoms produce both red light at 630 nm and green light at 558 nm. Excited nitrogen molecules produce a series of emission bands that are quite intense between 650 and 680 nm. Ionized nitrogen molecules emit substantial light between 391 and 470 nm. Which of the radiations predominates at any given time, and the general shape and size of the aurora itself depends on a complex set of factors involving the condition of the solar wind, the atmosphere, and the earth's magnetic field.

9.6 IN CONCLUSION

The natural phenomena we have discussed in this chapter are all relatively common and well known. Each can be understood using the basic principles developed throughout the text. There are of course countless other color phenomena, some common and some quite rare, that we could investigate. But then, all the mystery shouldn't be dispelled so easily. The interested student, armed with an understanding of the basic principles of light and color, might wish to investigate a few of these interesting occurrences personally before turning to textbooks for the answers.

PROBLEMS AND EXERCISES

1. Do you expect the light coming from a rainbow to be polarized? Explain.
2. Why don't class 1 rays that are reflected from water droplets produce a rainbow?
3. Suppose the raindrops producing a rainbow were somehow replaced by tiny glass beads. How would this affect the appearance of the rainbow?
4. Refer to Figure 9.14. Suppose an observer looking at reflected rays 1 and 2 sees that particular spot of the oil as magenta. Now imagine an observer below the surface of the water at position *A*. If this observer looks up at the same spot of oil, what color will be perceived?
5. In Rayleigh scattering, how much more efficiently is light of 420 nm scattered than light of 650 nm?
6. Let us imagine a beam of light consisting of 100 units each of light of wavelength 450 nm, 550 nm, and 650 nm. Suppose this beam penetrated the earth's atmosphere and conditions are such that, for each mile the light travels through the atmosphere, the *percentages* of 450 nm, 550 nm, and 650 nm scattered out of the beam because of Rayleigh scattering are 4.9, 2.2, and 1.1, respectively. Calculate how many of the original 100 units of each wavelength of light remain after the beam has traveled 2 miles, 4 miles, and 8 miles through the atmosphere.

APPENDIX A
CALCULATION OF THE CIE TRISTIMULUS VALUES

In Chapter 3 we described the CIE colorimetry system in some detail. The procedure for calculating the tristimulus values X, Y, and Z for a given colored surface or filter was outlined. Although conceptually simple, the actual calculation can be quite tedious. Briefly let us review what must be done. Figure A-1 shows the three CIE color matching functions for the imaginary primaries $\bar{x}$, $\bar{y}$, and $\bar{z}$. Figure A-2 shows the reflectance curve of some hypothetical surface. Finally, Figure A-3 shows the spectral content of CIE Standard Illuminant *C*, which is supposed to represent light from the sky (that is, the illumination typical of open shade outdoors). To compute the tristimulus values of the surface of Figure A-2 when illuminated by Standard Source *C*, we proceed as follows. First we divide the visible spectrum into a number of equal size segments. The more segments that are used, the more accurate will be the calculation. Let us imagine that the 300 nm of the visible spectrum has been divided into 10 equal segments each 30 nm wide. Thus the first segment goes from 400 to 430 nm, the second from 430 to 460 nm, and so on. Next we notice what wavelengths are in the exact center of each segment. Since there are 10 segments there will be 10 such wavelengths. For example, the center of the first segment is clearly at 415 nm, the center of the second is at 445 nm, and so on. These 10 center wavelengths are important, for they will form the basis of our actual calculations.

To compute the X tristimulus value we begin with the first of our 10 central wavelengths, 415 nm. On Figure A-1 we read the value of $\bar{x}$ at 415 nm. On Figure A-3 we read the intensity of Illuminant *C* at 415 nm. We multiply these two numbers. Next, on Figure A-2 we read the value of the reflectance at 415 nm and multiply this number by the result of our previous multiplication. Write this number down as the *first* of 10 numbers that will be needed to compute X. We now repeat the above procedure entirely using the next of our 10 wavelengths, 445 nm. The result of this calculation is written down as the *second* of the 10 numbers needed to calculate X. We continue to repeat the above procedure for each of the remaining eight central wavelengths, 475 nm, 505 nm, 535 nm, 565 nm, 595 nm, 625 nm, 655 nm, and 685 nm. When we finish we shall have 10 numbers that are *added together* to give X.

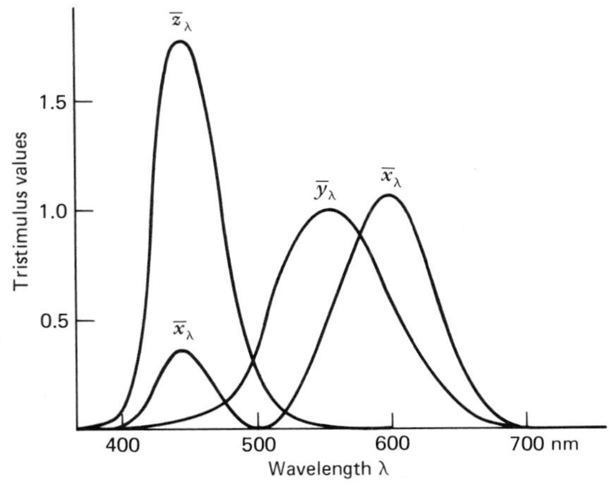

Figure A-1 The CIE color matching functions.

To compute the Y tristimulus value the same procedure as above is used except that on Figure A-1 we read values from the $\bar{y}$ curve instead of the $\bar{x}$ curve. The Z tristimulus value is found by using the $\bar{z}$ curve.

The calculations outlined above undoubtedly seem rather cumbersome and time-consuming. Fortunately, a scheme has been developed that greatly simplifies the matter. Figure A-4*a* is a computing form that can easily be used to compute the value of X for a surface with a known reflectance curve. Notice on the figure that 10 wavelengths are listed (the 10 central wavelengths discussed above), and that next to each wavelength is a double scale. The upper scale

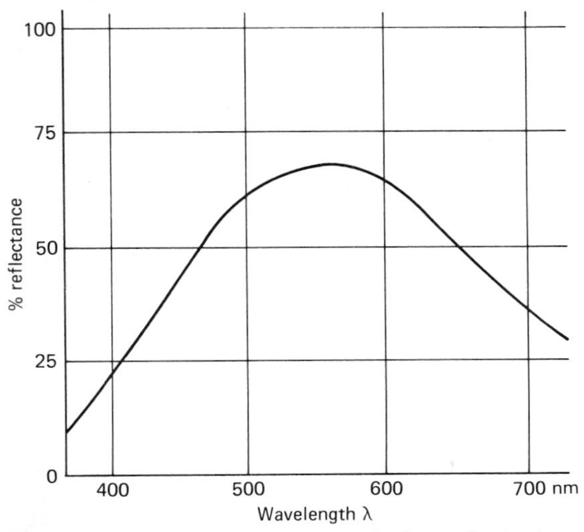

Figure A-2 The reflectance curve of a hypothetical surface.

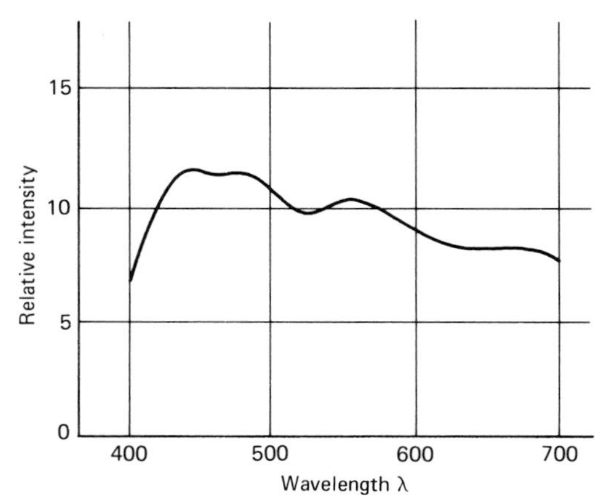

Figure A-3 The spectral content of Standard Source *C*.

goes from zero to 100. The bottom scale goes from zero to some maximum value, different for each wavelength. To use the scales we proceed as follows. From the reflectance curve (Figure A-2) read the reflectance percentage at 415 nm. Find this percentage on the *top* scale next to 415 nm on the computing form (Figure A-4*a*). Make a mark at this percentage. This mark will fall at a definite place on the *lower* scale following 415 nm. Read the number from the lower scale and write it in the blank space provided to the right of the scale. Now repeat this procedure for each wavelength listed on the computing form. When this has been done you will have a column of 10 numbers, one in each blank space on the right of the computing form. Simply add these 10 numbers to get X.

The computing forms shown in Figure A-4*b* and A-4*c* are used to find Y and Z in the same fashion as Figure A-4*a* was used to find X. The numbers X, Y, and Z are the tristimulus values of the surface whose reflectance curve is given by Figure A-2. These tristimulus values assume that the surface is illuminated by standard source *C*, and the computing forms (Figures A-4*a, b,* and *c*) are constructed with this source in mind. If the illumination were changed to standard source *A*, for example, then a new set of computing forms would be required (computing forms for Standard Illuminants *A, B,* and *C* are available commercially, from Bausch and Lomb, for example).

Once the tristimulus values are known, the color coordinates can be found from:

$$x = \frac{X}{X + Y + Z} \qquad y = \frac{Y}{X + Y + Z}$$

Along with the color coordinates x and y it is necessary to specify *Y*, which gives the effective reflectance of the surface. A perfectly white surface will have a Y value of 100, while a perfectly black surface will have a Y value of zero. It

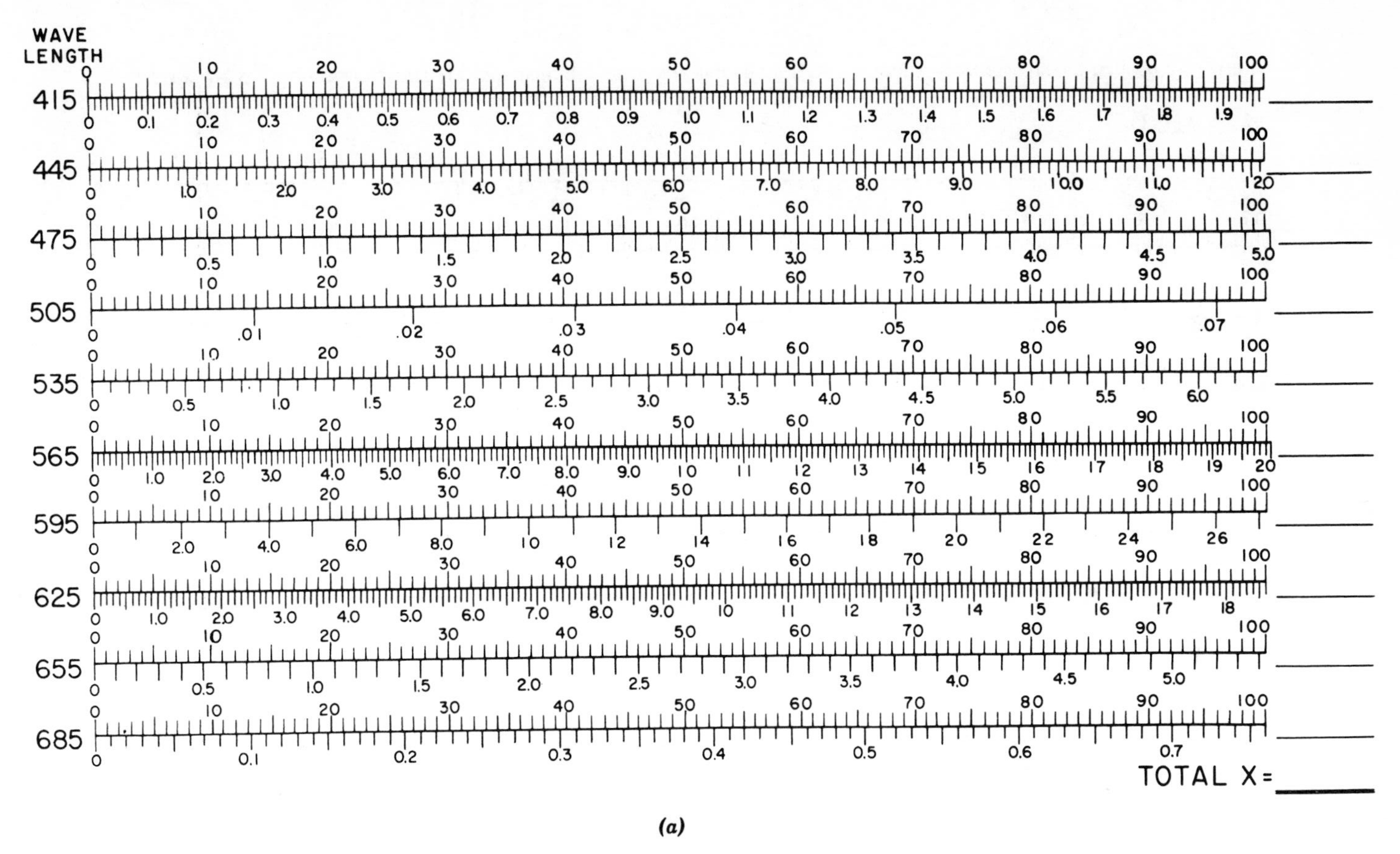

(a)

Figure A-4*a* Computing form for calculating X, assuming illumination by Standard Source *C*. (From Bausch and Lomb)

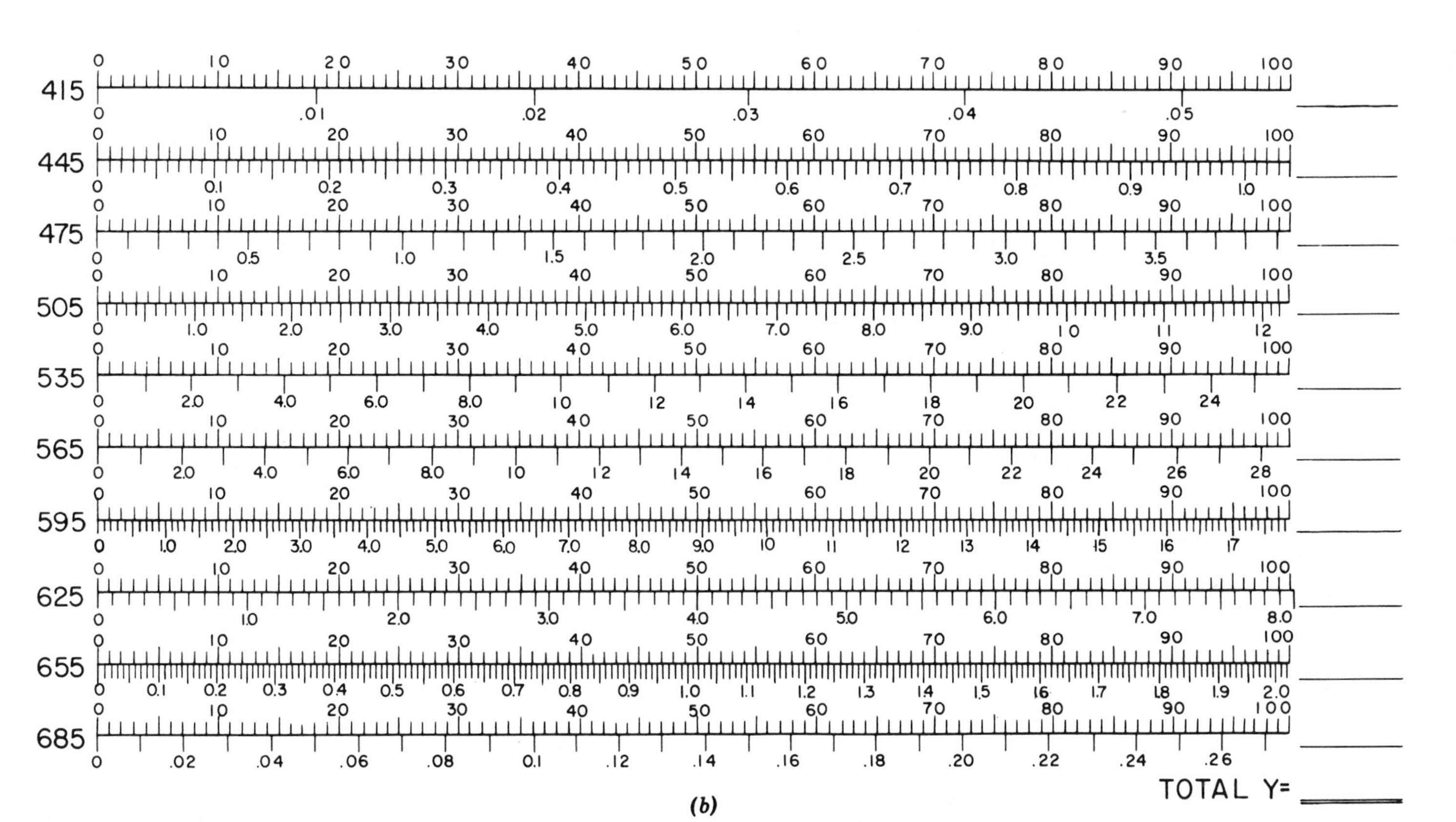

(b)

Figure A-4*b* Computing form for calculating Y assuming illumination by Standard Source *C*. (From Bausch and Lomb)

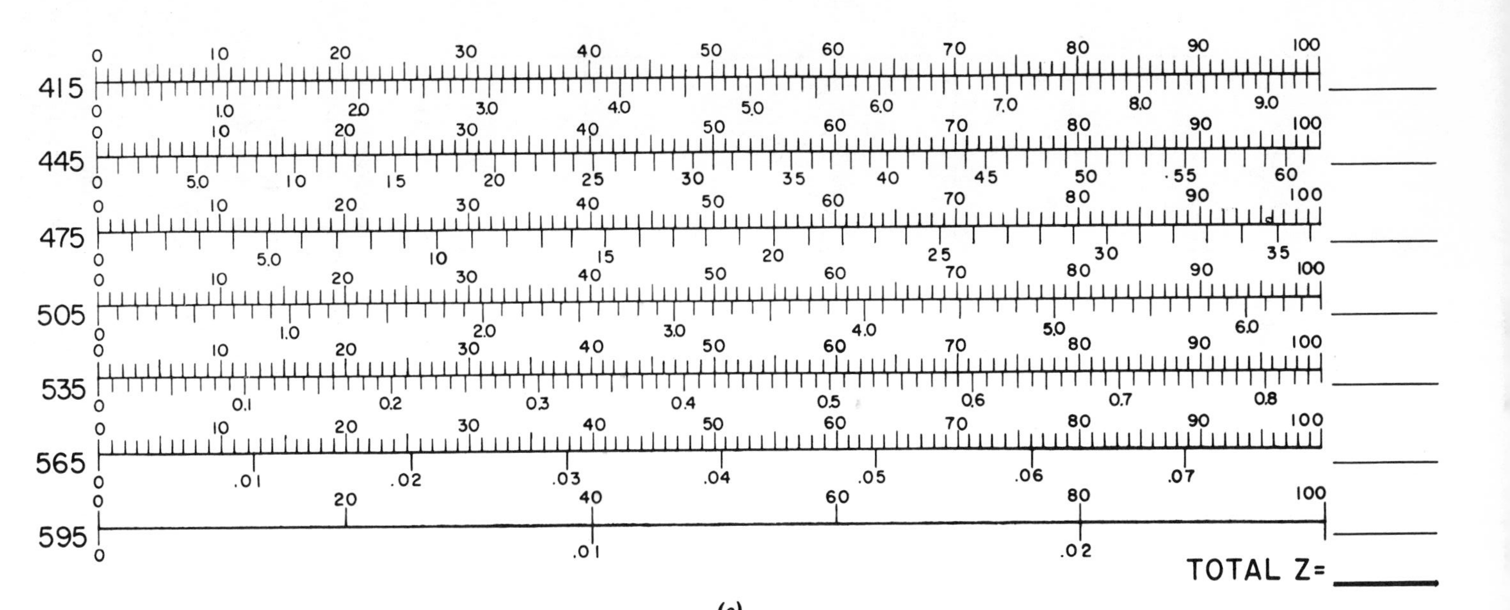

(c)

Figure A-4*c* Computing form for calculating Z assuming illumination by Standard Source C. (From Bausch and Lomb)

X = ________ Y = ________ Z = ________

$x = \frac{X}{X + Y + Z}$ $y = \frac{Y}{X + Y + Z}$

x = ________ y = ________

From Figure A-5:

λ_D = ________

P = ________

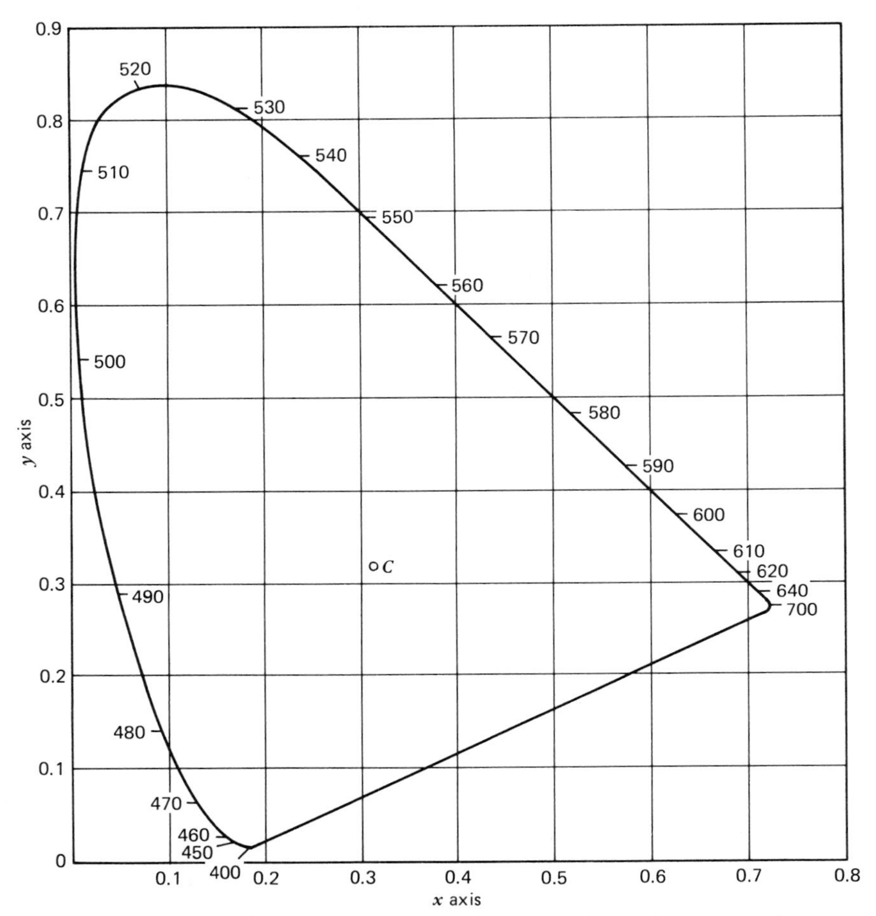

Figure A-5 CIE Chromaticity Diagram for use in determining λ_D and p.

is entirely possible for two different surfaces to have the *same* values of x and y, but very different values of Y. The color coordinates x and y specify the color, while Y measures the lightness of the surface.

With the aid of Figure A-5, the color coordinates can be used to find the dominant wavelength and purity of the color in question. A full discussion of this procedure can be found in Section 3.2.

The calculations above assume that you have a reflectance curve for the surface of interest. If such a curve has not been measured, then it will be necessary to make the measurements yourself. However, measurements need only be made at those wavelengths listed on the computing forms. A number of instruments are available for making these measurements. Appendix B describes how such instruments, known as spectrophotometers, operate and what their limitations are.

APPENDIX B THE SPECTROPHOTOMETER

Throughout the text we have referred to spectral curves of various kinds. Such curves were used to represent the output of different light sources, the transmittance of filters, or the reflectance of surfaces. In order to arrive at these kinds of curves it is necessary to use a device called a spectrophotometer. There are different ways to construct a spectrophotometer, but they all have much in common. In each case there will be a light source, an optical system for image formation, a dispersive element such as a prism or grating to separate the wavelengths, and a light detector of some kind.

A schematic diagram of a typical spectrophotometer is shown in Figure B-1. Let us follow the light through the device step by step to see how the apparatus operates. First, we have the light source. If the spectral characteristics of the light source itself are to be measured, then the light from the source will be allowed to pass directly through the spectrophotometer to the detector. If transmittance or reflectance measurements are to be made, the source will be a white light source. In the case of transmittance measurements, it is necessary to make two readings. One reading is made using the source by itself. A second reading is made with the filter to be measured inserted between the detector and the exit slit. The transmittance (for the particular wavelength being examined) is computed by comparing the two readings. A similar kind of comparison of readings must be done for reflectance measurements. In this case, however, the light from the exit slit must be rerouted so that it illuminates either a white surface or the color surface of interest before entering the detector.

Turning our attention back to the light source, we see that after the light leaves the source it passes through the entrance slit (*S*1). This slit serves as a brightly illuminated and sharply defined object that will be imaged by the optical system of the spectrophotometer. Slit *S*1 is located at the focal point of lens *L*1 and therefore the light leaving *L*1 consists of a parallel beam of light. This light passes through the prism (or diffraction grating, which also spreads out the different wavelengths) and is separated into the different colors of the spectrum. Each color is diffracted through a slightly different angle. The second lens, *L*2, simply focuses the light into several sharp images of the entrance slit, one image

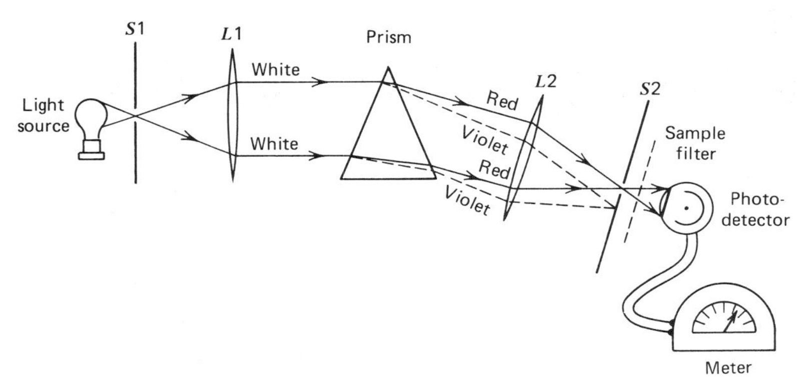

Figure B-1 Basic elements of a prism spectrophotometer.

for each color in the spectrum. These images run together to form a continuous spectrum. The exit slit, *S2*, allows only a very narrow band of color (a nearly pure wavelength) to pass on to the detector, *D*. By mechanically rotating the prism, we can shift the spectrum back and forth across the exit slit so that different wavelengths can be allowed to reach the detector. Usually, the mechanical knob that rotates the prism is calibrated directly in the wavelengths of the spectrum. Thus setting the knob at a particular wavelength ensures that this wavelength is indeed the one that is passing through the exit slit and into the detector.

Most practical spectrophotometers used in color analysis are not really designed to measure the spectral characteristics of different light sources. First, these instruments usually have their own built-in white light source. Second, the phototubes used as detectors have very different sensitivities to wavelengths in different parts of the spectrum. Thus, unless the response characteristics of the phototube are known accurately, it would be impossible to measure the absolute intensity of a light source at different wavelengths. On the other hand, this difficulty does not arise in making transmittance or reflectance measurements. As noted earlier, a transmittance measurement simply involves a comparison of detector readings with a filter present versus absent. The same light source and wavelength setting are used in both readings. Thus the different detector sensitivity at *different* wavelengths does not come into play. Likewise, reflectance measurements involve comparing detector readings from a white surface to those of the colored sample. Again, each pair of readings is made at the same wavelength so that no problem with detector sensitivity arises.

In the color industry, the arrangement shown in Figure B-1 is known as forward optics. That is, light is dispersed into different colors before passing through or reflecting from a sample filter or surface. This can lead to incorrect results when the sample is fluorescent. Consider a reflectance sample that contains a fluorescent dye that is able to absorb light at the blue end of the spec-

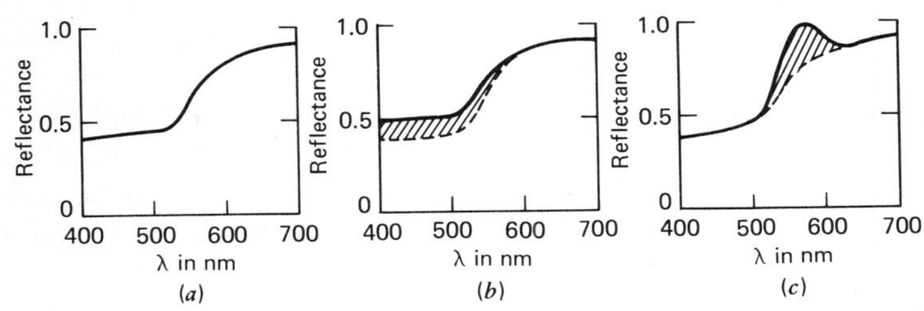

Figure B-2 (*a*) True reflectance of a sample containing a fluorescent dye — fluorescent light is not included. (*b*) Reflectance as measured by forward optics. Fluorescent light (shaded area) is incorrectly measured as occurring at short wavelengths. (*c*) True reflectance of fluorescent sample including fluorescent light as measured by reverse optics. Fluorescent light is the shaded peak.

trum and reemit this light in a band around 580 nm. Suppose we attempt to measure the reflectance of this sample at, say, 450 nm, so that light of this wavelength illuminates the sample. Now the following will happen. A certain amount of the 450 nm light, say 40%, will simply *reflect* from the sample. Thus the true reflectance of the sample at 450 nm is 40%. An additional amount of the incident light will be absorbed and not reradiated, say 50%. But now suppose the remaining 10% of the incident 450 nm light is absorbed by the fluorescent dye and reradiated near 580 nm. This 580 nm light will combine with the *reflected* 450 nm light and *both* will enter the detector. The detector cannot distinguish between light of different wavelengths, and thus responds to the 580 nm light as well as the 450 nm light. The result is an overly large detector response and thus a spuriously large apparent reflectance at 450 nm. This effect will occur at every wavelength that can be absorbed and reradiated by the fluorescent dye. Thus the true reflectance of the sample might be as shown in Figure B-2*a*, but the reflectance as measured by the forward optics arrangement could turn out as in Figure B-2*b*.

For many manufacturers of colored materials neither of the curves of Figures B-2*a* or B-2*b* represent what they really need to know. The true reflectance curve of Figure B-2*a* does not account for the fluorescence at all, while the curve of Figure B-2*b* incorrectly distributes the 580 nm fluorescent light to the shorter wavelength end of the spectrum. Most manufacturers need to know the following: given that a surface is illuminated by a certain source, what exactly is the spectrum of light that "reflects" (including any fluorescence) from the surface. To make this type of measurement we must do two things. First, in Figure B-1 the light source must be the same as the one that will illuminate the surface under the conditions of interest. Second, the surface (or filter in the case of transmittance measurements) must be introduced into the spectrophotometer *before* the light is dispersed by the prism. For example, this could be done between the source and the entrance slit. This arrangement is called reverse optics in the color industry, and ensures that the light measured at the detector is the actual distribution of intensities that leave the surface. In the fluorescent example discussed above, the resulting spectral curve would look like Figure B-3*c*, and would show a pronounced peak at 580 nm. This curve would accurately represent the reflectance of the sample (compared to the reflectance of a perfectly white standard) when illuminated by the particular light source in the spectrophotometer.

APPENDIX C
SINE FUNCTIONS

Angle	Sine	Angle	Sine	Angle	Sine
0.0	0.000				
0.5	0.009	19.0	0.326	37.5	0.609
1.0	0.017	19.5	0.334	38.0	0.616
1.5	0.026	20.0	0.342	38.5	0.622
2.0	0.035	20.5	0.350	39.0	0.629
2.5	0.044	21.0	0.358	39.5	0.636
3.0	0.052	21.5	0.366	40.0	0.643
3.5	0.061	22.0	0.375	40.5	0.649
4.0	0.070	22.5	0.383	41.0	0.656
4.5	0.078	23.0	0.391	41.5	0.663
5.0	0.087	23.5	0.399	42.0	0.669
5.5	0.096	24.0	0.407	42.5	0.676
6.0	0.104	24.5	0.415	43.0	0.682
6.5	0.113	25.0	0.423	43.5	0.688
7.0	0.122	25.5	0.431	44.0	0.695
7.5	0.131	26.0	0.438	44.5	0.701
8.0	0.139	26.5	0.446	45.0	0.707
8.5	0.148	27.0	0.454	45.5	0.713
9.0	0.156	27.5	0.462	46.0	0.719
9.5	0.165	28.0	0.470	46.5	0.725
10.0	0.174	28.5	0.477	47.0	0.731
10.5	0.182	29.0	0.485	47.5	0.737
11.0	0.191	29.5	0.492	48.0	0.743
11.5	0.199	30.0	0.500	48.5	0.749
12.0	0.208	30.5	0.508	49.0	0.755
12.5	0.216	31.0	0.515	49.5	0.760
13.0	0.225	31.5	0.522	50.0	0.766
13.5	0.233	32.0	0.530	50.5	0.772
14.0	0.242	32.5	0.537	51.0	0.777
14.5	0.250	33.0	0.545	51.5	0.783
15.0	0.259	33.5	0.552	52.0	0.788
15.5	0.267	34.0	0.559	52.5	0.793
16.0	0.276	34.5	0.566	53.0	0.799
16.5	0.284	35.0	0.574	53.5	0.804
17.0	0.292	35.5	0.581	54.0	0.809
17.5	0.301	36.0	0.588	54.5	0.814
18.0	0.309	36.5	0.595	55.0	0.819
18.5	0.317	37.0	0.602	55.5	0.824

Angle	Sine	Angle	Sine	Angle	Sine
56.0	0.829	67.5	0.924	79.0	0.982
56.5	0.834	68.0	0.927	79.5	0.983
57.0	0.839	68.5	0.930	80.0	0.985
57.5	0.843	69.0	0.934	80.5	0.986
58.0	0.848	69.5	0.937	81.0	0.988
58.5	0.853	70.0	0.940	81.5	0.989
59.0	0.857	70.5	0.943	82.0	0.990
59.5	0.862	71.0	0.946	82.5	0.991
60.0	0.866	71.5	0.948	83.0	0.992
60.5	0.870	72.0	0.951	83.5	0.994
61.0	0.875	72.5	0.954	84.0	0.994
61.5	0.879	73.0	0.956	84.5	0.995
62.0	0.883	73.5	0.959	85.0	0.996
62.5	0.887	74.0	0.961	85.5	0.997
63.0	0.891	74.5	0.964	86.0	0.998
63.5	0.895	75.0	0.966	86.5	0.998
64.0	0.899	75.5	0.968	87.0	0.999
64.5	0.903	76.0	0.970	87.5	0.999
65.0	0.906	76.5	0.972	88.0	0.999
65.5	0.910	77.0	0.974	88.5	1.000
66.0	0.914	77.5	0.976	89.0	1.000
66.5	0.917	78.0	0.978	89.5	1.000
67.0	0.921	78.5	0.980	90.0	1.000

APPENDIX D
ANNOTATED BIBLIOGRAPHY

GENERAL

KODAK 1962

Kodak Publication No. E-74, *Color as Seen and Photographed.* Eastman Kodak Company, Rochester, New York, 1962

In addition to a good discussion of the principles of photography, this book also has considerable material on light, color, and the psychophysical aspects of color. Extensive illustrations are provided.

RAINWATER 1971

Clarence Rainwater, *Light and Color,* Golden Press, New York, 1971.

This little book is filled with color illustrations and covers, on a very elementary level, nearly all the topics treated by us in the present text.

TIME-LIFE 1966

Conrad G. Mueller, Mae Rudolph and the Editors of LIFE, *Light and Vision,* Time, New York, 1966.

This is a very readable book dealing primarily with the basic properties of light and vision. It contains many outstanding illustrations, and touches on subjects such as color printing, colorimetry, and color in art.

COLOR AND COLOR VISION

BEGBIE 1973

G. Hugh Begbie, *Seeing and the Eye,* Anchor Press/Doubleday, Garden City, New York, 1973.

Begbie has written a good popular level book that covers basic visual phenomena in a fair amount of detail.

BILLMEYER 1981

Fred W. Billmeyer, Jr. and Max Saltzman, *Principles of Color Technology,* 2nd ed., Wiley, New York, 1981.

This book begins with an introduction to light and vision. It then goes on to describe the principles of colorimetry in great detail. The last half of the book is devoted to colorants, industrial coloring materials, and problems in color technology.

CORNSWEET 1970

Tom N. Cornsweet, *Visual Perception,* Academic Press, New York, 1970.

The various aspects of visual perception are discussed in some detail in this book. The level is definitely above BEGBIE, but should be accessible to most students.

EVANS 1948

Ralph M. Evans, *An Introduction to Color,* Wiley, New York, 1948.

An outstanding introduction to color vision, this book is well illustrated and very thorough. Unfortunately, it is out of print at the moment.

HURVICH 1981

Leo M. Hurvich, *Color Vision,* Sinaner Associates, Sunderland, Massachusetts, 1981.

This book provides an up-to-date discussion of color vision from an opponent theory point of view. Hurvich, one of the leading scientists in the field, has produced an outstanding work.

MUNSELL 1963

A. H. Munsell, *A Color Notation,* Munsell Color Company, Baltimore, Maryland, 1936–63

A complete description of the origins and operation of the Munsell color system is given in this book.

SHEPPARD 1968

Joseph J. Sheppard Jr., *Human Color Perception,* American Elsevier, New York, 1968.

The unique emphasis of this book is the attention it gives to the experiments that underlie the various theories of color vision.

WASSERMAN 1978

Gerald S. Wasserman, *Color Vision: An Historical Introduction,* Wiley, New York, 1978.

Wasserman's book is a good account of what is known about color vision, but its particular strength is the excellent historical review of the subject that begins the book.

WYSZECKI 1967

Gunter Wyszecki and W. S. Stiles, *Color Science,* Wiley, New York, 1967.

This book is not so much a textbook as a detailed reference for those interested in various aspects of color. It includes an enormous quantity of numerical data on colorimetry, filters, light sources, and many other subjects as well as detailed discussions of most of the important topics related to color science.

OPTICS

JENKINS 1976

Francis A. Jenkins and Harvey B. White, *Fundamentals of Optics,* 4th ed., McGraw-Hill, New York, 1976.

This is a fairly advanced text that covers geometrical and physical optics in great detail.

MILLER 1977

Franklin Miller Jr., *College Physics,* 4th ed., Harcourt, Brace Jovanovich, New York, 1977.

There are many standard algebra-trigonometry level college physics texts that include large sections devoted to optics. Miller's book is an example of such a text that has a good discussion of optics.

NEWTON 1730

Sir Isaac Newton, *Opticks,* Dover Publications, New York, 1952 (Reprint of 4th ed., London, 1730).

Here is the monumental text by Newton where the principles of color are first given in a more or less modern form. The reading is somewhat awkward because of the eighteenth century style, but it is not too difficult, and is well worth the effort.

SEARS 1976

Francis W. Sears, Mark W. Zemansky, and Hugh D. Young, *University Physics,* Addison-Wesley, Reading, Massachusetts, 1976.

Several calculus level general physics texts exist that have large sections devoted to geometrical, wave, and physical optics. This is one such text.

PHOTOGRAPHY

DAVIS 1975

Phil Davis, *Photography,* 2nd ed., William C. Brown, Dubuque, Iowa, 1975.

This is a good general review of the basic principles of photography including a discussion of light, color, lenses, the camera, and practical aspects of photography.

KODAK 1979

The editors of Kodak, *The Joy of Photography,* Addison-Wesley, Reading, Massachusetts, 1979.
The principles of photography are covered on a popular level in this book.

NATURAL COLOR PHENOMENA

GREENLER 1989

Robert Greenler, *Rainbows, Halos, and Glories,* Cambridge University Press, 1980.
This is a beautifully illustrated book on the subjects given in the title.

MINNAERT 1954

M. Minnaert, *The Nature of Light and Color in the Open Air,* Dover Publications, New York, 1954.

Here is an exhaustive treatment of all kinds of natural color phenomena such as rainbows, halos, glories, mirages, the blue sky, fog effects, and much more.

INDEX